TOXICITY REDUCTION IN INDUSTRIAL EFFLUENTS

TOXICITY REDUCTION IN INDUSTRIAL EFFLUENTS

Edited by

Perry W. Lankford
W. Wesley Eckenfelder, Jr.

 VAN NOSTRAND REINHOLD _____ New York

Copyright © 1990 by Van Nostrand Reinhold

Library of Congress Catalog Card Number 90-41625
ISBN 0-442-00234-3

Printed in the United States of America

Van Nostrand Reinhold
115 Fifth Avenue
New York, New York 10003

Van Nostrand Reinhold International Company Limited
11 New Fetter Lane
London EC4P 4EE, England

Van Nostrand Reinhold
480 La Trobe Street
Melbourne, Victoria 3000, Australia

Nelson Canada
1120 Birchmount Road
Scarborough, Ontario M1K 5G4, Canada

16 15 14 13 12 11 10 9 8 7 6 5 4 3 2 1

Library of Congress Cataloging-in-Publication Data

Toxicity reduction in industrial effluents / Perry W. Lankford, W.
    Wesley Eckenfelder, Jr., editors.
        p.   cm.
    Includes bibliographical references.
    ISBN 0-442-00234-3
    1. Factory and trade waste—Purification. 2. Factory and trade
waste—Toxicology.   I. Lankford, Perry W.   II. Eckenfelder, W.
Wesley (William Wesley), 1926–   .
    TD897.T69   1990
    628.1′683—dc20
                                                        90-41625
                                                        CIP

Contributors

Perry W. Lankford, P.E., Editor
Executive Vice President
ECKENFELDER INC.

W. Wesley Eckenfelder, Jr., Editor
Senior Technical Director
ECKENFELDER, INC.

Billy G. Isom
Director, Aquatic Toxicology and Ecology Division
ECKENFELDER, INC.

W. Scott Hall
Aquatic Toxicologist
ECKENFELDER, INC.

Dr. Richard J. Mirenda
Senior Aquatic Toxicologist
ECKENFELDER, INC.

Dr. Richard E. Speece
Centennial Professor of Environmental Engineering
Vanderbilt University

Dr. Yerachmiel Argaman
Professor, Environmental and Water Resources Engineering
Technion-Israel Institute of Technology

Dr. Alan Bowers
Associate Professor, Civil and Environmental Engineering
Vanderbilt University

Jeffrey L. Pintenich
Director, Waste Management Division
ECKENFELDER, INC.

Kevin D. Torrens
Assistant Director, Wastewater Management Division
ECKENFELDER, INC.

Contents

Preface

Wastewater treatment technology is undergoing a profound transformation as a result of the fundamental changes in regulations and permit requirements. Established design procedures and criteria which have served the industry well for decades are no longer useful. Toxicity reduction requirements have forced reconsideration of design standards and caused practicing environmental engineers to seek additional training in the biological sciences. Formal academic programs have not traditionally provided the cross-training between biologists and engineers which is necessary to address these issues. This book describes not only the process of identifying the toxicity problem, but also the treatment technologies which are applicable to reduction or elimination of toxicity. The information provided in this book is a compilation of the experience of ECK-ENFELDER INC. in serving the environmental needs of major industry, and the experience of the individual contributors in research and consultation.

Acknowledgment

This book is the result of tremendous efforts, not only by the editors and contributors, but by numerous reviewers, proofreaders, and word processors. Some of those who are not acknowledged elsewhere and who put in long, inconvenient hours in the book's preparation are:

Teresa G. Summers
Bonnie G. Branner
Jackie L. Thomas
Nerissa Shelton
Teresa O. Williams
Kimberly J. Neumair
Dee P. Biggert

Richard L. Williamson, Jr.
Dr. John L. Musterman
Gregory W. Pulliam
Stephan R. Tate
T. Houston Flippin
Dr. Ajit Ghorpade

Without the support and efforts of these individuals, this book would still be just another good idea.

TOXICITY REDUCTION IN INDUSTRIAL EFFLUENTS

1

Toxicity in Industrial Wastewater

Perry W. Lankford

INTRODUCTION

In 1977, Congress amended the Federal Water Pollution Control Act, more commonly known as the Clean Water Act (CWA), to deal with the growing concern over toxicity. This concern was primarily focused on individual toxic substances. In March 1984, the United States Environmental Protection Agency (U.S. EPA) issued a policy statement encouraging the use of aquatic toxicity limitations in the National Pollutant Discharge Elimination System (NPDES) permitting process. More recently, the reauthorization of the Clean Water Act in 1987 specifically addressed toxics management as a priority, and required the states to control the discharge of all priority pollutants under Section 304(1). In addition to the requirement to reduce or eliminate the discharge of toxics, many industrial plants are facing a requirement to conduct a toxicity reduction evaluation (TRE). Complying with these requirements can pose significant scientific and engineering challenges. This book is designed to make the reader more knowledgeable on the major issues.

The generally unknown origin of effluent aquatic toxicity, as well as the complex makeup and varying toxicants in individual wastestreams, frequently requires a toxicity reduction program to begin at the discharge and proceed upstream to the source. The key toxicant(s) and the actual toxic mechanism are normally not apparent from traditional characterization data and may even consist of refractory organics or byproducts generated in the biological treatment processes. Due to the complex interaction between toxicants, test organisms, and ambient characteristics, a toxicity theshold concentration for specific compounds cannot be universally applied. A simple example of this is that a heavy metal may be quite toxic at very low levels in low-hardness dilution water and be relatively nontoxic in a wastewater because of increased levels of hardness. This phenomenon is further described in Chapter 2.

In the past, NPDES permit requirements primarily related to the "conventional pollutants," such as total suspended solids or biochemical oxygen demand, for which there were rational design and operational criteria. More recently, limits on specific organics were added to permits; treatment of these was addressed through the development of predictive models. The latest addition to the problems facing wastewater treatment systems is the requirement to reduce aquatic toxicity, for which no predictive models exist. Individuals charged with the responsibility for complying with these new requirements must have an understanding of wastewater treatment technology, aquatic toxicity measurement methods, and the relationship between the two. Formal cross-training of this type for engineers and biologists has existed only to a limited extent to date. It is the intent of this book to fulfill some of the potential user's basic needs of cross-training. It is not, however, the intent of the book to serve as an in-depth text on each of the technical areas covered; references are cited in each chapter to meet that need.

An overview of the growing need for this cross-training is provided in Chapter 1. The discussion includes the regulatory authorities which may bring about a toxicity limit to a discharger under Federal regulations and state programs. In addition, the fundamental concepts of aquatic toxicology and toxicity reduction are described to provide the user with an introduction to the subject. Various approaches to toxicity reduction are also described. These tie together the later chapters into a logical series of considerations for determination of the best solution to the problem.

Chapter 2 provides an in-depth description of aquatic toxicity tests and procedures. An understanding of the use and limitations of the various tests is an essential first step in order to define a toxicity problem and evaluate the available data. Although definitive procedures are not provided in this chapter, comparisons and applications are described which will serve the reader well.

Toxicity identification evaluations (TIEs), especially the procedures for sample fractionation, are described in Chapter 3. Such procedures are an essential element of toxicity reduction, as described in Chapter 1. It is necessary to understand the toxicity tests described in Chapter 2 in order to utilize Chapter 3. The purpose of the TIE procedures described is to identify the specific toxic components or class of components in the sample which causes most or all of the toxicity. The end result of the TIE can be the identification of a specific cause of toxicity, or more likely, the identification of one or more technologies, which can result in a reduction or elimination of toxicity. This second, and more common, result is the principal difference and advantage of the procedure described here over other published techniques.

Another important element of toxicity reduction is the identification and treatment of the wastewater source or sources contributing the toxicity. Since most wastewater streams are made up of many component sources, the process of testing each one can be tedious and expensive. In addition, since the toxicity of concern is that following treatment of the wastewater, not prior to treatment, there is a need to analyze each source as it would appear after treatment, not in its untreated form. These principles, as well as applicable source treatment technologies, are described in Chapter 4.

The toxicity identification evaluation effort described in Chapter 3 can result in the identification of either a toxic component or one or more treatment technologies which reduces or eliminates the toxicity. One of the components evaluated is heavy metals. In numerous instances of toxicity, heavy metals are found to be the problem. Chapter 5 provides data on the treated levels of various metals which might be required and the technologies which might achieve those levels.

Through proper design techniques, many toxic organics can be reduced to nontoxic levels through aerobic biological treatment. Chapter 6 describes these techniques, the achievable levels, and the biological processes best suited to this application. This chapter builds on the basics described in the toxicity identification work of Chapter 3 and the source testing work of Chapter 4.

Many organics which are found to be toxic and nonbiodegradable or poorly degradable under aerobic conditions will degrade to end-products under anaerobic conditions. In these cases, anaerobic pretreatment can effectively reduce toxicity and render the wastewater amenable to subsequent aerobic biological treatment. Chapter 7 provides the background, design concepts and supporting data for this technology.

Stripping of volatile priority pollutants and some inorganics from a waste stream is a component of the toxicity identification effort of Chapter 3. Although volatile organics have not been found to be as toxic as the heavy metals and other organics, these compounds are sometimes found to contribute to effluent toxicity. In other cases, stripping is needed as a pretreatment step to improve the effectiveness of subsequent treatment processes. Chapter 8 provides descriptions of various stripping processes, design procedures, and operating data.

Activated carbon is often used for reduction of toxicity, particularly because of the nonspecific nature of adsorption on this material. A wide variety of organics and most heavy metals will adsorb to some degree, making the process a good choice where the cause of toxicity is unknown and/or variable. Activated carbon is one of the techniques utilized in the toxicity identification work described in Chapter 3. The two principal alternative carbon adsorption processes involve use of granular or powdered carbon, respectively. Granular carbon is normally applied in column contactors and can be designed to be quite efficient; this technology and its performance are described in Chapter 9. Powdered carbon is typically applied in mixed slurry contactors, which while not always as efficient as granular, are superior in some cases and usually less expensive; this technology is described in Chapter 10.

Chemical oxidation is a promising technology for treatment or pretreatment of a variety of organics and inorganics. It is included as one of the test procedures in the toxicity identification evaluation described in Chapter 3. In most cases, the primary objectives of chemical oxidation are detoxification and rendering the organics more biodegradable in subsequent treatment steps. Chemical oxidation is also a key pretreatment process in metals removal as described in Chapter 5. Chapter 11 provides a description of the optional oxidants, the systems involved and the performance expectations of chemical oxidation.

Sludge handling and disposal are critical aspects of toxics management. Although the fate of specific compounds may not be predictable, alternatives can be evaluated based upon the origin of a sludge. Chapter 12 provides an in-depth description of the regulations and technical alternatives for handling and disposal of sludges generated in the treatment of toxic waste waters.

At various stages of a toxicity reduction evaluation, it can be necessary or desirable to narrow the number of alternatives considered by comparing their costs. Chapter 13 describes how to prepare an evaluation of relative costs at various points in a project, depending on the desired or required accuracy. This chapter also provides tabulated cost factors for comparing all of the process technologies described in the book.

A glossary and an index are provided following Chapter 13 to allow the reader to obtain definitions or references for unfamiliar terms.

FUNDAMENTAL CONCEPTS

Discharge of substances in amounts potentially harmful to humans has long been forbidden, though in recent years those requirements have become more stringent. More recently, the limitations on toxicity are being placed on effluents to avoid toxicity to the aquatic environment. This is determined by tests performed on test organisms such as small fish or water fleas. Acute toxicity is an observation of mortality of the test organism; chronic toxicity is an observation of some effect less severe than mortality, such as diminished growth or reproduction. Toxicity can be expressed in one of two ways: either the concentration ($\mu g/l$) of a specific constituent which is found to be toxic, or as a dilution (percent) of the sample which is found to be toxic. The latter expression, called whole effluent toxicity, is increasingly utilized in wastewater management because it in effect sums the toxic effects of all constituents present. A high value for whole effluent toxicity, say 80 percent (a sample made up of 80 percent effluent and 20 percent dilution water was found to be toxic), indicates that the effluent is not very toxic. A low value, say 1 percent (a sample made up of

1 percent effluent and 99 percent dilution water was found to be toxic), indicates that the effluent is highly toxic.

The terms most commonly used to denote toxicity are the LC_{50} for acute and the LOEC and NOEC (same as LOEL/NOEL) for chronic. The LC_{50} is defined as the concentration which is lethal to 50 percent of the test organisms. The LOEC (LOEL) is defined as the lowest observed effect concentration (or level), or the lowest amount of effluent added to a sample which resulted in the observed effect; the effect being growth or reproduction differences. The NOEC (NOEL) is the no observed effect concentration (or level) or the highest amount of effluent added to a sample where no effect was observed.

Almost every compound or material is toxic to some organism at some concentration; for example, common salt is toxic to freshwater organisms, lack of salt is toxic to marine organisms. Materials on which some organisms depend or flourish are toxic to other organisms. Therefore, universal statements on toxicity are not possible and are dangerous. The toxicity of some organic and inorganic priority pollutants can vary by as many as five orders of magnitude ($100,000 \times$) depending on the organism selected for testing. Typically, regulatory authorities recommend or require that the most sensitive applicable organism from among the common test organisms be used for testing. Determination of which organism is most sensitive may require multiple tests with a number of alternative organisms.

The tests used to measure toxicity vary depending on the organism being tested. The temperature, duration, salinity, sample renewal, and handling techniques can differ dramatically from one to another. In addition, several different data analysis techniques can be used which may produce differing results.

Toxicity reduction consists of determination of a specific cause or general characteristic of the toxicity and then devising a means of reduction or elimination. Since historical concerns were limited predominantly to conventional pollutants, effluents from some well designed and well run treatment systems have been found to be quite toxic. Nearly all of these toxicity conditions relate to either a metal or other inorganic which can be removed when identified or changed to a less toxic form, or to a synthetic organic which can be destroyed or altered to a nontoxic state.

When numerous components and contaminants are combined into a single stream, as is the case in wastewater management, it is necessary to consider whether the toxicity of the combined sample is greater (synergistic) or less (antagonistic) than the sum of the toxicities of the components. Although these phenomena have been observed in some cases, most effluents have been found to exhibit toxicity equal to the sum of the toxicities of each component (additive). Toxicity here is measured in toxic units (TU), defined as the inverse of the whole effluent toxicity times 100, or the concentration of the toxic constituent divided by the LC_{50} concentration of the toxic constituent. An advantage of the whole effluent toxicity approach over controls based on single toxicants is that it automatically takes additivity, synergy, etc. into account.

REGULATORY AUTHORITIES

All Federal or state discharge permitting programs under the Clean Water Act included a basis for toxics control through either instream numerical limits or by extrapolation from narrative criteria, sometimes called "free froms." Instream numerical limits consist of a regulatory level of a specific parameter, above which the water would presumably be harmful to its inhabitants or to human health. "Free froms" consist of general quality descriptions such as "the waters of the State shall be free from substances in concentrations or combinations toxic to humans, wildlife or aquatic life."

On March 9, 1984, the U.S. EPA issued a policy statement which asserted that permit levels on toxicity could be imposed and that toxicity reduction evaluations (TREs) could be required. This policy strongly encouraged states to force dischargers suspected of having an excessively toxic discharge to perform TREs without necessarily having toxicity as a permit limit. The U.S. EPA has recently proposed to make the essential features of its recommendations henceforth be binding regulations.

Since 1984, the various state and Federal regulations governing individual toxic substances have been significantly strengthened. In addition to the Clean Water Act described previously, control of toxicity in a discharge stream might be required under one or more of the following authorities:

1. Federal effluent limitations guidelines. As a part of BAT (Best Available Technology) controls, limitations on individual toxic substances are being very rapidly added as permit conditions. Where a plant has only a single substance above legal limits, a treatment system design can be designed fairly simply to address that parameter. However, in many cases, a plant will have numerous substances to reduce, and this can be just as complex a problem to solve as that of a plant displaying high whole effluent toxicity.
2. State water quality standards for individual substances. A few states already have substantial chemical-specific controls, and Section 304(1) of the 1987 amendments to the Clean Water Act (CWA) requires all states to do so. Many plants will be forced to reduce individual toxic substances which have not been required to do so under effluent limitations guidelines.
3. State water quality standards for whole effluent toxicity. These standards vary considerably from state to state as described in subsequent paragraphs.

4. Federal authorities other than the Clean Water Act. Reduction of toxicants is increasingly being required under consent orders under the Toxic Substances Control Act (TSCA), or more commonly under the Federal Insecticides, Fungicide and Rodenticide Act (FIFRA). In addition, elimination of the toxicity of an effluent could be required for any wastewater which is determined to be a hazardous waste under the Resource Conservation and Recovery Act (RCRA). Toxicity reduction of discharges may also be required for corrective action at contaminated sites under RCRA and the Comprehensive Environmental Response Compensation and Liability Act (CERCLA, more commonly known as Superfund).
5. Ocean Discharge Criteria. Although not a significant factor to this point, the U.S. EPA and the states may begin taking the toxicity aspects of the Ocean Discharge Criteria under the Clean Water Act far more seriously.

States have generally selected one of two options, or a combination of them, to satisfy the programmatic requirements of the Clean Water Act. One option is to limit instream concentrations of specific chemicals to below their toxic levels through the use of instream application factors and safety factors. The second option is to limit levels of whole effluent toxicity where specific chemicals could not be identified. The first option is currently predominant because it is generally easier to test, and results in permit limits similar to the technology-based limits contained in Federal effluent limitations guidelines. The second approach is rapidly being adopted as a supplement to the first.

The programs of the various states differ considerably because of varying local water quality conditions, the provisions of state legislation, and preferences of the enforcement agencies. Table 1-1 provides a summary of the toxics management programs of eight jurisdictions which have taken dramatically different paths. California's approach has been to break the state into regions. Each region can set its own guidelines and can require toxicity testing on up to twelve different species or alternatively may not impose any toxicity limit at all in a permit. At least in the recent era of toxics management, Pennsylvania began with a chemical specific approach while its neighbor, New Jersey, emphasized a whole effluent approach with a single minimum toxicity requirement in permits. Both Pennsylvania and New Jersey have since moved to a dual full-fledged approach requiring both chemical specific and whole effluent toxicity limits. Virginia began with a whole effluent approach, but has not immediately imposed limits, requiring TREs instead. Proposed U.S. EPA regulations will require all states to adopt the dual strategy.

Some in industry would like to view whole effluent toxicity as a single-parameter monitoring method which obviates the need for the growing list of parameters on most NPDES permits. There is also some support for this view by regulators for reasons of convenience. However, the U.S. EPA Antibacksliding Regulation requires that a modified permit be at least as stringent as that which

Table 1-1. Examples of Toxics Control Strategies.

	Administration	Status	Strategy	Biomonitoring-Based Permit Requirement	Protocol	Species
U.S. EPA Guidance	—	1987 WQA and Permit Writer's Guide	Individual and whole effluent	—	Acute or chronic	—
California	Regions	Proposed or in development	Whole effluent	Monitoring and toxicity limits	Acute and chronic	Multispecies, up to six for each
Louisiana	DEQ	No regultions; U.S. EPA issuing permits; program just starting	Whole effluent and chemical specific	U.S. EPA has set limits	U.S. EPA acute and chronic methods	U.S. EPA recommended species; developing methods for brackish water species
Missouri	DNR	Promulgated for individual substances	Chemical-specific	No limits; some toxicity tests	Acute primarily	U.S. EPA species but some 7- or 8-day chronics
New Jersey	DEP	Promulgated	Whole effluent	Toxicity limit	Acute and chronic	One of those allowed by State law
Pennsylvania	DER	Promulgated	Whole effluent and individual	Few toxicity limits currently	Acute	U.S. EPA species
Puerto Rico	U.S. EPA Region II	None	Whole effluent	Monitoring	Acute, flow-through	Mysid, sheepshead, silversides
South Carolina	DNEC	No regulations; in development	Whole effluent testing	Monitoring primarily	Acute, instream stream assessment, and mini-chronic	Ceriodaphnia
Virginia	WCB	In review	Whole effluent	Monitoring and TRE's	Acute and chronic	EPA recommendation plus indigenous

it replaces, making it difficult for parameters to be removed once established. More important, the technology based effluent limitation guidelines applicable to many industries require limitations on many individual pollutants. Hence it is doubtful that the addition of a toxicity limit to a permit will result in the elimination of specific chemical limits. Moreover, while such a strategy would protect the aquatic environment, it would not necessarily protect human health. A number of known or suspected carcinogens have little or no aquatic toxicity. In any case, this issue was put to rest by Section 308(d) of the Water Quality Act of 1987, which required states to promulgate specific numerical water quality criteria for all priority pollutants.

One of the most widely varying aspects of the state programs is the definition, requirements for, and use of toxicity reduction evaluations (TREs), a requirement originally portrayed to be far simpler than it has proven to be. In some cases, such as Pennsylvania and Virginia, a TRE is the only automatic requirement for a discharge found to be toxic. This approach presumes that a TRE is an action-oriented program which will include implementation of solutions, not simply identification of them. In other cases, such as New Jersey, a TRE is required only to document progress by the discharger toward a solution to the problem.

There has been no EPA-mandated definition of the content or scope of a TRE, allowing each state to provide its own definition. This is, in some ways, fortunate in that each state can tailor its program to meet local conditions, and experience can be gained from the successes and difficulties of various approaches. However, the differences between states has caused, and will continue to cause, significant uncertainty and guesswork on the part of industry, particularly multi-plant companies who strive for consistency between plants under the same corporate leadership.

In 1988, the U.S. EPA began the process of developing TRE guidance for state agencies and permittees. This involved selective input from the U.S. EPA staff, state representatives, industry, laboratories and consultants. The resulting document, published in 1989, provides a generic procedure following a flow diagram similar to Fig. 1-1. This approach follows three parallel pathways of evaluation:

1. Housekeeping
2. Raw materials
3. Treatment improvements

These three pathways converge at the toxicity identification stage, where a causative agent approach and/or a treatment approach are identified which will resolve the problem. Although it is a thorough, comprehensive approach, it can result in very high costs, since all routes are explored concurrently rather than in the order of lowest cost.

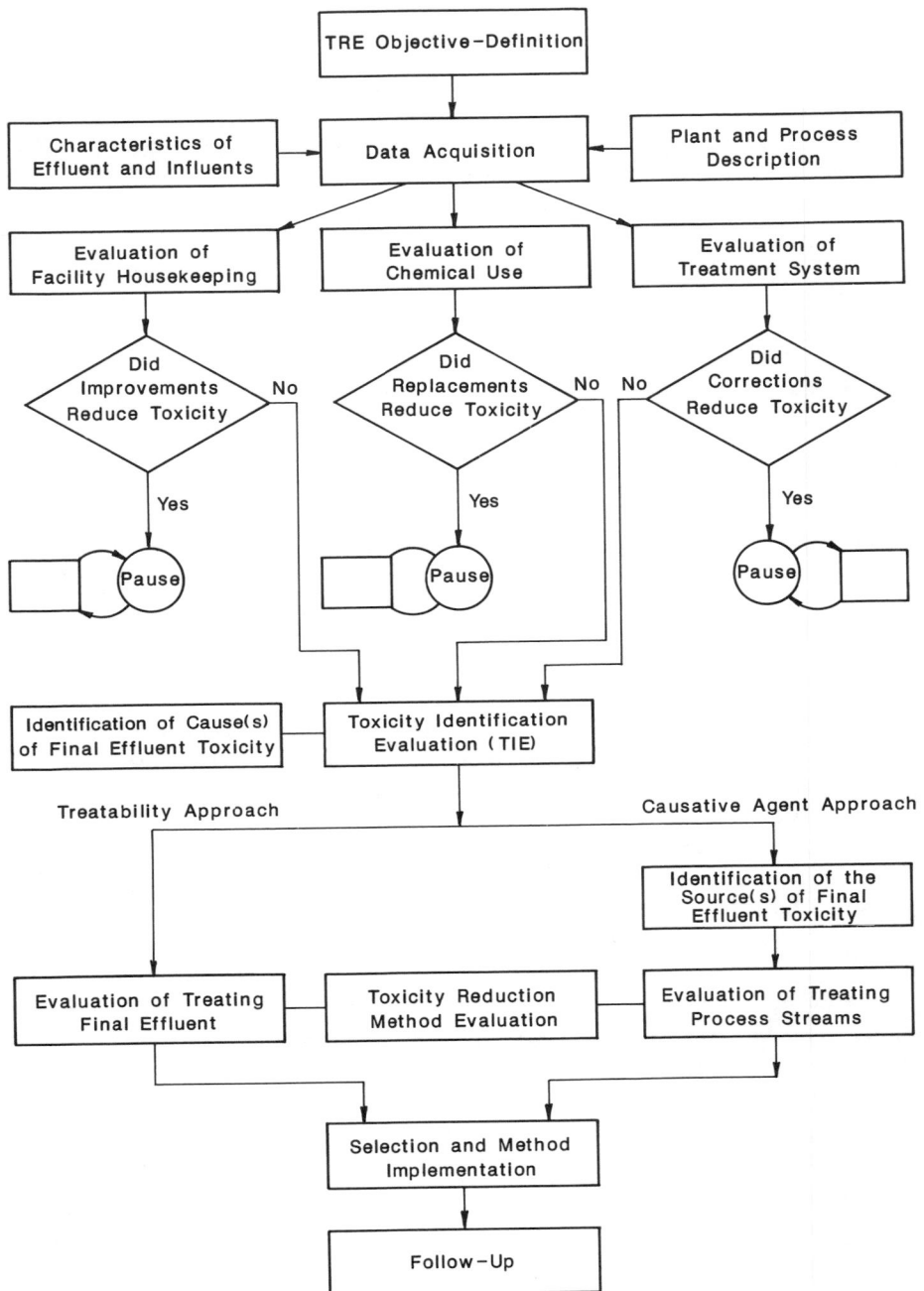

Fig. 1-1. U.S. EPA industrial toxicity reduction evaluation flow diagram (U.S. EPA 1989).

Other guidance is available from some state agencies. In its 1986 *Toxics Management Strategy Guidance Manual*, Pennsylvania provided valuable guidance and information for industrial TREs. Fig. 1-2 presents the flow diagram presented in the Pennsylvania manual. This approach proceeds stepwise, with an engineering analysis followed by evaluation of alternatives beginning with in-plant controls and concluding with treatment required. A shortcoming of the Pennsylvania approach, however, is that it assumes that the toxicity observed in the discharge can be directly connected to one or more raw materials and/or products. In fact, toxicity is often the result of reaction products, and can sometimes arise from completely unexpected sources (such as cleaning agents or even the biological treatment process itself).

TOXICITY REDUCTION EVALUATIONS

Just as each state program has different features, each industrial facility has different features as well. No definitive TRE protocol can be universally applicable. The problems are different, the goals are different, and the plant components are different. Efforts to design a single TRE approach for all dischargers are fundamentally flawed, and efforts to follow such approaches result in unnecessary expense, and often a less than optimal outcome.

The first step in the process for a specific facility must be to tailor a protocol to the problem and needs that exist. However, there are some common features of all plans which should be considered and included to the extent applicable.

In all cases, professional judgment is needed to design the optimal control strategy for a given facility. Frequently, in order to avoid significant waste of time and money, toxicity reduction is best approached in three sequential phases, as presented in Fig. 1-3. The first phase is data validation, in which the effluent toxicity problem should be verified through additional testing. The components and considerations to be included in phase 1 (data validation) are:

- Test additional samples
- Establish regular test program
- Establish QA/QC program
- Test duplicates
- Test standard reference toxicants
- Send samples to multiple labs
- Establish variability of data
- Verify health of organisms
- Check applicability of organisms
- Validate surrogate test or species

Because of the inaccuracy and variability of the bioassay test, as described in Chapter 2, it takes numerous tests to be confident that the extent of toxicity has

```
┌─────────────────────────────────────────────────┐
│                    Phase 1                      │
│                                                 │
│         Process Engineering Analysis            │
│                                                 │
│      (a) – Materials Inventory                  │
│      (b) – Examination of Relevant Literature   │
│      (c) – Comparison of Effluent With Raw      │
│            Material Inventory                    │
└─────────────────────────────────────────────────┘
┌─────────────────────────────────────────────────┐
│                    Phase 2                      │
│                                                 │
│          Pollutant Control Evaluation           │
│                                                 │
│      (a) – Inplant Controls                     │
│      (b) – Resource Recovery                    │
│      (c) – Good Housekeeping Measures           │
│      (d) – Alternative Waste Management         │
│            Practices                            │
│      (e) – End-of-Pipe Treatment                │
└─────────────────────────────────────────────────┘
┌─────────────────────────────────────────────────┐
│                    Phase 3                      │
│                                                 │
│       Toxics Reduction Evaluation Report        │
│                                                 │
│      (a) – Identify Pollutant Sources           │
│      (b) – Review Control Options and           │
│            Effectiveness                        │
│      (c) – Select Control Option                │
│      (d) – Implementation Schedule              │
└─────────────────────────────────────────────────┘
┌─────────────────────────────────────────────────┐
│     Review Of The TRE Report By The State       │
└─────────────────────────────────────────────────┘
┌─────────────────────────────────────────────────┐
│       Modification Of The NPDES Permit          │
│        To Reflect TRE Report Findings           │
└─────────────────────────────────────────────────┘
┌─────────────────────────────────────────────────┐
│      Implementation Of Control Measures         │
└─────────────────────────────────────────────────┘
```

Fig. 1-2. Pennsylvania industrial toxics reduction evaluation flow diagram. (*Source: Commonwealth of Pennsylvania 1986.*)

12

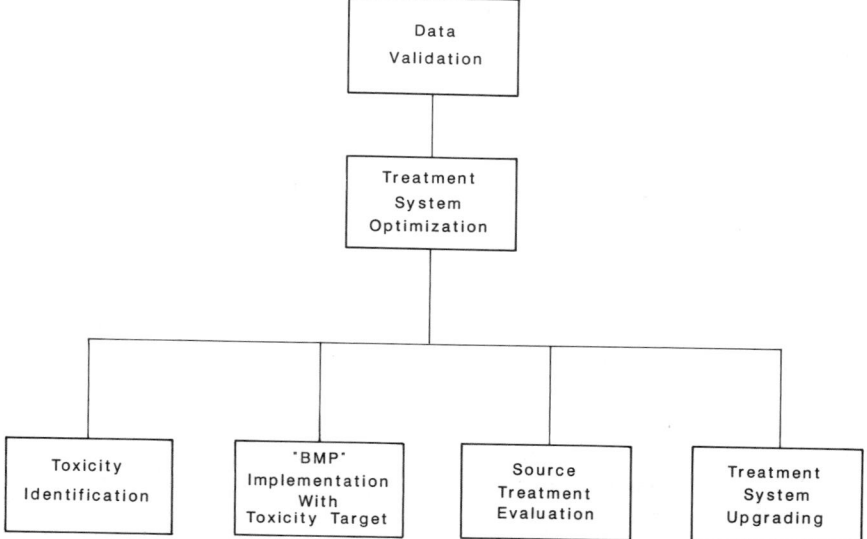

Fig. 1-3. Toxicity reduction logic.

been defined. Reliance on only a few data points can be misleading. The existence or extent of toxicity in a discharge can only be concluded from a statistically valid data base. A period of three to six months or more can be required for the data validation phase.

The second phase is existing treatment system optimization. The performance of most treatment systems, industrial or public, can be improved through operational modifications. This generally applies as much to toxicity as it does to conventional parameters such as BOD or solids. In most cases, extremely toxic effluents cannot be completely resolved through optimization of existing treatment, but in nearly all cases some improvement can be expected. Generally, time and money should not be expended on investigation of upstream causes or improvements (unless the cause is already suspected) until the existing end-of-pipe treatment system is evaluated and utilized to the maximum possible extent. Some of the components and considerations for this phase are:

- Adjust conditions to maximize system performance
- Adjust conditions to attain an active, healthy, diverse biomass
- Adjust solids retention time (SRT) to the optimal range
- Improve metals and solids removal
- Minimize potentially toxic additives such as chlorine, nutrients, polymers, etc.
- Closely monitor system material balance
- Model existing treatment system in bench or pilot scale

The optimization of the existing treatment system can require several months or longer to complete.

Some examples of system optimization steps which can result in improved performance in an existing system are:

- Addition of an increased dosage of powdered carbon
- Change in the type of powdered carbon used in a PACT® system
- Increase in solids retention time in an activated sludge system

As shown in Chapter 10, the performance of an activated sludge system can be expected to generally improve through addition of increasing amounts of powdered carbon. TOC, color, metals and toxicity will all follow exponential decay trends as more powdered carbon is added. Chapter 10 also explains that use of an alternative carbon can dramatically improve performance. In one case, the required dosage of different qualities of carbon varied from 31 mg/l to 230 mg/l to achieve identical toxicity reduction results.

When concerned with a specific organic compound, activated sludge system performance is directly affected by the solids retention time (SRT) or sludge age as described in Chapter 6. For example, a change in the SRT at one plant from 4.5 days to 21 days reduced the nonylphenol ethoxylated surfactant (NPEO) from 1 mg/l to 0.2 mg/l and increased the whole effluent LC_{50} from 20 percent to 60 percent.

The third phase of a toxicity reduction project, normally not to be started until completion of the second phase, consists of four parallel, concurrent efforts. These are the following:

- Toxicity identification
- Best management practices (BMP) implementation
- Source treatment
- Treatment system upgrading

In most cases, these four efforts should be conducted in parallel with close coordination of resources and results. However, it is essential to bear in mind the need for flexibility and judgment. For example, in a particular case, an experienced investigator might realize that plant disposal practices may be contributing significantly to toxicity. In such a case, a rigorous BMP plan (described more thoroughly below) might be sufficient to deal with this problem, eliminating far more expensive stages.

The toxicity identification step consists of the components presented below, in order:

- Inventory and characterize sources
- Compare levels with AQUIRE or QSAR data
- Perform sample fractionation steps as required

- Utilize GC/MS or other analytical technique
- Verify through long-term analysis

Data bases such as AQUIRE and QSAR which can assist in this effort are described in Chapter 2. Some form of sample fractionation procedure such as that depicted in Chapter 3 should be employed. Such procedures require a few weeks per sample to complete. This must be repeated regularly to verify that the results are consistent over the long term because frequently the identified toxicant varies with the plant production schedule.

The best management practices (BMP) implementation step in phase 3 consists of the following components:

- Inventory raw materials, products, and additives
- Determine losses and uses of all materials
- Compare losses with available AQUIRE and QSAR data
- Generate toxicity data on materials as necessary
- Institute housekeeping practices on chemical use, dumps, washouts, cleaning chemicals
- Decrease process losses (improve efficiency)

This step is principally concerned with housekeeping, loss reduction, and raw material substitution. Information on the chemicals used is typically available in the published data bases. Once initiated this effort typically becomes a part of the standard plant procedures and as such is never actually completed, it continues to function indefinitely, and should be upgraded and improved with time and experience.

The source treatment evaluation step in the third phase consists of components such as those identified below:

- Inventory and characterize toxic source streams
- Determine applicable processes based on character
- Determine process performance based on toxicity and other parameters

This effort, which is described in detail in Chapter 4, consists of source inventory development, identification of alternative processes, and performance testing. Some of the processes which are applicable to source treatment are chemical oxidation, wet air oxidation, resin adsorption, stripping, and membrane processes. The length of time required for this step is dependent on the number and type of processes considered.

The components of the last step in the third phase, treatment system upgrading, will depend on the problem. Among the key steps to be considered are:

- Review toxicity identification data for direction
- Maximize performance of the existing system

- Add metals removal capabilities
- Modify the system to achieve a more efficient biological process
- Add a carbon adsorption step—intermittent or continual
- Evaluate other end-of-pipe treatment techniques

These would be evaluated in parallel with the other steps in Phase 3 so that the most cost effective combination is selected. The duration of this effort is dependent on the alternative considered.

Chapter 10 describes the improvement experienced by converting an existing activated sludge system to PACT®. The median toxicity (LC_{50}) in one example improved from 9 percent to nearly 80 percent through the addition of 250 mg/l of powdered carbon.

Another beneficial improvement can be the reduction of ammonia toxicity. Free ammonia (NH_3) is toxic at a level (LC_{50}) of 0.05 mg/l to 0.2 mg/l. However, the amount of NH_3/NH_4^+ in the free ammonia form is largely dependent on the pH of the test sample. Fig. 1-4 presents the relationship of effluent toxicity in percent by volume to: free ammonia; total ammonia at pH 6.5; total ammonia at pH 7.0; and, total ammonia at pH 7.5. As can be seen, at an effluent pH of 7.5, 24 mg/l of total ammonia results in toxicity (LC_{50}), but at a pH of 6.5, the total ammonia level resulting in an LC_{50} increases to 300 mg/l. Therefore, depending on the effluent pH, ammonia removal through stripping or nitrification may or may not improve effluent toxicity.

Complete toxicity reduction evaluations can be very lengthy undertakings. Even the simplest investigation typically requires three to six months to com-

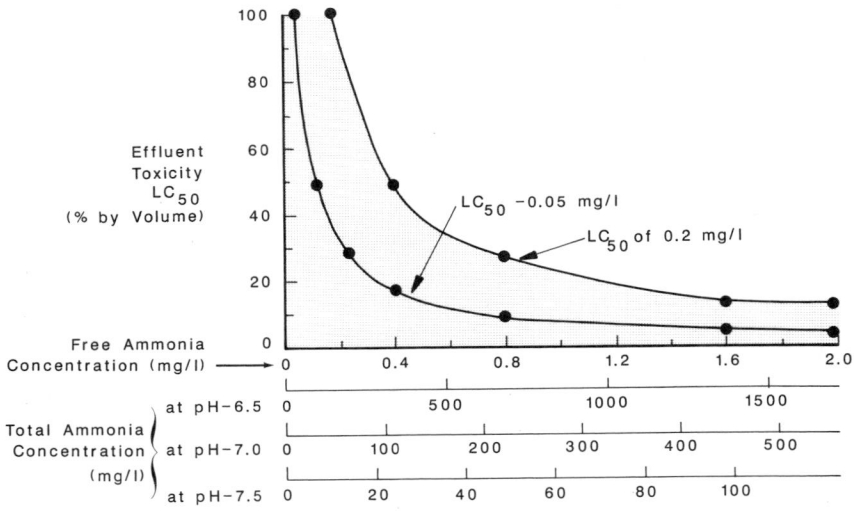

Fig. 1-4. Toxicity of ammonia versus pH.

plete. Complex TREs can easily extend to eighteen to twenty-four months, with a median duration in the area of twelve months.

REFERENCES

Commonwealth of Pennsylvania. 1986. Revised 1989. Toxics Management Strategy, Appendix C. Pennsylvania Department of Environmental Resources, Bureau of Water Quality Management. Sections C-1 through C-30.

U.S. EPA. April 1989. Generalized Methodology for Conducting Industrial Toxicity Reduction Evaluations. EPA/600/2-88/070.

2

Aquatic Toxicity Testing

Billy G. Isom

INTRODUCTION

In recent years the terms *toxicity testing*, *bioassay*, and *biomonitoring* have often been used as if they were synonyomous when applied to wastewater effluents. In fact, the terms are somewhat different. The U.S. EPA (1985) defines toxicity testing as "the means to determine the toxicity of a chemical or effluent using living organisms. A toxicity test measures the degree of response of an exposed test organism to a specific chemical or effluent." A bioassay is technically a "determination of the relative strength of a substance (such as a drug) by comparing its effect on a test organism with that of a standard preparation" (Anonymous 1986a). Biomonitoring is a much broader concept that includes all of the biological techniques available for determining the biological integrity of waters, including toxicity testing. Weber (1980) provided an excellent review of federal and state biomonitoring programs including the statutory and scientific bases.

The standard technique for toxicity determination of a wastewater is the toxicity test. There are many forms and methodologies for a toxicity test, the two most commonly used today are the *subchronic* and the *acute*. The subchronic test is utilized to determine and monitor the concentration at which a discharge has an effect on survival, growth, and reproduction of test organisms. The term *subchronic* is used to distinguish a four to eight day test from a full chronic test which lasts one or more life cycles. Depending on the organism, such full chronic tests range from a month to more than a year. Tests of such long duration are impractical for regulatory programs. The acute toxicity test is a short-term survivial determination which involves exposure of a selected test organism, such as fish or water fleas, to a known dilution or concentration of sample for a specific time period, typically 48 to 96 hours, but occasionally as short as 24 hours.

BACKGROUND

This paper deals exclusively with aquatic toxicity testing. These tests are used to determine the toxicity of both individual chemicals and complex effluents to aquatic organisms. Other, sometimes similar techniques are used to determine toxicity to humans. Rand and Petrocelli (1985) provide a comprehensive review of aquatic toxicology, which is defined as "the qualitative and quantitative study of the adverse or toxic effects of chemicals and other anthropogenic materials or xenobiotics on aquatic organisms."

The theoretical basis for aquatic toxicology evolved from studies in pharmacology and the toxicity of substances to humans, domestic animals, and terrestrial biota in general. The response of animals to known doses of substances is the basis of numerous scientific disciplines including pharmacology, biochemistry, physiology, pathology, and toxicology. Dose-response relationships for many chemicals are widely known by virtue of measured dose applications to the skin, eyes, feeding, injection, or by ingestion.

In aquatic toxicology the critical measurement is concentration-response, rather than dose-response. Concentration responses are observed when testing effluent toxicity and when chemicals are mixed with water or contained in effluents; the effluent concentration or the chemical concentration is known or may be measured in the water, but the dose received by the organism is an unknown quantity. "Therefore, the response observed is more correctly correlated with the exposure concentration, hence the term concentration-response relationship" (Rand and Petrocelli 1985). This is less of a distraction than it initially appears to be, since in both laboratory testing and in environmental exposure, the most likely mode of toxic uptake is by the organisms' gills or other exposed tissues.

The relationship between exposure to effluents or chemical concentrations (dose) and the magnitude of observed effects (response) is best described by concentration-response curves. Like dose-response curves, concentration-response curves are basically of two types: one where no response is observed until some minimum concentration is reached and the other where no threshold is observed. For chemicals of the latter type, there are no concentrations free of risk. An example might be certain carcinogens. It is obvious that acute and chronic aquatic toxicity testing only addresses effluents where there are observed threshold effects such as shown in Fig. 2-1.

In addition to protocols available from the open literature, there are five primary sources of standardized protocols for conducting toxicity tests:

- The American Society for Testing Materials publishes peer reviewed toxicity testing protocols in Volume 11.04 (ASTM 1989) which are developed by Subcommittee E47.01 on Aquatic Toxicology.

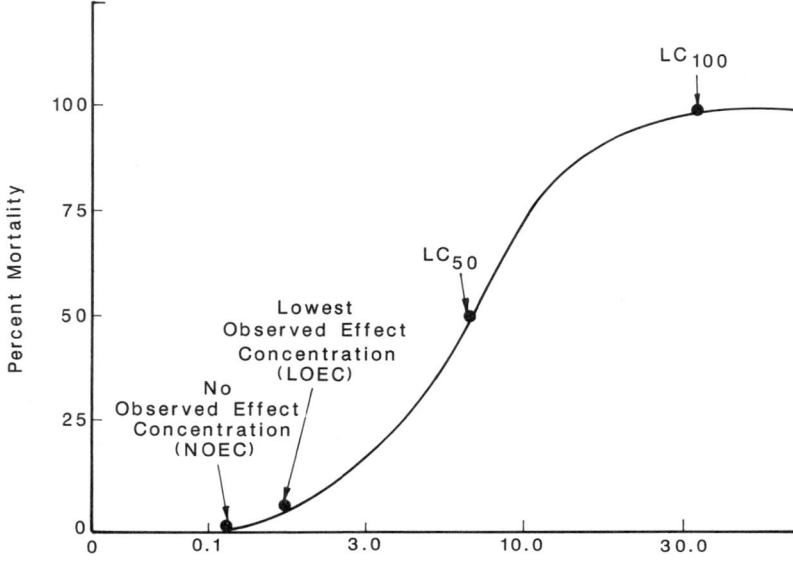

Fig. 2-1. Threshold concentration-response curve.

- Standard Methods for Examination of Water and Wastewater (APHA, AWWA, WPCF 1989) has extensive toxicity testing methods for aquatic organisms in Part 8000.
- U.S. EPA formalized current toxicity testing methods for whole effluent testing of freshwater and marine organisms in 1985, 1988, and 1989 (Peltier and Weber 1985; Horning and Weber 1985; Weber et al. 1988; Weber et al. 1989).
- U.S. EPA regulations for testing aquatic toxicity of individual chemical substances under the Toxic Substances Control Act.
- Many states have their own modified toxicity testing protocols which are not generally standardized or peer reviewed.

ELEMENTS OF TOXICITY TESTING

The four major elements of a toxicity test for a whole effluent include:

- Obtaining a representative effluent or water sample(s)
- Procuring or culturing a healthy test organism
- Applying the proper experimental design
- Calculating the endpoint

Sample Procurement

Regulatory permits usually identify the methods of choice for collecting representative samples. Appendix B of U.S. EPA (1985) discusses sampling. The same document indicates that the overall effluent variability uncertainty factor may be as high as 100 where chronic toxicity is a concern but no chronic data are available. Therefore, taking of a representative effluent or water sample is very important. Some states require 6 hour composite sample or grab samples to be taken over 24 hours with independent tests conducted on each fraction. Other states require that samples be composited over 24 hours. A third group of states requires flow-weighted proportional samples be taken over 24 hours. The difference between these methods is that a spill or perturbation would be more likely to be detected in the first example, but with toxicity attenuated in the other examples (Weber et al. 1989). If the water treatment facility has equalization basins that hold effluents 24 hours or more before or after treatment, then 24 hours composites would be representative of the effluent. The greatest risk of not obtaining a representative sample is a single grab. In all cases, the toxicity test results can be no better than the samples provided.

Healthy Organisms

The quality of the test organisms is another factor that greatly affects test results, especially in chronic testing, since in these tests both reproduction and growth can be concerns. Transported organisms, in all cases, undergo some stress that should not be a concern with organisms cultured on site. Even though controls are used, the chances of an aborted test due to excess control mortalities is much greater with use of transported organisms, especially in chronic or embryo-larval tests. When conducting acute tests, test organisms can usually be held for a period sufficiently long to assure quality.

Another factor in using transported test organisms is that quality assurance/quality control (QA/QC) is much more expensive, since standard reference toxicants must be used to test the sensitivity of each "batch" of new test organisms. In contrast, laboratory cultured organisms can be tested for sensitivity monthly or biweekly. As a result, the cost of extra organisms and the cost of conducting the extra QA/QC tests can be saved. Test organisms of known genetic history and sensitivity that have been cultured on site can provide better toxicity test results (Peltier and Weber 1985; Weber et al. 1989). Since some TRE approaches require many different toxicity tests, the cost differential can become significant, even for large industrial plants.

Experimental Design

As mentioned elsewhere, another factor that can affect toxicity test results is application of the test or test design. Ideally, the test should be applied completely at random to meet the assumptions of many parametric statistical tests. No matter which statistical procedure is used, bias will be reduced by randomization of rows, containers, position in the row, application of organisms, and addition of the sample. More details on randomization or experimental design can be found in any of the methods referenced above, especially Weber et al. (1989).

Endpoint Calculation

Another major element of toxicity testing is the calculation of test endpoints. Selection and proper use of appropriate methods for calculating endpoints for acute and chronic testing is discussed below, in the references cited above, and in Stephan (1977).

The endpoint or effect for determining acute toxicity to aquatic organisms is usually death. The mean lethal concentration, that is, the concentration estimated to kill 50 percent of the test organisms, expressed as an LC_{50}, is usually the value calculated in animal tests (Stephan 1977). Because of the way it is expressed, the lower the LC_{50}, the more toxic the effluent. The effects concentration, or EC_{50}, is usually calculated in tests with algae and plants, since death is not easily detected. Data are applied in the same way as for LC_{50}.

Acute toxicity limitations or numeric criteria have been established in some states. NPDES permits are expressed in numerous ways based on an individual state's perceived needs. Some states require that an effluent have acute LC_{50}s for fish and invertebrates not to exceed 50 percent, and others not to exceed 80 percent, for example. That is, at least half the organisms must survive a mixture of 50 percent effluent/50 percent dilution water or 80 percent effluent/20 percent dilution water, respectively. Still, other states simply ignore the theoretical basis for estimating toxicity and require that an effluent not have more than 10 percent mortality to fish or invertebrates in 100 percent effluent during a prescribed test period, usually 24 to 96 hours. Failure of such tests usually requires additional definitive testing.

Such absolute toxicity limits lack a rational basis, and are inconsistent with the purposes of state water quality standards, which provide the legal authority for toxicity limits. For these reasons, U.S. EPA and most states are in agreement that the critical toxicity limit for protecting the aquatic environment is assurance that the instream waste concentration (IWC) is equal to or less than the no-observed-effect concentration, or NOEC, which is determined by chronic testing. The NOEC is ''the highest measured continuous concentration of an effluent or toxicant that causes no observed effect on a test organism'' (U.S. EPA 1985).

The *Toxicology Handbook* (anonymous 1986b) defines chronic as ''occurring over a long period of time, either continuously or intermittently; used to describe ongoing exposures and effects that develop after a long exposure.''

Weber et al. (1988) and Weber et al. (1989) make clear that the current NPDES chronic tests are short-term methods for estimating chronic toxicity. The U.S. EPA (1980) issued a policy statement that early life stage (ELS) test data could be used for establishing water quality criteria where full life cycle test data were not available. Apparently, the technical basis for this decision was the work of McKim (1977) and Macek and Sleight (1977). McKim (1977) recommended use of ELS test data for establishment of water quality criteria. Macek and Sleight (1977) determined that for fish, critical life stage tests provided estimates of chronically safe toxicant concentrations similar to those derived from full life cycle toxicity tests. They noted that "In all cases, estimates of specific application factors derived from critical life stage studies provided information equal or greater in predictive value than information derived from the use of an arbitrary general application factor."

Chronic tests produce a number of endpoints. The most significant endpoint is the NOEC (no-observed-effect concentration) noted above. For some purposes, U.S. EPA considers the "safe" concentration to be the NOEC or NOEL (no-observed-effect level). For other purposes, it uses the "chronic value," which is the geometric mean of the NOEC and LOEC, which is the lowest concentration at which an effect is observed. Use of the NOEC instead of the chronic value in permitting is extremely conservative.

Application of the NOEC in permitting is extremely conservative since U.S. EPA recommends the use of 1Q5 or 1Q10 design flow (1Q5 is the lowest average daily flow for one day in any given five-year period and 1Q10 is the lowest average daily flow for one day in any given ten-year period). For the criteria maximum concentration (CMC), the highest instream concentration of a toxicant to which organisms can be exposed for a brief period without mortality, and 1Q10 or 7Q10 for a criteria continuous concentration (CCC), the highest instream concentration of a toxicant or an effluent to which organisms can be exposed indefinitely. U.S. EPA recommends that the 1Q10 be used as the hydrologically based "design flow for calculating dilution for the CMC and that 7Q10 be used as the design flow for calculating dilution for the CCC." In addition, concentrations in excess of the CMC and the CCC are allowed only once in three years, just as for chemical-specific criteria (U.S. EPA 1987).

Tebo (Peltier and Weber 1985) described the procedure for estimating acute and chronic toxicity of effluents in receiving waters based on LC_{50} data alone. Two significant features of the estimate are an application factor for acute data and the use of low flows such as the 7Q10. The calculated instream value must protect aquatic life in rivers and streams from acute lethality. Chronic NOECs are calculated from LC_{50} data by applying a factor of 20 ($LC_{50}/20$) for non-persistent and 100 ($LC_{50}/100$) for potentially persistent wastes. Species sensitivity is factored into both acute and chronic estimates by adding a safety factor of one order of magnitude. While the use of such application factors for regulatory purposes is overly conservative, it has the advantage of avoiding the additional costs of chronic testing.

Where mixing zones are allowed by the states (as is the case for most states), values can be calculated to protect lakes, impoundments, and estuaries. If no mixing zone is allowed, then obviously the IWC effectively is end-of-the-pipe. U.S. EPA's 1983 policy on mixing zones states that "any mixing zone should be free from materials which cause acute toxicity to aquatic life" (U.S. EPA 1985).

SELECTION OF ACUTE OR CHRONIC TESTS

Different organisms and life stages (such as adults, juveniles and eggs) are selected by the permitting agency based on whether the effluent and the receiving stream are fresh or salt water, and the relative sensitivity of different species to the effluent (Table 2-1). Each of the organisms may exhibit a different toxicity threshold to the same or similar compounds, as shown in Table 2-2. In addition, there is a relatively large variability in test results for any one test species as a result of biological factors.

Toxicity test results typically exhibit considerable variability when tests are run in duplicate or triplicate because of such factors as the species of the organism, test conditions, and number of organisms used in the test. The variability of the test is also related to the lab conducting the test, with rather large vari-

Table 2-1. Standard Bioassay Tests and Organisms.

Type of Bioassay Test	Freshwater Organism	Marine Organism
Subchronic	Algae (*Selenastrum capricorutum*),[a] *Ceriodaphnia dubia*,[d] fathead minnow (*Pimephales promelas*)[b,c]	Sheepshead minnow,[b,c] inland silverside,[b] *Mysidopsis bahia*,[e] sea urchin,[f] *Champia parvula*[g]
Acute	Bluegill (*Lepomis macrochirus*),[h] *Ceriodaphnia dubia*,[i] *Daphnia magna*,[i] *Daphnia pulex*,[i] fathead minnow (*Pimephales promelas*),[h] rainbow trout (*Salmo gairdneri*)[h]	Atlantic silverside (*Menidia menidia*), grass shrimp (*Palaemonetes pugio*),[i] inland silverside (*Menidia beryllina*),[i] Mysids (*Mysidopsis bahia*), sheepshead minnow (*Cyprinodon variegatus*)[i]

[a] Four-day static algal growth test.
[b] Seven-day, static renewal, larval survival and growth test.
[c] Eight- or nine-day, static, embryo-larval survival and teratogenicity test.
[d] Seven-day, three-brood, static renewal, survival and reproduction test.
[e] Seven-day, static renewal, survival, growth, and fecundity test.
[f] Eighty-minute gamete fertilization test.
[g] Seven- to nine-day alga sexual reproduction test.
[h] Ninety-six hour static, static renewal, or flow-through test.
[i] Forty-eight hour static test.

Table 2-2.   Acute Toxicity of Selected Compounds (96-hr LC$_{50}$)[a].

	Units	Fathead Minnow	*Daphnia*	Rainbow Trout
Organics[b]				
Benzene	mg/l	42.70	35.20	38.70
Carbon tetrachloride	mg/l	17.30	15.20	14.50
Chlorobenzene	mg/l	13.20	11.60	11.10
1,1-Dichloroethane	mg/l	120.00	96.40	113.00
1,1,2-Trichloroethane	mg/l	88.70	72.60	81.10
2-Chlorophenol	mg/l	21.60	18.60	18.40
1,4-Dichlorobenzene	mg/l	3.72	3.46	2.89
1,2-Dichlorobenzene	mg/l	87.40	71.10	80.50
2,4-Dinitrophenol	mg/l	5.81	5.35	4.56
4,6-Dinitro-*o*-cresol	mg/l	2.79	2.65	2.10
Pentachlorophenol	μg/l	170.00	—	—
Ethylbenzene	mg/l	11.00	9.97	9.47
Methylene chloride	mg/l	326.00	249.00	325.00
Toluene	mg/l	31.00	26.00	27.40
Trichloroethylene	mg/l	55.40	46.20	49.50
Phenol	mg/l	39.60	33.00	35.40
1,4-Dinitrobenzene	mg/l	1.68	1.61	1.24
2,4,6-Trichlorophenol	mg/l	5.91	5.45	4.62
2,4-Dichlorophenol	mg/l	9.27	8.35	7.49
Naphthalene	mg/l	5.57	5.07	4.44
Nitrobenzene	mg/l	118.00	95.40	110.00
1,1,2,2-Tetrachloroethane	mg/l	31.10	26.70	26.70
Metals[b]				
Arsenic	μg/l	15,600	5.278	13,340
Chromium, hexavalent	μg/l	43,600	6,400	69,000
Cadmium	μg/l	38.2	0.29	0.04
Copper	μg/l	3.29	0.43	1.02
Lead	μg/l	158.00	4.02	158.00
Mercury	μg/l	—	5.00	249.00
Nickel	μg/l	440.00	54.00	—
Selenium	μg/l	1,460.00	710.00	10,200
Silver	μg/l	0.012	0.00192	0.023
Zinc	μg/l	169.00	8.89	26.20
Inorganics				
Unionized ammonia (Total Ammonia)[c]				
pH 7.0	mg/l		0.093 (23)	0.093 (23)
pH 8.5	mg/l		0.260 (6.8)	0.260 (6.8)

[a]Estimates of 96 hour LC$_{50}$ in mg/l for common aquatic test organisms based on the primary mode of action and structure–activity relationship.
[b]From EPA/Montana State QSAR (Quantitative Structure/Activity Relationship) system.
[c]Highly variable depending on pH and Temperature (*Federal Register* Volume 50, No. 185, Monday, July 29, 1985, pp. 30784 to 30786). Data represent criteria to protect aquatic life at pH 7.0 and 20°C and pH 8.5 and 20°C, one hour average, mg/l.

ability observed when multiple labs conduct replicate tests. Results of replicate bioassay tests for several test organisms at different laboratories using standard reference toxicants are presented in Table 2-3 (Broderius 1983; Schimel 1981). From the results of the coefficient of variation analysis, it is apparent that test variability is dependent upon the toxicant present and species used, thus indicating greater sensitivity of certain organisms to specific compounds.

Figure 2-2 presents bioassay confidence limits for a series of *Mysidopsis bahia* bioassay tests conducted on a batch chemical production facility wastewater plant effluent. It can be seen that the precision of the test results decreases significantly as the LC_{50} increases. This is due to the nature of the test, which involves statistical evaluation of test data for determination of LC_{50} concentration based on organism mortality. At high LC_{50} values, there is a low mortality rate which, if the test is conducted with only 5–10 organisms, can result in a wide range of actual LC_{50} concentrations. If, however, there is a high mortality,

Table 2-3. Precision of acute Toxicity Tests Using Reference Toxicants.

| | Reference Toxicants[a] | | | |
| | Silver Nitrate | | Silver Nitrate | |
Test Organism	N	CV	N	CV
1. Fathead minnow (*Pimephales promelas*),[b] freshwater fin fish				
96-hour static test	10	52	12	38
96-hour flow-through test	10	40	12	46
2. Rainbow trout (*Salmo gairdneri*),[b] freshwater fin fish				
96-hour static test	10	64	12	50
96-hour flow-through test	10	32	12	43
3. *Daphnia magna*,[b] freshwater zooplankton				
48-hour static test	8	71	12	51
4. *Acartia tonsa*,[c] Marine zooplankton				
96-hour static test	5	42	6	82
5. *Mysidopsis bahia*,[c] Marine crustacean				
96-hour static test	6	27	5	62
96-hour flow-through test	6	22	6	58
6. Sheepshead minnow (*Cyprinodon variegatus*,[c] Marine fin fish				
96-hour static test	4	35	6	37
96-hour flow-through test	5	50	6	46

[a]N is the number of data points used in calculation. CV is coefficient of variation = standard deviation × 100/mean.
[b]Adapted from Broderius (1983).
[c]Adapted from Schimel (1981).

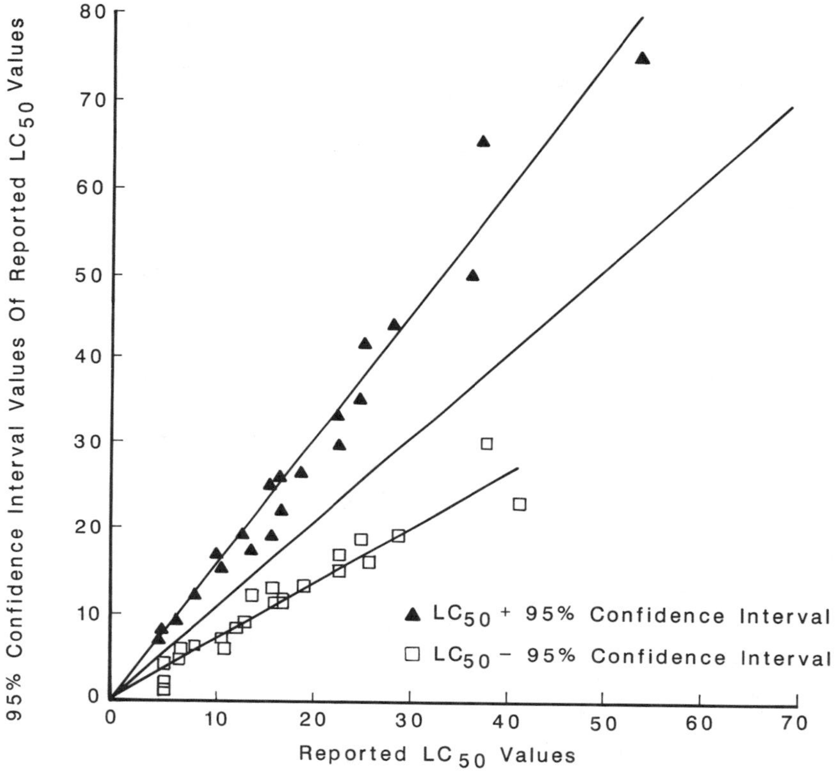

Fig. 2-2. Mysid shrimp bioassay 95 percent confidence limits.

the results are more accurate since a greater percentage of the organisms were affected by the sample, thus providing a statistically more accurate representation of actual toxicity. It is obvious there are many factors to consider in selecting and applying toxicity tests.

U.S. EPA (1987) notes that "There is generally no reason to mix two types of monitoring for the same outfall." Choice of test type, acute or chronic, is determined by the more limiting long-term average toxicity results. "Permit limits are then derived directly from whichever performance level is more restrictive" (U.S. EPA 1987). Standard toxicity tests and test organisms are shown in Table 2-1, as noted above.

The basic rule is to use the test that will provide a quantifiable response. That is, if there is true acute toxicity then acute tests should be used to the point where toxicity reduction efforts result in no acute toxicity. If there is no acute toxicity then chronic tests should be conducted. If there is chronic toxicity, and calculations indicate that there would or could be receiving water toxicity, then toxicity reduction evaluations should be used to achieve toxicity removal as discussed in other chapters. This would be followed by additional chronic testing to assure that there is no instream toxicity at some conservative stream flow value such as 7Q10 ("lowest average daily flow during any consecutive seven days in any ten-year period," Tebo) (Peltier and Weber 1985). Even in those cases where there is a basis for concern with both acute and chronic effects, a suite of acute and chronic tests would allow establishment of an acute/chronic ratio which should obviate the need for continuing both test types.

ACUTE TESTS

There are three types of acute toxicity tests that are readily applicable to wastewater effluents. They are the range-finding, screening, and definitivie acute tests. Peltier and Weber (1985) list recommended species for acute testing which includes both freshwater and marine organisms (also see Table 2-1).

Range-Finding Test

A range-finding test may be used when the characteristics of an effluent are essentially unknown. This test is rarely used for regulatory purposes, although a few NPDES permits specify the use of range-finding tests.

A range-finding test is a simple test where five or more organisms are exposed to three to five widely distributed effluent concentrations, such as 1 percent, 10 percent, 50 percent, 100 percent effluent, and a single control. Test duration is usually 8 to 24 hours. If receiving water is used for dilution, then both receiving water and culture or laboratory water controls should be included in this test. Some receiving waters have residual toxicity which could result in unacceptable control mortalities as well as affect effluent toxicity results. Range-finding tests are used to obtain better results in subsequent definitive tests since the dilutions tested can be targeted around the range-finding results rather than using 0 to 100 percent effluent.

Screening Test

A screening test is usually conducted with a single 100 percent effluent concentration. Duration of the test is usually 24 hours. Time to death should be noted if acutely toxic in less than 24 hours. No control is required. If toxicity is observed, then a definitive test may be required.

Definitive Test

Acute definitive tests are distinguished by their duration, usually 48 hours for invertebrates and 96 hours for fish, and the use of multiple dilutions. At least five dilutions are applied in a geometric series (X, $2X$, $4X$, $8X$, $16X$, . . .), for example, 6.25 percent, 12.5 percent, 25 percent, 50 percent, 100 percent effluent, and appropriate controls. If range-finding test data are available, those data may be used to better define the dilution series around the target value. Test concentrations and controls are conducted in duplicate or triplicate with at least 20 to 30 organisms per concentration. Although replication is not absolutely essential to calculate an LC_{50}, it is desirable to do so (Peltier and Weber 1985) and some states so require. In addition, it is desirable to apply the experiment randomly, that is, rows and positions are assigned randomly and organisms and test solutions are applied to containers randomly. These procedures reduce the bias that may occur without randomization. In addition, parametric statistical evaluation procedures like the probit assume randomicity and independence of observation (Finney 1971).

The end point for determining acute toxicity to aquatic organisms is usually death. The concentration that is estimated to kill 50 percent of the test organisms, or LC_{50}, is usually the value calculated as noted elsewhere.

CHRONIC TESTS

U.S. EPA (Horning and Weber 1985; Weber et al. 1989) currently has four freshwater chronic tests for wastewater effluents. They include the fathead minnow (*Pimephales promelas*) larval survival and growth test, the *Ceriodaphnia dubia* survival and reproduction (three brood) test, the four-day algal (*Selenastrum capricornutum*) growth test, and the fathead minnow embryo-larval survival and teratogenicity test.

U.S. EPA toxicity testing methods for marine and estuarine organisms (Weber et al. 1988) include the sheepshead minnow (*Cyprinodon variegatus*) larval survival and growth test; the inland silverside (*Menidia beryllina*) survival and growth test; the estaurine mysid (*Mysidopsis bahia*) survival, growth, and reproduction test; the sea urchin (*Arbacia punctulata*) sperm cell test; the marine macroalga (*Champia parvula*) sexual reproduction test; and the sheepshead minnow embryo-larval survival and teratogenicity test. The authority for promulgating these test procedures is embodied in Section 304(h) of the Clean Water Act, according to Weber et al. (1988).

CALCULATING ENDPOINTS

Determination of Acute Endpoints

The alternative means of determining the acute toxicity test endpoints are the following:

- Straight-line graphical interpolation
- Moving average interpolation
- Probit method

- Binomial method
- Litchfield–Wilcoxin abbreviated method
- Trimmed Spearman–Karber method

Each method has its particular advantages and disadvantages and is described in greater detail in the references cited previously.

Acute test endpoints ($LC_{50}s$ or $EC_{50}s$) can be quickly and easily calculated by the graphical (log concentration versus percent mortality) method. Walsh et al. (1987) used this method, the moving average, probit analysis, and the binomial method to compare results from 187 algal toxicity tests. EC_{50} values were essentially identical for all four methods. They concluded that since straight-line interpolation is a simple and rapid method, it may be used instead of the more complex computer-based methods. However, the graphical method does not provide the 95 percent confidence interval for the LC_{50} estimate which is required by most permits.

The moving-average angle method is recommended for use in calculating $LC_{50}s$ and associated 95 percent confidence intervals for two reasons: the method permits calculation of the 95 percent confidence limit where no partial mortalities occur, and it is easily calculated with an engineering pocket calculator with log, antilog, inverse sine, and inverse tangent functions.

The probit method requires two partial mortalities and is perhaps the most often used method for LC_{50} calculations. The probit method assumes randomness and independence of observations in obtaining data. The method was essentially developed for analyzing data from concentration-response studies.

Other methods that are available for calculating $LC_{50}s$ and their associated confidence intervals are the binomial, the Litchfield–Wilcoxin, and the trimmed Spearman–Karber. The binomial method does not require partial kills (Peltier and Weber 1985). The Litchfield–Wilcoxin method requires that partial mortalities occur in the test. The trimmed Spearman-Karber method requires that toxicant concentrations cover the range of 0 and 100 percent kill. This is a disadvantage of the method unless some assumed data are used.

Computer programs for calculating aquatic toxicity endpoints can be obtained from Computer Sciences Corporation, Room B-12, 26 West Martin Luther King Drive, Cincinnati, Ohio 45268. Table 2-4 is an example calculation of 96 hour $LC_{50}s$ utilizing the binominal, moving average, and probit methods. All three methods produced similar results with this set of data.

Determination of Chronic Endpoints

The most often used methods for calculating chronic test endpoints are:

- Fisher's exact test
- Probit analysis

Table 2-4. Acute Endpoint Calculation Using Various Procedures.

Effluent Concentration	Number of Organisms Exposed	Number Dead
100	20	20
50	20	16
25	20	7
12.5	20	1
6.25	20	0

Binomial method	LC_{50} 31.26 (12.5–50 CL)
Moving average method	LC_{50} 30.05 (24.4–37.7 CL)
Probit method	LC_{50} 30.65 (24.98–37.8 CL)
	LC_1 9.2 (4.66–13.04 CL)

- Dunnett's procedure
- Bonferroni's T-test
- Steel's many-one rank test

Fisher's exact test is used to determine the NOEC and LOEC endpoint estimates for the *Ceriodaphnia* survival and reproduction test. It provides a conservative test assuming the independence of responses from a binomial population (Weber et al. 1989). Probit analysis is used to calculate the LC_1, LC_5, LC_{10}, and LC_{50}, in the same test. The total number dead at a given concentration is the response (Weber et al. 1989).

As noted previously, the NOEC is the ''safe'' concentration derived from the chronic tests. The lowest-observed-effect concentration or LOEC is also often determined. The chronic value (ChV) is the geometric mean of these two values. Probit analysis can be used to calculate, LC_{50}, LC_{10}, LC_5, and LC_1 values and associated 95 percent confidence intervals for chronic data sets, as noted above.

Dunnett's procedure and Bonferroni's T-test make the assumption that observations are normally distributed and that the variance of the observations for all concentrations and the control are homogeneous (Weber et al. 1988). These assumptions should be tested if one of these procedures is considered. If the test for homogeneity fails, then a nonparametric test such as Steel's many-one rank, the Wilcoxon rank sum test, or some other nonparametric test should be used (Weber et al. 1989).

Weber et al. (1989) provided excellent guidance on the selection of statistical tests to be used with each type of chronic test. Fig. 2-3 is an example of a flow chart for statistical analysis of fathead minnow larval growth data.

STANDARD REFERENCE TOXICANTS

Horning and Weber (1985) and Weber et al. (1989) discuss the use of standard reference toxicants for QA/QC. Standard reference toxicants should be used

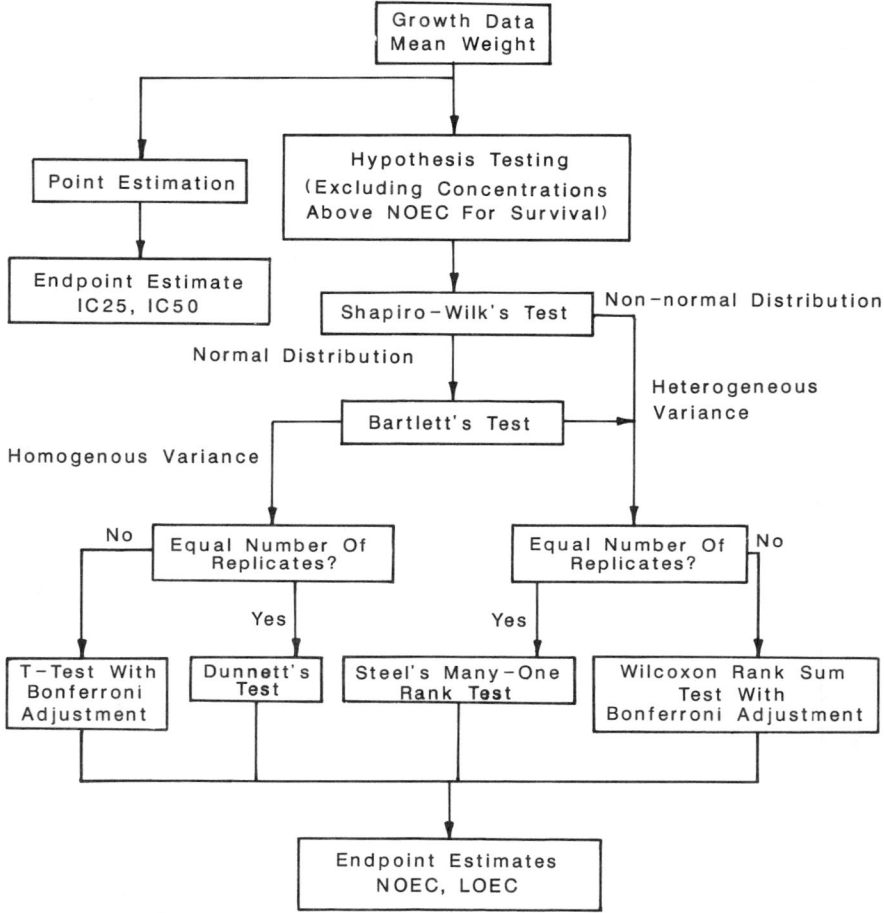

Fig. 2-3. Flow chart for statistical analysis of fathead minnow larval growth data. (*Source:* Weber et al. 1989.)

periodically as required to assure that organism sensitivity is within the expected toxicity range (see Fig. 2-4). If the toxicity value for a standard reference toxicant does not fall in the expected range, the sensitivity of the organism, and thus the test results, are suspect. Test organisms cultured on site should be tested for sensitivity to the standard reference toxicant at least once each month. A major disadvantage of using purchased organisms is that each batch or purchased organisms should be tested upon receipt.

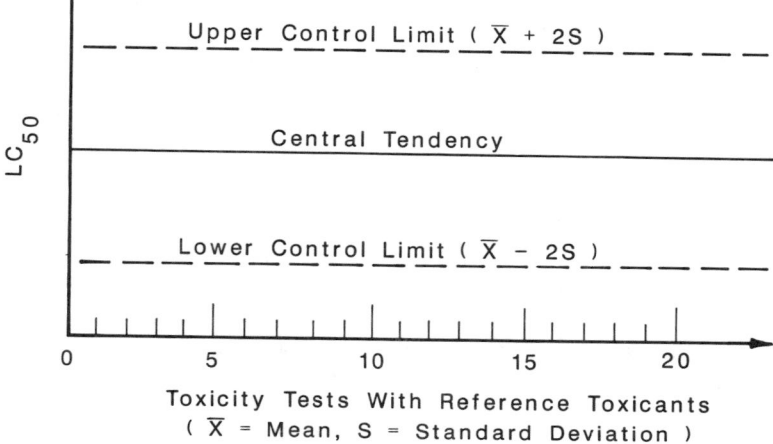

Fig. 2-4. Standard reference toxicant control chart, (*Adapted from:* Horning and Weber 1985.)

WATER QUALITY PARAMETERS

It is very important to collect water quality data at the beginning, during, and at termination of toxicity tests. Peltier and Weber (1985) recommend that if the pH of effluent falls outside the range of 6.0 to 9.0, pH adjusted controls should be included in the test. The U.S. EPA criterion for pH is 6.5 to 9.0 for protection and propagation of freshwater aquatic life (U.S. EPA 1986).

Weber et al. (1989) recommends in Method 1000.0 that dissolved oxygen (DO) concentrations should not fall below 40 percent of air saturation. DO saturation at the test temperature of 25°C and 0 chlorinity is 8.26 mg/l. Forty percent of this value is 3.3 mg/l. The U.S. EPA (1986) criterion for early life stages of aquatic biota is a 7 day mean of 6.0 mg/l and a 1 day minimum of 5.0. Based on the water quality criteria, the 7 day mean DO saturation should be about 70 percent with a 1 day minimum of no less than 60 percent.

ASTM (1989) recognizes the need for a higher DO than U.S. EPA Method 1000.0, of no less than 60 to 100 percent saturation (E724-89, E1241-88).

Growth of animals, especially in a growth test, can be affected by less than adequate DO. In addition, effects of low DO and many toxicants are additive since they affect many of the same enzyme systems.

The DO and pH should be measured at the beginning and end of each 24 hour test period in one replicate at high, medium, and low concentrations and the control. Conductivity, alkalinity, and hardness should be measured in 100 percent effluent, receiving water, and in the controls as a minimum.

REFERENCES

Anonymous. 1986a. *Webster's Ninth New Collegiate Dictionary*. Springfield, Massachusetts: Merriam-Webster Inc.

Anonymous. 1986b. *Toxicology Handbook*. Rockville, Maryland: Government Institutes, Inc.

APHA, AWWA, WPCF. 1989 *Standard Methods for the Examination of Water and Wastewater*, 17th Ed.

ASTM. 1989. *1989 Annual Book of ASTM Standards*. Philadelphia. Vol. 11.04.

Broderius, S.J., 1983. *Analysis of an Interlaboratory Comparative Study of Acute Toxicity Tests with Freshwater Organisms*. Duluth: U.S. EPA. 54 pages.

Finney, D.J. 1971. *Probit Analysis*, 3rd Edition. New York: Cambridge University Press.

Horning, W.B. and C.I. Weber. 1985. *Short-Term Methods for Estimating the Chronic Toxicity of Effluents and Receiving Waters to Freshwater Organisms*. EPA/600/4-85/014. The second edition of this document was released in June 1989 (EPA 600/4-89-001).

Macek, K.J. and B.H. Sleight. 1977. *Utility of Toxicity Tests with Embryos and Fry of Fish in Evaluating Hazards Associates with Chronic Toxicity of Chemicals to Fishes*. ASTM STP 634. Philadelphia.

McKim, J.M. 1977. Evaluation of Tests with the Early Life Stages of Fish for Predicting Long-Term Toxicity. *J. Fisheries Research Board* **34**:1148–1154.

Peltier, W.H. and C.I. Weber. 1985. Methods for Measuring the Acute Toxicity of Effluents to Freshwater and Marine Organisms. EPA/600/4-85/013. U.S. EPA Cincinnati, Ohio. 216 pages.

Rand, G.M. and S.R. Petrocelli, 1985. *Fundamentals of Aquatic Toxicology*. Washington, D.C.: Hemisphere Publishing.

Schimel, S.C., 1981. *Results Interlaboratory Comparison—Acute Toxicity Tests Using Estuarine Animals*. Gulf Breeze, Florida: U.S. EPA.

Stephan, C.E., 1977. Methods for Calculating an LC_{50}. In *Aquatic Toxicology and Hazard Evaluation*. ASTM STP 634. Philadelphia.

U.S. EPA. 1980. Appendix B—Guidelines for Deriving Water Quality Criteria for the Protection of Aquatic Life and its Uses. *Federal Register* **45**(231).

U.S. EPA. 1985. *Technical Support Document for Water quality-Based Toxics Control*. EPA 440/4-85-032.

U.S. EPA. 1986. *Quality Criteria for Water 1986*. EPA 440/5-86-001.

U.S. EPA. 1987. *Permit Writer's Guide to Water Quality-Based Permitting for Toxic Pollutants*. EPA 440/4-87-005.

Walsh, G.E. et al. 1987. Comparison of the EC_{50}s of Algal Toxicity Tests Calculated by Four Methods. *J. Environ. Toxicology and Chemistry* **6**(793).

Weber, C.I. 1980. Federal and State Biomonitoring Programs. In *Biological Monitoring for Environmental Effects*, Lexington, Massachusetts: D.C. Health and Co., pp. 25–52.

Weber, C.I. et al. 1988. *Short-Term Methods for Estimating the Chronic Toxicity of Effluents and Receiving Waters to Marine and Estuarine Organisms*. EPA/600/4-87/028.

Weber, C.I. et al. 1989. *Short-Term Methods for Estimating the Chronic Toxicity of Effluents and Receiving Waters to Freshwater Organisms*. EPA/600/4-89/001.

3

Toxicity Identification Evaluations

W. Scott Hall and Richard J. Mirenda

INTRODUCTION

Inherent to nearly all toxicity reduction evaluations (TRE) is the process of conducting a toxicity identification evaluation (TIE). A TIE attempts to *identify* the constituent(s) or group of constituents responsible for toxicity, but does not necessarily develop a means to remove effluent toxicity. The increased emphasis on reducing effluent toxicity has resulted in the need to develop cost-effective toxicity identification protocols. Many toxicity identification protocols make use of effluent fractionation procedures designed to characterize the source(s) of toxicity in an effluent. This chapter presents a generalized protocol for conducting toxicity identification evaluations which can be used to identify the likely source(s) of and sometimes solutions to effluent toxicity.

One of the first efforts to develop toxicity identification procedures was a draft, unpublished U.S. EPA document which first appeared in 1980. This procedure used physical and chemical effluent fractionation methods to separate different components in a complex effluent. These components were tested for toxicity and compared to untreated effluent in an attempt to isolate the source(s) of toxicity. Walsh and Garnas (1983) demonstrated that effluent fractionation procedures could be used to isolate the source of toxicity in freshwater and saltwater effluents. This paper was important because it was the first to demonstrate the utility of fractionation procedures, and showed that these procedures could be used on a variety of effluents. Generalized effluent fractionation and/or toxicity identification procedures have been presented in the U.S. EPA acute toxicity test manual (U.S. EPA 1985a) as well as in the Technical Support Document (U.S. EPA 1985b), and the Permit Writer's Guide (U.S. EPA 1987). Lankford, et al. (1987) also reviewed the state-of-the-art in toxicity identification procedures and proposed specific methods to identify the source(s) of toxicity in complex effluents. Fava et al. (1987) also released a report outlining toxicity identification (and toxicity reduction) strategies for industrial effluents.

The U.S. EPA recently released *Methods for Aquatic Toxicity Identification Evaluations—Phase I Toxicity Characterization Procedures* (U.S. EPA 1987). This "Phase I Document" provides specific laboratory procedures for conducting effluent fractionations. These procedures identify the physical/chemical nature of the causative toxicant(s). Two (2) companion documents (U.S. EPA 1989a; U.S. EPA 1989b), outline methods for specifically identifying causative toxicants and confirming the identity of these toxicants (Phase II, and Phase III Documents, respectively). These documents serve as guidelines for conducting effluent fractionations. The utility of these procedures was quickly recognized, and many laboratories now use these in similar fractionation procedures. Gasith et al. (1988) published results of effluent fractionations of freshwater effluents, showing that fractionation steps other than those used in the U.S. EPA protocols could be successfully used in conjunction with the U.S. EPA procedures. Most recently, Lankford, et al. (1988) discussed approaches to toxicity identification and effluent fractionation, and presented an approach to identifying the source of toxicity which characterizes waste streams based on their toxicity, biodegradability, key chemical components, and flow proportion.

SCOPE OF A TOXICITY IDENTIFICATION EVALUATION

The process of conducting a TIE can be quite complex and often must consider the discharge flow characteristics, chemical make-up, variability, and relative toxicity of an effluent. TIEs are often quite interdisciplinary in nature, and can require engineering, chemical, and toxicological expertise. An effective toxicity identification protocol must be comprehensive yet flexible. Such a program is presented in Fig. 3-1, and typically includes the following steps:

1. Review of all historical analytical and toxicological data. The problem should be validated through repeated tests, the most sensitive species identified, and a baseline established of effluent toxicity and analytical parameters.
2. Evaluation of effluent constituents. Depending on the existing data, detailed analytical evaluations should be conducted of constituents present in the effluent. This step may also include continuous monitoring of analytical parameters in conjunction with subsequent toxicity identification efforts.
3. Evaluation of single streams. Toxicity testing and chemical analyses should be conducted on contributing wastestreams.
4. Fractionation of effluent. Effluent fractionation procedures should be used to identify the physical/chemical nature or categorical source of final effluent toxicity or of individual wastestream toxicity.

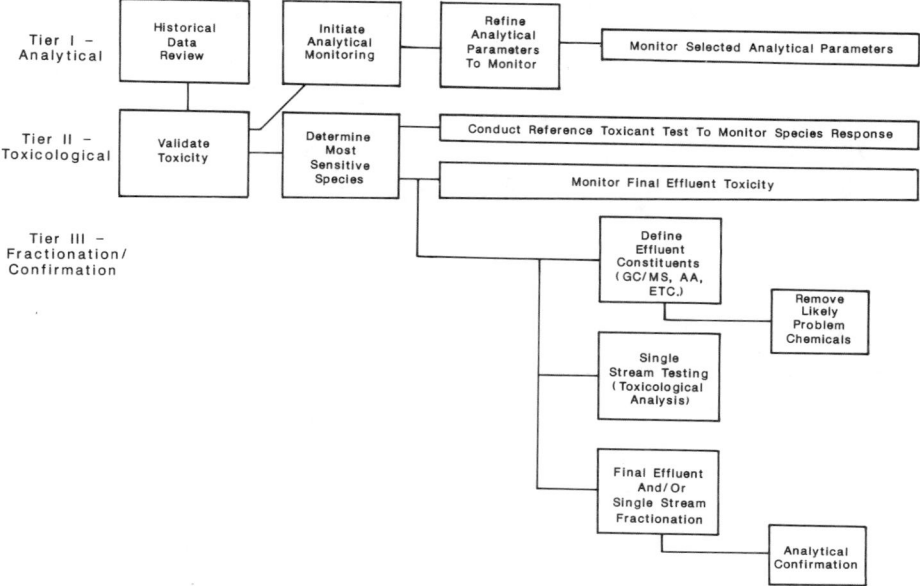

Fig. 3-1. Flow diagram for typical toxicity identification evaluation.

5. Performance of analytical confirmation. Detailed analytical work should be conducted in conjunction with effluent fractionation, possibly including analyses of constituents removed from the effluent by a particular fractionation step.

Although not necessarily conducted in the above order, this series of steps provides a conceptual framework from which to work when conducting a TIE. It cannot be stressed enough that this process is highly dynamic in nature, and must remain flexible. Certain steps may need to be emphasized more than others, or possibly deleted, depending on the situation. Some of these steps may of course be conducted simultaneously. As illustrated in Fig. 3-1, other steps may need to be conducted throughout the course of the TIE (e.g., monitoring certain analytical parameters while other toxicity identification efforts are being conducted). Depending on the scope and goal of the TIE, the true causative toxicant may or may not need to be identified. A general or categorical identification of the toxicant may be all that is required, and the "weight of the evidence" can be the deciding factor.

BASELINE DATA COLLECTION AND HISTORICAL DATA REVIEW

Prior to the initiation of the study, all historical analytical and toxicological data on the effluent should be reviewed for possible clues as to which parameters might relate to the observed toxicity. If the extent of the toxicity problem is not already documented, repeated tests should be conducted to determine the extent of the problem. Defining the magnitude and variability of toxicity in an effluent is critical for making subsequent decisions.

BASELINE DATA COLLECTION

The actual baseline data collection phase consists of a period in which background or preliminary toxicity and chemistry data are collected on the discharge. A monitoring program including parameters believed to relate to effluent toxicity should be initiated in this phase.

Most Sensitive Species

Early in the baseline data collection phase it is necessary to determine the species most sensitive to the effluent. This will insure that the "worst case" is determined initially and save the need to retest if the regulatory agency later tests with a more sensitive species. Organisms from at least two and preferably three taxonomic groups should be tested (most often a fish, an invertebrate, and an algal species), with special attention to the test species used by the regulatory agency. Regardless of the species tested, multiple comparisons will be necessary to determine the most sensitive species.

Microtox® Assay

It may be advisable at this stage to compare the correlation of Microtox® assays with toxicity tests using the required species. Microtox® tests are very rarely used as permit parameters, but may offer a more cost-effective means of monitoring toxicity if it can be shown to correlate with conventional aquatic species data. However, the correlation of Microtox® and aquatic species data is highly wastewater specific. Although Microtox® tests are most often used to predict the toxicity of contributing waste streams to a biological treatment process, the test sometimes also correlates to aquatic toxicity.

Relative Toxicity

After determination of the most sensitive species, the magnitude and variability of the effluent's toxicity must be documented. Knowing the relative toxicity of the effluent (especially when coupled with analytical information on the effluent) can provide clues as to the source(s) of toxicity. Documenting the variability of effluent toxicity is important in recognizing "true" changes in effluent toxicity when altering or "fine tuning" conditions. In complicated cases, the use of two test species may show changes in the relative sensitivity of each species through time. When the relative sensitivity of the test organisms changes through time it is usually a good indication that different toxicants are responsible for effluent toxicity at different times. The toxicity tests should be continued until both the magnitude and variability of toxicity are known, and until an indication has been obtained as to which species is most sensitive.

Sampling

In order to verify the extent of the problem, a sampling and testing program should be designed which involves frequent collection of samples representative of the discharge. Flow-proportional sampling should be used whenever practical. It may be prudent to monitor throughout the course of the TIE any chemical parameters suspected of possibly influencing toxicity. This will serve not only as part of the baseline data set, but contribute to the "weight of the evidence" in making subsequent decisions. Throughout the study, all sampling and testing protocols should be identical in order to insure comparability.

Reference Toxicant Tests

Finally, reference toxicant tests (described in Chapter 2) should be initiated following selection of the most sensitive species. Reference toxicant tests serve as "positive controls" to document both the sensitivity of the species and the consistency of the organism's response. Reference toxicant tests must be conducted throughout the course of the study in order to insure the comparability of toxicity data generated through time. If a certain group of chemicals is suspected of causing toxicity (e.g., metals, dyes, surfactants, pesticides), a chemical representative of that group should be selected as the reference toxicant. If this is not the case, a U.S. EPA reference toxicant is appropriate.

EFFLUENT CONSTITUENT EVALUATION

The evaluation of constituents present in the final effluent may provide an indication of the possible sources and causes of toxicity when evaluated in conjunction with the industrial survey procedures described in Chapter 4. Chemical analyses for potentially toxic constituents (e.g., heavy metals, GC/MS scan) followed by a literature search for toxicity data on these constituents can be quite useful. GC/MS scans are normally beneficial only if the specific pollutant identified is produced or used at the facility. The toxic heavy metal scans are more often found to be useful. If specific toxic chemicals are identified at this stage, caution should be used in developing conclusions from literature data. The toxicity of many chemicals is influenced by water quality characteristics, for example:

- Increased water hardness reduces the toxicity of many heavy metals
- Suspended solids may reduce the toxicity of sorptive chemicals
- Increased pH increases the proportion of the more toxic unionized form of ammonia

Pure chemical toxicity tests on suspected toxins and comparison of results to concentrations present in the effluent can be useful in well defined effluents with a limited number of toxic chemicals. Laboratory synthesis of the sample with certain chemicals or wastestreams deleted can also be useful in these cases. However, if this approach is taken, particular consideration must be given to any subsequent treatment processes a wastestream or chemical undergoes which are not accounted for in the design of the synthesis experiment.

Information on chemical additives (polymers, antifoams, biocides, etc.), raw materials, and chemicals present in cleaning agents can point to unsuspected sources of toxicity. Caution should be used when evaluating the available toxicity data on these constituents. For example, the toxicity of polymers in the wastewater matrix is likely to be much lower than their toxicity under laboratory test conditions. The suspended solids present in the wastewater which are bound by the polymers likely renders them less toxic. In addition to the previously mentioned toxicants (heavy metals, organic priority pollutants, chemical additives), other common sources of effluent toxicity which may need to be evaluated include ammonia, chlorine, high levels of dissolved solids, unidentified organics in the form of TOC or COD, surfactants, and process byproducts.

SINGLE STREAM EVALUATION

Assuming none of the above procedures provides any substantial evidence as to the source of toxicity, it may be advisable to proceed with single stream toxicity testing and analytical characterization. Fractionation of the effluent may also be considered at this point and may serve to direct the single stream evaluation. Monitoring analytical parameters of some individual streams to correlate with observed final effluent toxicity, mass loading to the final effluent, etc., may also be useful in this phase. A key component of single stream toxicity testing is the simultaneous monitoring of final effluent toxicity. This will allow the correlation of final effluent toxicity to the relative toxicity of single streams.

Although useful, one should be careful when conducting toxicity tests on single waste streams. Their relative toxicity and contribution to the final effluent volume must be considered. Since single streams are often very concentrated, they are likely to exhibit a high degree of toxicity. Also, due to their highly concentrated nature there may be a problem of extremes in pH, dissolved oxygen, and dissolved solids. All of these factors can result in toxicity, and knowing the cause of toxicity observed in single stream tests is critical to interpreting the wastestream's likely contribution to final effluent toxicity. Some examples follow:

- Extremes in pH are often buffered or diluted out prior to discharge.
- Because of dilution alone, dissolved solids toxicity is much less likely to be a problem than toxicity due to heavy metals.
- Since some heavy metals are toxic in trace (ppb) amounts, dilution alone may not be sufficient to reduce their levels to below toxic concentrations in the final effluent. Analytical work on the most toxic streams will provide information on likely sources of toxicity. A good approach to single stream evaluations is provided by Lankford et al. (1988). In this approach, streams are classified according to relative toxicity, biodegradability, percent of total flow, and key components in the stream.

EFFLUENT FRACTIONATION

The identification of key toxic components in the discharge or in individual waste streams may require the fractionation of samples through chemical and/ or physical means, as presented in Fig. 3-2. The goal of each fractionation step is to remove toxicity due to a specific group of chemicals and compare these results to the toxicity present in the unaltered sample. A thorough plan should be developed prior to beginning this phase which addresses the specific conditions at the facility. Special attention should be given to collecting representative samples of the final effluent if considerable variability was observed in previous steps.

The effluent fractionation procedure outlined below makes use of physical and chemical treatments which remove toxicity associated with specific chemical groups. Comparison of the toxicity of treated and untreated samples can yield information on the cause or source of toxicity and insight on possible treatment options.

The procedure is a partial modification of that described in the "Phase I Document" (U.S. EPA 1988). In some ways, the procedure has been simplified for logistical and practical reasons; for example, the initial toxicity test is performed parallel with rather than prior to the fractionation. In other ways, the procedure has been expanded to include treatments which are not used routinely in the U.S. EPA procedure and suggests a more definitive toxicity test protocol than that suggested by U.S. EPA.

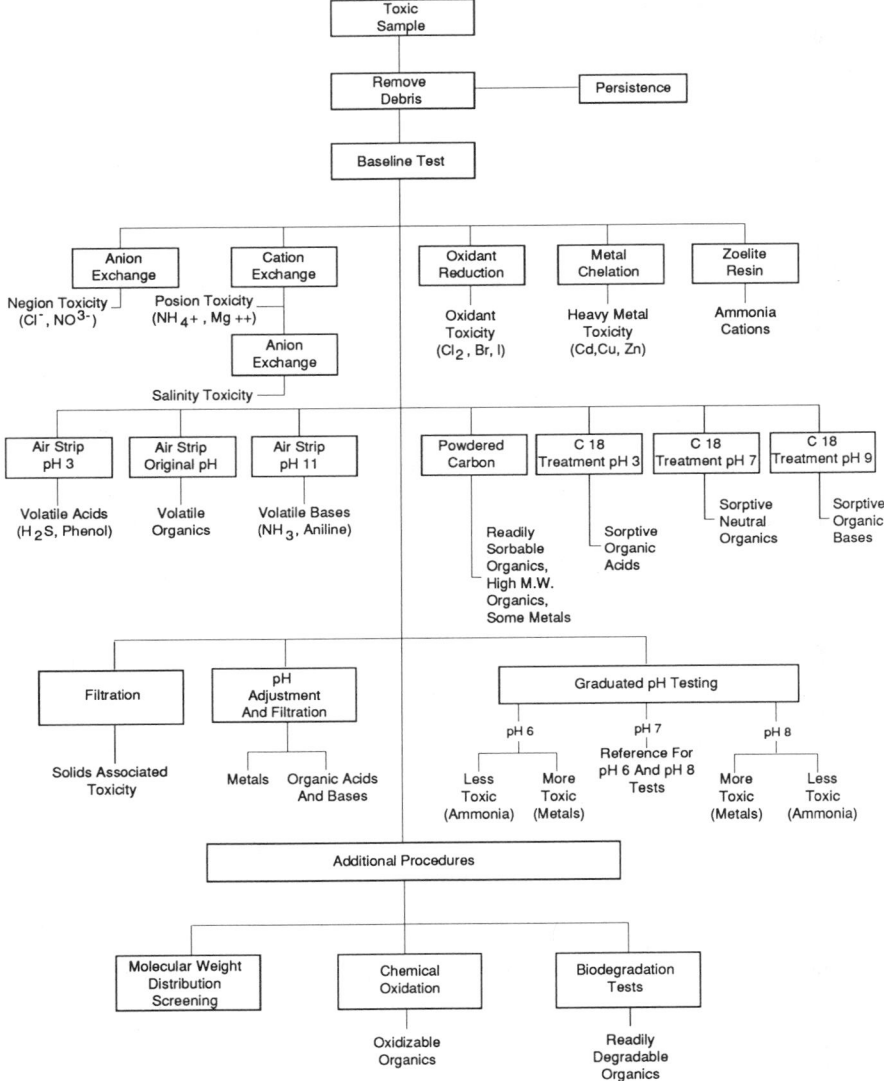

Fig. 3-2. Flow diagram for comprehensive effluent fractionation.

The number of times the fractionation procedure is repeated depends on the reproducibility of results. At a minimum, the procedure should be conducted twice on each effluent. Rerunning certain fractionation steps and not others is appropriate as patterns develop.

A narrative description of the considerations and steps involved in effluent fractionation is provided in the following paragraphs. General considerations are followed by more specific recommendations, followed by the steps involved in the routine laboratory fractionation procedure. Finally, other fractionation steps are described which are not routinely used, but which may be useful in certain cases. Each procedure should be evaluated prior to use in order to verify its likely applicability.

FRACTIONATION PROCEDURES

Quality Assurance Considerations

Critical to any effluent fractionation procedure is the consideration of quality assurance and proper experimental controls on the treatments. Important considerations include:

- Use of system blanks to identify toxic artifacts. Controls should be maintained to identify "false negatives," and care should be taken to avoid "false positives." For example, a "false negative" occurs when it appears that the treatment did not remove toxicity, when in fact the treatment both removed and caused toxicity. By running the same treatment on control waters, this situation can usually be avoided. An example of a "false positive" is the reduction in toxicity through filtration due to the effect of organic binders in the filters on dissolved organics in the effluent. This would lead the investigator to believe that toxicity was associated with the solids removed during filtration, when in fact it was due to a dissolved constituent which had been bound by the organic binders on the filters. Due to matrix effects on the toxicity of chemicals added to water, it cannot be guaranteed that a toxic or nontoxic treatment of the control waters results in the same toxic or nontoxic effects in the effluent.
- Reference toxicant (positive control) tests must be run throughout the study to insure that test organism sensitivity has not changed over time. Differences in test organism sensitivity between animals from different sources can lead to erroneous conclusions.
- To insure precision, adequate replicates should be used on samples for chemical analysis, including the use of "spikes" or standard additions to assess accuracy.
- Sample analysis should be used to correlate the amount of a constituent present when different levels of toxicity are observed, in order to establish true cause-and-effect relationships whenever possible.
- The use of "laboratory controls" in the form of laboratory culture water controls insures that any toxicity observed in upstream dilution water controls is in fact due to the unacceptable nature of the dilution water, and not due to an unhealthy stock of test organisms.

Proper effluent sampling and handling techniques should be used and chain of custody records maintained. This includes such matters as observing proper holding period limits (36 hours for nonpersistent and 72 hours for persistent toxicity). Water quality conditions should be determined at the start of the tests (pH, dissolved oxygen, conductivity, hardness, etc.) and all tests should be conducted at the same temperature.

Toxicity Test Procedure

The recommended toxicity test procedure essentially follows U.S. EPA protocol 600/4-85/013 for acute toxicity tests (U.S. EPA 1985a). Following effluent fractionation, static acute (48 or 96 hour) nonrenewal toxicity tests with invertebrates (*Daphnia pulex*, *D. magna*, or *Ceriodaphnia dubia* for freshwater, *Mysidopsis bahia* for saltwater) or young fish (fathead minnow or sheepshead minnow in fresh and saltwater, respectively) are used to evaluate changes in toxicity. Other appropriate species may also be used as the situation warrants. Generally, five effluent concentrations and a control are used for each toxicity test. One replicate of 50 to 200 ml total volume with seven organisms is used for each control and effluent concentration. LC_{50} values and 95 percent confidence limits are calculated for each treated sample and compared to the LC_{50} value. This procedure is preferred over timed lethality tests or tests with fewer dilutions for the following reasons:

- Information is obtained on effluent toxicity over a wide range of concentrations
- Data are generated with 95 percent confidence limits
- Time is saved by not having to rerun tests after a screening phase
- Tests are simplified by avoiding time-lethality observations while conducting other procedures

Initial Sample Handling

If any large debris is present in the sample, it should be removed by filtering the sample through a large mesh (300 to 500 micrometer) Nitex® net. Some samples may contain solids not removed by large-mesh screening. It is advisable, especially in early stages of work, to filter these samples as described later under "Filtration." Results of all other tests are considered after accounting for the effects of filtration alone.

The sample should then be warmed to the test temperature of $20° \pm 1°C$ in a water bath.

Ammonia and residual chlorine analyses should be conducted on the untreated effluent. Simple, inexpensive colorimetric test kits are appropriate for ammonia and chlorine analysis at this stage of the study.

Next, the dissolved oxygen concentration and sample pH at the test temperature ($\pm 1\,^\circ$C) should be determined. If dissolved oxygen is less than 40 percent saturation, a sufficient volume of the original sample should be set aside to run an additional toxicity test on unaltered effluent. The original effluent sample should be aerated to increase the dissolved oxygen concentration to the desired level. Results of all tests must be interpreted in light of the possibility that some component of the sample may have been volatilized during reaeration.

If the effluent pH is less than 6.0 or greater than 9.0, a portion of the sample should be used to set up an additional toxicity test with "pH unadjusted" effluent. The remainder of the sample should have the pH adjusted to an acceptable value with $0.1N$ NaOH or $0.1N$ HCl. The volume of acid or base used should be recorded, and limited to less than 10 percent by volume to avoid extensive sample dilution. The results of subsequent fractionation steps must be interpreted in light of the results of the "pH adjusted" and "pH unadjusted" toxicity tests on untreated effluent.

Controls (dilution water carried through identical treatments) are tested for each treatment evaluated.

Toxicity Tests on Unaltered Sample

Toxicity tests on the unaltered effluent sample provide the basis for comparison with all treated samples to determine which treatments reduced or removed toxicity. If desired, untreated effluent may be tested for toxicity under light and dark conditions to establish the importance of photolysis in altering toxicity. "Light" tests are run under light conditions of 16 hours light and 8 hours dark, while "dark" tests are run under continuously dark conditions. Generally, only the "light" phase tests are run. Water chemistry performed at the start of testing includes pH, dissolved oxygen, conductivity or salinity, alkalinity, and hardness in the controls and 100 percent untreated effluent. As a rule, additional water chemistry is not done on the treated samples described below except to readjust pH where noted. Tests with treated and untreated effluent should be set up simultaneously to eliminate time as a variable.

Toxicity Persistence

The persistence of effluent toxicity can be evaluated through exposure to direct light for 24 to 48 hours at room temperature. A glass container containing the necessary volume for a toxicity test is set aside at room temperature in sunlight or under laboratory lights. A toxicity test is conducted at the end of this period. Routine water chemistry analyses (as stated previously) are conducted on the control and undiluted sample. Although not designed to identify toxicity due to any specific chemical group, results of the persistence evaluation may be used to support data generated from other fractionation steps. The dissolved oxygen level should be checked prior to preparing test dilutions. If the sample has a high oxygen demand, it will need to be aerated prior to testing. In this event, results of the persistence test must consider the results of the air stripping tests described below.

Filtration

Filtration through a one-micrometer glass fiber filter will indicate whether effluent toxicity is related to the soluble or insoluble phase of the sample. After filtration into a glass vacuum flask, toxicity tests are conducted on the filtrate. The filters and associated materials should be saved for possible future analytical work. If toxicity is reduced, the filtered material may be resuspended in control water to verify that it was solids removal and not some phenomenon such as organic binders on the filter which removed toxicity. In rare cases involving highly volatile chemicals, filtration may falsely indicate the mechanism of removal due to "degassing" as the sample passes into the filtration flask. Additional filtration at pH 3 or pH 11 is done when a precipitate is noticed during pH adjustment for other tests (see below). Care should be taken to analyze blanks for toxicity contributed by the test media.

Precipitation Testing

Whenever a precipitate is observed as a result of pH adjustment in other tests, a precipitation/filtration test should be conducted. Using NaOH or HCl, the pH of a fresh sample should be adjusted to the pH observed to induce precipitation (e.g. pH 11) and allowed to equilibrate for one hour. The sample is then filtered through a one micrometer glass fiber filter to remove the precipitate. The sample pH is readjusted to the original pH and tested for toxicity. In pure chemical tests with cadmium, this procedure at pH 11 resulted in a substantial reduction in toxicity. A similar precipitation procedure utilizing calcium carbonate ($CaCO_3$) was also successful in reducing cadmium toxicity.

Ion Exchange

Toxicity due to inorganic materials can be evaluated using cation and anion exchange resins individually or in combination. These treatments can result in nearly complete removal of potentially toxic inorganic ions as well as some charged organic molecules. The use of either cation or anion exchange can provide some indication as to the precise identity of the causative toxicant. However, the products of exchange must be carefully evaluated for introduction of toxicity. In certain cases, toxicity in the control waters has been observed following ion exchange.

A column treatment technique is used. The volume of sample treated per mass of resin is highly dependent on the resin exchange capacity and characteristics of the sample. Following sample treatment the pH must be reset to the original sample pH, since extremes of pH are seen after treatment as a result of H^+ and OH^- exchanges. Concentrated acids and bases may be needed to readjust pH to avoid exceeding sample dilution by the 10 percent limit. In some cases, it may be necessary to add ions back to the treated water if, for example, alkalinity and hardness or conductivity are low following ion exchange. Samples which are toxic due to very high levels of dissolved solids have been successfully treated by a sequential treatment of cation followed by anion exchange.

Sodium Thiosulfate Addition

Residual chemical oxidants in a sample can be highly toxic to the test organisms. This includes chlorine and chloramines used for disinfection, and ozone and hydrogen peroxide used in sludge conditioning. A batch reduction of these oxidants using a reducing agent such as sodium thiosulfate will remove the residual oxidant and any associated toxicity. Toxicity tests with sodium thiosulfate treated effluent are conducted along with controls containing the same amount of sodium thiosulfate. In previous tests, controls treated with up to 160 mg/1 of sodium thiosulfate were not toxic to *D. pulex*, *Ceriodaphnia dubia*, or fathead minnows.

EDTA Addition

The toxicity of cationic metals (excluding mercury) can be determined by chelation of samples using EDTA and evaluating the change in toxicity. However, EDTA can be toxic at certain concentrations, depending on water hardness. Tests should be performed using low, medium, and high dosages in initial testing. These dosages would typically be 7.5, 37, and 75 mg/1 EDTA at a water hardness of 100 mg/1 $CaCO_3$. For subsequent testing, only one level of EDTA is used based on results of initial tests. A single batch treatment of the effluent using 75 mg/1 of EDTA has been used successfully.

In cases where the water hardness of the sample is higher than that of the control for which EDTA toxicity data are available, one of two approaches can be used to establish the amount of EDTA added to the effluent. The EPA Phase I document (U.S. EPA 1988) provides EDTA toxicity data as a function of water hardness, and can be used as a guide. Alternatively, the amount of EDTA required to reach the hardness titration end point can be determined, and from this the high, medium, and low dose scenario described above may be implemented.

Air Stripping

Vigorous batch air stripping at acidic pH, the original sample pH, and basic pH can result in complete or partial removal of compounds which are volatile at any of the pH conditions as well as pH-extractable organics. Typically, samples at the original pH, pH 3, and pH 11 are aerated for a period of at least one hour, although periods of up to twenty-four hours may sometimes be necessary. The pH is adjusted to pH 3 or pH 11 with NaOH or HCl as required, again limiting sample dilution to 10 percent or less. Following air stripping, the pH of the samples is readjusted to the original pH. A 30 minute equilibration period is recommended, then the sample pH is rechecked and readjusted if necessary.

In some cases where air stripping removes toxicity, it may be advisable to reevaluate the sample using nitrogen gas (N_2) as the stripping gas. This procedure evaluates the possibility that the toxic component(s) were oxidized rather than volatilized during the air stripping step. If toxicity is markedly reduced under pH 11 air stripping conditions, ammonia may be a candidate for toxicity in the untreated sample. Subsequent or concurrent treatment with zeolite may clarify this situation.

A few notes of caution must be made concerning air and N_2 stripping procedures. First, the pH of the unadjusted sample often changes as a result of the vigorous air stripping procedure. If this occurs, the pH should be readjusted to the original sample pH. If N_2 stripping is used, dissolved oxygen is often purged from the sample. If this occurs, reaeration of the sample is necessary prior to testing. In addition, air flotation or foaming may occur during air stripping, particularly if surfactants are present. In these cases, removal of the toxicants can result from deposition onto the container walls, misleading the investigator as to the true mechanism of toxicity removal. This problem can be avoided by rinsing any deposited material back into the sample with control water prior to use of the sample in toxicity tests.

Zeolite

Batch air stripping at a basic pH may result in complete or partial removal of ammonia as well as volatile organics. If volatile organics are a candidate for toxicity, an alternative ammonia removal technique can be used utilizing zeolite. Expanded stripping tests, zeolite tests, and additional analytical data may be needed to confirm ammonia as the source of toxicity.

Zeolite treatment can be conducted in a column apparatus. The pH must be adjusted following treatment, since the sample pH is generally increased as a result of zeolite treatment. A control must also be run simultaneously with the treated effluent. Caution must be employed when evaluating the results of zeolite treatment because ammonia is not the only constituent removed. Zeolite may also remove other molecules of the size range trapped by the particular pore size of the zeolite used.

Solid Phase Extraction

The solid phase extraction procedures utilize two sorptive media. Long chain C_{18} resin is used for selective removal of sorptive, hydrophobic compounds, and powdered activated carbon (PAC) is used for nonselective removal of toxicants. The mechanisms of removal can be different for these two media, with powdered carbon removing some compounds due to sorption and pore entrapment. In addition to sorptive removal of hydrophobic organics, the C_{18} resin also has some physical filtration ability due to the small pore size of the C_{18} resin. This filtration phenomenon may falsely indicate the true mechanism of toxics removal.

A nonpolar compound is one which does not ionize in the carrier medium and therefore has more affinity for an adsorptive medium than for the carrier medium. Through the use of resin adsorption procedures, nonpolar organics can sometimes be identified as toxicants. Since the nonpolar (sorptive) form of some chemicals changes with pH, samples are tested at pH 3, pH 9, and the original sample pH. The respective samples are pulled across the C_{18} resin by a vacuum and analyzed for toxicity following the readjustment of pH. The C_{18} columns are previously conditioned with methanol and distilled water at the respective treatment pH following the manufacturer's instructions. Control water should be simultaneously tested for toxicity at all pH conditions. These columns will remove organics from solutions so long as the mass to be removed does not exceed 5 to 8 percent of the mass of the sorbent. Beyond this point, exchange sites are exhausted, and a new column must be used. Nonpolar organics sorbed to C_{18} may be eluted with methanol and reintroduced to control water and tested for toxicity as suggested by U.S. EPA (1989a). However, a methanol control must be run and the toxicity of the carrier determined so as not to exceed threshold levels of carrier toxicity. Elution of sorbed toxicants from C_{18} columns may also be used for analytical identification using GC/MS or high pressure liquid chromatography (HPLC) procedures.

Powdered activated carbon treatment is a batch treatment of the sample at the original pH with high doses of carbon. The sample/carbon solution is stirred on a stir plate for at least one hour. The sample is then filtered using a one micrometer glass fiber filter and the filtrate is tested for toxicity. A control with the same amount of carbon is tested with each sample. It may be necessary to readjust the pH of the samples following carbon treatment. Carbon treatment has removed toxicity in many of the effluents evaluated due to the nonselective nature of powdered carbon.

Graduated pH

The graduated pH tests essentially follow the procedures outlined by U.S. EPA (U.S. EPA 1988). The graduated pH test generally consists of adjusting the undiluted sample to a pH of 6.0, 7.0, and 8.0, using $0.1 N$ HCl or NaOH. The pH of the dilution water is also adjusted to the same three pH values, and used as control and dilution water in the respective graduated pH tests. Tests are conducted in glass beakers filled to the rim and covered with laboratory film to minimize gas exchange. Final solution pH is recorded at the end of the testing period, and results interpreted in light of the stability of the test solution pH. This approach has been useful for observing changes in effluent toxicity with changes in pH, suggesting that ammonia or metals may be involved in the toxicity. Increasing the pH results in an increase in the proportion of ammonia in the unionized form (NH_3). Since unionized ammonia is the more toxic form, increased toxicity as a result of increased pH makes ammonia a likely causative toxicant candidate. This information may be useful in conjunction with results from the air stripping and zeolite tests. Increased toxicity associated with a decrease in pH may indicate metals are causing or contributing to toxicity. Decreases in pH result in a higher solubility of most metals, thereby increasing metal toxicity.

Chemical Oxidation

In industrial applications, chemical oxidation of a wastewater is often used to oxidize pollutants to end-products or to intermediate products that are more readily biodegradable or more readily removed by adsorption. Chemical oxidation can also directly reduce the toxicity of many chemicals. Treatment of a wastewater in this manner can reduce both the toxicity and organic content of the wastewater. Common oxidants used are chlorine, ozone, hydrogen peroxide, and potassium permanganate. Chemical oxidation may be markedly dependent on pH and the presence of catalysts.

In an expanded fractionation scenario, chemical oxidation of residual organics in the effluent may provide some indication of the cause of effluent toxicity. A simple batch treatment of the effluent with hydrogen peroxide (H_2O_2) and an iron catalyst may be used, followed by sodium thiosulfate reduction of any residual H_2O_2. Since none of the routinely applied fractionation steps removes soluble, polar organics in their original form, H_2O_2 oxidation provides a means to evaluate the possibility of polar organics being the cause of toxicity.

Molecular Weight Distribution

Evaluation of the molecular weight distribution of both influent and effluent samples may be beneficial in determining the possible toxic compounds present in a wastestream. Evaluation of the influent molecular weight distribution, and toxicity determination of each range of molecular weight can often narrow the list of suspected toxic compounds. Treatment of compounds known to fall in the toxic molecular weight range can then be studied. The molecular weight distribution testing and separation may be accomplished using a series of molecular sieves. Toxicity due to filter leachates should also be evaluated by filtering and testing control waters. Although molecular weight distributions are not routinely carried out on fractionated samples, they may provide a means of evaluating the effluent in greater depth following initial studies.

Biodegradation

This procedure is not routinely used in a fractionation protocol because most effluents have already undergone some biological treatment. However, controlled biological treatment of effluent samples can result in almost complete oxidation of the degradable portion of the organic constituent. Subsequent testing can then assess the toxicity associated with the nondegradable components as well as the reduction in toxicity attainable by biological treatment. Care must be taken to avoid desorption of toxic components or generation of toxic end products from the biological mass. Several control tests must be run in parallel to determine whether either of these is occurring. Batch degradation tests can be run in one liter glass beakers with continuous aeration and stirring for at least twenty-four hours. However, proper microbial acclimation must be achieved before results of these batch treatments can be utilized.

Indirect Determination of Polar Organics

A polar compound is one which readily ionizes in the carrier medium, such as an acid or base, a salt, or an organic or inorganic radical. Polar compounds do not readily adsorb (although certain compounds will adsorb on carbon to some extent) and are not volatile in their ionic, soluble form. Thus, none of the treatments described above directly removes toxicity caused by polar organics. However, toxicity due to polar organics can be evaluated indirectly by evaluating the amount of organic carbon remaining after a sorptive exchange process. Following a filtration or resin adsorption procedure, the residual organics (as measured by TOC), can be assumed to be polar organics. Any remaining toxicity associated with TOC is thus presumed to be due to polar organics. Identification of specific compounds by GC/MS, HPLC or other techniques should be attempted, as well as a determination of the molecular weight distribution. Data gathered from this procedure may also be used to compare with the results of the C_{18} treatments or those obtained from the hydrogen peroxide oxidation procedure.

ANALYTICAL CONFIRMATION

Whenever possible, analytical confirmation should be conducted during some phase of the effluent fractionation study. This confirmation will add to the "weight of the evidence" when making final and possibly costly treatment decisions. Analytical procedures appropriate to the above protocol are discussed below.

Basically, two approaches may be taken. One approach is to monitor analytical parameters to correlate with toxicity removal as previously discussed, and the other is to conduct a before and after treatment evaluation of the effluent to look at constituents removed. The U.S. EPA "Phase II Document" (U.S. EPA 1989a), outlines various methods for analytical evaluation of the treated effluent. One of the most notable is to conduct methanol elutions of the C_{18} columns following effluent treatment. This process uses various concentrations of methanol to elute chemicals of different hydrophobic capacities. Eluted samples can be reevaluated in toxicity tests or submitted for GC/MS evaluations. The before/after treatment approach can also be used for constituents identifiable using HPLC, GC/MS, and atomic absorption techniques. However, any appropriate analytical technique may be applied for a given fractionation step to evaluate constituent removal. With the exception of EDTA treatments and the graduated pH tests, all of the previously discussed fractionation steps can be used to evaluate the removal of a given constituent following effluent treatment. The variability and reproducibility of results must be verified as with all aspects of the study.

FRACTIONATION CASE STUDIES

The results of selected case studies from a variety of industrial effluent fractionations are presented and discussed below. The case studies were selected to illustrate the selective use of different fractionation treatments on various effluents, and the use of different test species. An effort was made to select results which illustrate the ambiguity inherent in some studies, while other results were selected to illustrate more definitive findings. A special effort was also made to select results which demonstrate the special considerations previously discussed for some fractionation steps.

All of the studies were carried out with invertebrates less than 24 hrs old or with fish species two to three weeks old. All of the tests were conducted at 20°C, utilizing the fractionation procedures previously described. LC_{50} values and 95 percent confidence limits were generated using the binomial procedure. All LC_{50} values for invertebrates are 48 hour values, and all values for fish species are 96 hour values. The values are expressed as percent effluent, and the criterion used to establish statistically significant differences in LC_{50} values was that 95 percent confidence intervals did not overlap (APHA, AWWA, and WPCF 1980).

Case Study No. 1

Table 3-1 presents results of a study with *Ceriodaphnia dubia* in which the fractionation of a moderately toxic effluent ($LC_{50} \approx 35$ percent) resulted in an increase in LC_{50} to >100 percent for several treatments. This is fairly common with effluents which are not markedly toxic.

Toxicity was reduced to the extent that an LC_{50} could not be calculated ($LC_{50} > 100$ percent) as a result of five different treatments. These treatments were:

- Nitrogen stripping at pH 11
- C_{18} treatment at the original sample pH
- Powdered carbon treatment
- Anion exchange
- Sodium thiosulfate addition

It should be noted that the confidence intervals of all of the data points overlapped. Due to the wide range of confidence intervals around LC_{50}s generated using the binomial method, it is very difficult to achieve statistical significance. However, the importance of the degree of toxicity removal by a treatment is obvious regardless of the lack of statistical significance. The importance of evaluating the trends in toxicity reduction as a result of fractionation steps has been discussed elsewhere (U.S. EPA 1989).

Other treatments which resulted in a clear increase in LC_{50} values included air stripping at pH 11 and cation exchange, which each raised the LC_{50} of untreated effluent from 35.4 percent to greater than 70 percent. Removal of toxicity by powdered activated carbon (PAC) and C_{18} at the original sample pH may indicate that the chemical(s) responsible for toxicity are sorptive at near neutral pH. The fact that anion exchange and sodium thiosulfate treatments both

Table 3-1. Case Study No. 1: Effluent Fractionation (*Ceriodaphnia dubia*).

Treatment	LC_{50} (95% confidence limits)		Treatment	LC_{50} (95% confidence limits)	
Untreated	35.4	(25 to 50)	Powdered carbon	>100	(0 to 100)
Persistence evaluation	35.4	(25 to 50)	Cation exchange	70.7	(0 to 100)
Air stripping			Anion exchange	>100	(50 to 100)
pH 3	35.4	(25 to 50)	Filtration	53.4	(6.3 to 100)
Orig. pH	35.4	(25 to 50)	EDTA addition	53.4	(0 to 100)
pH 11	84.4	(6.3 to 100)			
N_2 stripping			Sodium thiosulfate	>100	(0 to 100)
pH 3	26.7	(6.3 to 50)	C_{18}, orig. pH	>100	(25 to 100)
Orig. pH	42.3	(25 to 100)			
pH 11	>100	(6.3 to 100)			

removed toxicity indicates that an oxidant of some sort played a role in causing toxicity in the untreated sample. However, toxicity removal by nitrogen stripping at pH 11 and a large toxicity reduction by air stripping at pH 11 indicates that a base volatile compound was contributing to effluent toxicity. This is supported by the fact that toxicity was removed by cation resin, since a cationic organic compound would be uncharged at pH 11 and hence volatile. The success of both air and N_2 stripping at pH 11 verifies that the mechanism of toxicity removal was volatilization and not oxidation during the pH 11 air stripping tests. The clear success of the anion exchange resin contradicts results of the cation exchange treatments and of the pH 11 purging tests. Because the success of the anion exchange and sodium thiosulfate treatments indicates an anionic toxicant, and the success of cation exchange and high pH purging indicates a cationic toxicant, there were likely two different toxicants present in the untreated effluent. In order to resolve the actual sources of toxicity in this effluent, definitive analytical comparisons of before and after treatment samples would have to be conducted.

Case Study No. 2

Table 3-2 presents results of a fractionation study with *Daphnia pulex*. The effluent was moderately toxic and resulted in an LC_{50} of $\approx$ 52 percent.

Treatments increasing the LC_{50} to > 100 percent include:

- Air stripping at pH 3
- Air stripping at pH 11
- C_{18} treatment at pH 3
- Zeolite resin treatment

Removal of toxicity by air stripping at pH 3 and pH 11 indicates that two toxicants were likely present, each of which is readily volatile under different

Table 3-2. Case Study No. 2: Effluent Fractionation (*Daphnia pulex*)

Treatment	LC_{50} (95% confidence limits)		Treatment	LC_{50} (95% confidence limits)	
Untreated	51.9	(25 to 100)	Powdered carbon	57.9	(0 to 75)
Air stripping					
pH 3	>100	(25 to 100)	Cation exchange	72.2	(50 to 100)
Orig. pH	72.2	(50 to 100)	Anion exchange	72.2	(0 to 100)
pH 11	>100	(50 to 100)	Filtration	80.4	(0 to 100)
C_{18}					
pH 3	>100	(75 to 100)	EDTA addition	53.2	(0 to 100)
Orig. pH	61.2	(50 to 75)	Sodium thiosulfate	68.0	(50 to 100)
pH 9	64.8	(50 to 100)	Zeolite resin	>100	(0 to 100)

pH conditions. The fact that C_{18} treatment at pH 3 removed toxicity indicates that a chemical was present which is also readily sorbed under low pH conditions. An organic chemical unionized at pH 3 should both sorb to C_{18} resin and be volatile. The results are thus in good agreement in this regard, and indicate that an acid-extractable compound was at least partially responsible for toxicity. The success of pH 11 air stripping can only be explained by the presence of a second toxicant of similar toxicity. The marginal success of both the cation and anion exchange resins further indicates the presence of two toxicants. Due to the success of the zeolite resin treatment it is possible that ammonia is the other toxicant present in this effluent since ammonia would be removed by both zeolite and pH 11 air stripping.

Case Study No. 3

Table 3-3 presents results of an early fractionation study using sheepshead minnows. Only five organisms per concentration were exposed and 95 percent confidence intervals could not be obtained. However, this study is presented to demonstrate the use of a saltwater fish species, to illustrate how some treatments occasionally increase effluent toxicity, and to illustrate a case in which the mechanism of toxicity removal was found to be misleading with regard to the suspected toxicant(s).

An effluent of low toxicity ($LC_{50} \approx 64$ percent) was fractionated. However, pH 3 air stripping, C_{18} treatment at the original sample pH, sodium thiosulfate addition, powdered carbon treatment, and EDTA addition all increased effluent toxicity. These cases are rare, especially when increased toxicity occurs as a result of so many treatments, and the cause of these increases in toxicity is not known. Treatments increasing LC_{50}s to > 100 percent included:

- Air stripping at pH 11
- Cation exchange
- Anion exchange
- Filtration

Table 3-3. Case Study No. 3: Effluent Fractionation (*Cyprinodon variegatus*).

Treatment	LC_{50}	Treatment	LC_{50}
Untreated	64.1	Powdered carbon	7.5
Persistence evaluation	50.8	Cation exchange	>100
Air Stripping			
pH 3	27.4	Anion exchange	>100
Orig. pH	77.5	Filtration	>100
pH 11	>100	Sodium thiosulfate	7.5
C_8, orig. pH	11.5	EDTA addition	32.8

The success of the cation and anion exchange resins, coupled with the nature of the chemicals produced at this facility, indicated that some of the charged organic chemicals being produced were responsible for toxicity. It was also determined that deposition during aeration was responsible for the removal of toxicity during pH 11 air stripping, and that chemical volatilization was not the mechanism of toxicity removal. It was suspected from this that surfactants were the source of toxicity at the facility. This study demonstrated how the mode of toxicity removal can be falsely indicated as a result of physical deposition during air stripping.

Results such as those presented in Table 3-1, 3-2, and 3-3 typify the ambiguous results which may be obtained when effluents of low to moderate toxicity are fractionated. With effluents such as these, several fractionation steps often remove toxicity, making interpretation of results very difficult. Often, contradicting mechanisms seem to be removing toxicity. One reason for such ambiguity is that if effluent toxicity is relatively low, several treatments may remove a small amount of the toxicant which may be enough to remove toxicity in a given effluent. Another possibility is that toxic constituents may be acting synergistically in these cases. Together, the constituents exert some toxicity, but when one is removed the remaining constituent is less toxic. The true cause of these phenomena of course cannot be determined from the above data alone.

Case Study No. 4

Table 3-4 presents results of a fractionation study utilizing *D. pulex* and fathead minnows which compares toxicity removal for different species. The sample was very toxic to both *D. pulex* and the fathead minnow, with LC_{50}s of 0.12 and 1.2 percent, respectively.

Statistically significant increases in *D. pulex* LC_{50} were achieved by PAC, and by pH 11 precipitation/filtration techniques. Although not statistically significant, the largest increase in *D. pulex* LC_{50} was achieved by a high level addition of EDTA. Other clear increases in *D. pulex* LC_{50} values were achieved by C_{18} treatment at pH 9 and by cation exchange, with the lower 95 percent confidence intervals for these treatments just bordering the upper 95 percent confidence interval for the untreated effluent. Since pH 11 precipitation and filtration, high level EDTA treatment, and cation exchange all reduced toxicity, it is very likely that a heavy metal was responsible for toxicity to *D. pulex*. PAC can be expected to partially remove metals and the partial success of this treatment (although statistically significant) supports the hypothesis for metal toxicity. The success of the pH 9 C_{18} treatment also indicates metal toxicity since precipitation at pH 9 and physical removal (filtration) by the C_{18} resin bed may also remove metals. The success of the anion exchange resin does not necessarily contradict the metals toxicity theory. A metal containing charged

Table 3-4. Case Study No. 4: Effluent Fractionations (*Daphnia pulex* and *Pimephales promelas*).

Treatment	LC$_{50}$ (95% Confidence Limits)				Treatment	LC$_{50}$ (95% Confidence Limits)			
	D. pulex		*P. promelas*			*D. pulex*		*P. promelas*	
Untreated	0.12	(0 to 0.5)	1.2	(0.5 to 10)	Cation exchange	35.0	(0.5 to 100)	>100	(1 to 100)
Persistence evaluation	0.12	(0 to 0.5)	1.8	(0.1 to 10)	Anion exchange	10.9	(0 to 25)	>100	(≥100)[a]
Air stripping									
pH 3	0.12	(0 to 0.5)	3.2	(0.1 to 25)	Filtration	0.12	(0 to 0.5)	2.3	(0 to 10)
Orig. pH	0.12	(0 to 0.5)	2.3	(0.5 to 10)	EDTA addition	0.71	(0.5 to 1.0)	35	(10 to 100)
pH 11	<0.10	(NA)	2.3	(0.1 to 10)	Zeolite	5.7	(0.1 to 25)	>100	(≥100)[a]
C$_{18}$									
pH 3	<0.10	(NA)	2.3	(0.5 to 25)	pH 11 precip.	13.9	(1 to 25)[a]	>100	(≥100)[a]
Orig. pH	0.16	(0 to 0.5)	1.2	(0.5 to 10)	High EDTA	50.0	(0 to 100)	>100	(≥100)[a]
pH 9	15.8	(0.5 to 25)	>100	(0 to 100)	Sodium thiosulfate	0.20	(0 to 0.5)	1.2	(0.5 to 10)
Powdered carbon	3.2	(1 to 10)[a]	50	(0.5 to 100)					

[a]Statistically different from untreated effluent.

hydroxide forms, for example, may be removed by the anion exchange resin or be precipitated by the high pH resulting from this treatment.

The results of the fathead minnow fractionation with this same effluent also indicated that metals were the source of toxicity. The pH 9 C_{18} treatment, powdered carbon, cation exchange, anion exchange, zeolite, pH 11 precipitation and filtration, and high level EDTA treatment clearly reduced or removed effluent toxicity. Removal of toxicity ($LC_{50} > 100\%$) to the fathead minnow was achieved for all of these treatments except powdered carbon. Statistically significant increases in LC_{50} were achieved by the high EDTA treatment, zeolite resin, pH 11 precipitation/filtration, and anion exchange. Removal of metals toxicity to fathead minnows would be achieved by these treatments as previously discussed for *D. pulex*. The pH 11 precipitation/filtration process was the only treatment resulting in statistically significant increases in the LC_{50} for both species. This, coupled with the other results for both species (i.e., high EDTA treatments) solidifies the case for metal toxicity. Plant personnel had suspected copper as a likely source of toxicity, supporting the metal toxicity theory.

An interesting result of this case study was the observed difference in species response to the treated and untreated effluents. None of the treatments removed toxicity to *D. pulex* to the extent that an LC_{50} value could not be calculated, while six treatments accomplished this for the fathead minnow. This likely relates to the difference in sensitivity of *D. pulex* and *P. promelas* to heavy metals in general. Trace levels of heavy metals are often toxic to daphnids, while the fathead minnow is less sensitive to many metals. This finding illustrates the need to run tests with more than one species in preliminary fractionation studies.

Case Studies Summary

The case studies presented illustrate many of the previously discussed fractionation considerations. Most importantly, the results do not always clearly indicate the cause of toxicity. This is especially true for moderately or low toxicity effluents for which several treatments reduce toxicity. In these cases, the source of toxicity may be better defined by repeating fractionations until patterns develop. It cannot be emphasized enough that decisions on the source of toxicity should never be made based on the results of one fractionation study. Decisions on the source of toxicity and any control decisions should be based on multiple fractionations utilizing analytical support, and must consider specific conditions present at the facility.

The "all or nothing" pattern of toxicity removal was observed for many of the treatments employed. This pattern is very common and has been observed for many other effluents studied, particularly when the effluent is not markedly toxic. These case studies also emphasize the need to interpret the data with a knowledge of the true mechanisms removing toxicity by a given treatment. Mechanisms of removal are not always those typically thought of for some treatments (e.g., deposition during air stripping, physical filtration by C_{18}) and alterations of the form of the chemical can occur during some treatments (e.g., due to pH changes in ion exchange resin beds). Another observation which can be made as a result of these case studies is that trends or magnitude of change in LC_{50} values as a result of different treatments are valuable in determining the likely cause of toxicity. Statistical significance is not always necessary to gain valuable information, and in fact the largest reductions in toxicity due to a given treatment are often not statistically significant. Due to the variability inherent in biological data and the need for wide dilution ranges in preliminary tests, it will often be impossible to obtain statistical significance. Trends in the types of treatments removing toxicity and trends in the relative success of different treatments are often much more useful than statistical significance.

REFERENCES

APHA, AWWA, and WPCF. 1980. *Standard Methods for the Examination of Water and Wastewater.* Washington, DC: American Public Health Association.

Fava, J. A., D. Lindsay, W. H. Clement, R. Clark, S. Hansen, W. Rue, S. Moore, and P. W. Lankford. 1987. *Generalized Methodology for Conducting Industrial Toxicity Reduction Evaluations (TREs).* U.S. EPA Project Contract Number 68-03-3248. Cincinnati, OH: U.S. EPA.

Gasith, A., K. M. Jop, K. L. Dickson, T. F. Parkerton, and S. J. Kaczmarek. 1988. Protocol for the Identification of Toxic Fractions in Industrial Wastewater Effluents. In *Aquatic Toxicology and Hazard Assessment*, W. J. Adams, G. A. Chapman and W. G. Landis, eds., pp. 204–215. Philadelphia, PA: American Society for Testing and Materials.

Lankford, P. W., W. W. Eckenfelder, Jr., and K. D. Torrens. 1987. Technological Approaches to Toxicity Reduction in Municipal and Industrial Wastewaters. Presented at 1987 Annual Meeting, Virginia Water Pollution Control Association, Norfolk, VA.

Lankford, P. W., W. W. Eckenfelder, Jr., and K. D Torrens. 1988. Reducing Wastewater Toxicity. *Chemical Engineering*, **95**(16): 72–82.

U.S. EPA, 1985a. Methods for Measuring the Acute Toxicity of Effluents to Freshwater and Marine Organisms. EPA 600/4-85/013. Cincinnati, OH.

U.S. EPA, 1985b. Technical Support Document for Water Quality-Based Toxics Control. EPA 440/4-85/032. Washington, DC.

U.S. EPA, 1987. Permit Writer's Guide to Water Quality-Based Permitting for Toxics Control. EPA 440/4-87/005.

U.S. EPA, 1988. Methods for Aquatic Toxicity Identification Evaluations—Phase I Toxicity Characterization Procedures. EPA 600/3-88/034. Duluth, MN.

U.S. EPA, 1989a. Methods for Aquatic Toxicity Identification Evaluations—Phase II Toxicity Identification Procedures. EPA 600/3-88/035. Duluth, MN.

U.S. EPA, 1989b. Methods for Aquatic Toxicity Identification Evaluations—Phase III Toxicity Confirmation Procedures. EPA 600/3-88/036. Duluth, MN.

Walsh, G. E., and R. L. Garnas. 1983. Determination of Bioactivity of Chemical Fractions of Liquid Wastes Using Freshwater and Saltwater Algae and Crustaceans. *Env. Sci. Technol.* **17**: 180–182.

4

Source Identification, Testing, and Treatment

Perry W. Lankford

SOURCE SURVEYS

A source survey of an industrial plant can be an extremely valuable and powerful tool in incorporating source testing and treatment into an overall toxicity reduction effort. The intent and objective of the survey is to identify potentially toxic materials in the effluent, identify the source(s) of these materials, and consider alternative means of dealing with them.

There are many alternative ways of performing such a survey and the engineers most familiar with an industry or process will be the best suited to establishing and executing a plan. As a starting point, however, one logical and practical approach has been developed by the Pennsylvania Department of Environmental Resources in its *Toxics Management Strategy Guidance Manual* (Pennsylvania DER 1986), which was described in Chapter 1. The first phase in that approach is the process engineering analysis, which includes the essential ingredients of an industrial toxicity survey, the materials inventory, examination of relevant literature, and the comparison of the effluent with a raw materials inventory.

Materials Inventory

A water flow diagram should be developed for each process in the facility which generates a discharge. This diagram should identify the following as a minimum:

1. Sources of wastewater
2. Volume from each source

3. Possible contaminants in each source including raw materials and products

4. Flow of streams through treatment system(s), including indication of which streams are blended

5. Additives in the treatment system, including volume

6. Recycle streams with volume

7. Discharge(s) with volume

A complex chemical facility may require a flow diagram for each building or area.

For each flow diagram, a potential contaminant inventory should be prepared. This would involve a table identifying each raw material, product or additive which could enter the discharge stream. The table should be related back to the sources identified on the flow diagram and should include:

1. Contaminant (raw material, product, or additive)

2. Amount of each contaminant lost to effluent, or amount used and estimated loss, if losses are unavailable

3. Source of each contaminant

As an aid in preparation of these flow diagrams and contaminant inventories, sources such as handbooks and reference texts could be utilized. The U.S. EPA effluent guidelines documents for each industrial category provide some information on this subject. Table 4-1 provides a capsule summary of the basic contaminants to be considered in the organic chemicals industry. Appendix 2 provides similar tables for fifteen other industries. These, of course, are just starting points and are not accurate for any specific plant.

Table 4-1.   Organic Chemicals, Plastics, and Synthetic Fibers Industry Selected Discharge Contaminants. (U.S. EPA Treatability Manual EPA 600/2-82/001, 1981.)

Raw materials	Benzene, toluene, naphthalene, xylene, inorganic acids, phenol, formaldehyde, cellulose, methane, ethylene, propylene
Products	Dyes, biological stains, organic color pigments, acetates, nylon resins, synthetic resins, polyesters, rayon fibers, acrylic fibers, spandex
Potential toxicants	Acenaphthene, benzene, carbon tetrachloride, 1,2-dichloroethane, fluoranthene, naphthalene, phenol, pyrene, vinyl chloride, toluene, acetic acid, arsenic, silver, chloroethane

Examination of Relevant Literature

There are numerous sources of technical data in the literature on the presence, toxicity, and treatment of known pollutants. As discussed in Chapter 1, the U.S.

EPA and most state agencies have developed or collected data to assist in the permitting process and these data are also available to the permittee. Some examples of extremely useful data are:

1. U.S. EPA Categorical Industrial Effluent Guidelines
2. U.S. EPA *Quality Criteria for Water*, 1986
3. Data bases such as QSAR, AQUIRE, CETIS, OHM/TADS, and CHRIS
4. State manuals such as the Pennsylvania *Toxics Management Strategy Guidance Manual*

All of these will prove to be quite valuable in the planning and preliminary efforts of a program.

As an example of what can be assembled from these available sources, Table 4-2 was developed as a priority pollutant toxicity summary. This table provides the available data on the organic and inorganic priority pollutants relating to toxicity and treatment. The following information is provided:

1. Analytical method
2. Method detection limit
3. Chronic instream aquatic criteria
4. Acute instream aquatic criteria
5. Estimated treated effluent concentration (end-of-pipe treatment processes considered) from the Pennsylvania DER Manual

For a priority pollutant identified in the inventory or in the effluent, the potential impact on the receiving stream can be determined from the acute or chronic criteria and the stream dilution factor of the discharge. If it appears that the current level may be too high, the estimated treated effluent concentration level should be considered to determine if end-of-pipe treatment would be an applicable solution. If end-of-pipe treatment will not provide a technical or cost-effective solution, source treatment or process modification techniques will be necessary.

Comparison of Effluent with Raw Materials Inventory

The next step in the procedure is to compare the contaminants currently found in the discharge to the inventory tables previously prepared. The intent of this step is to narrow the ''hunt'' for possible solutions if the level of a contaminant is found to exceed acceptable levels.

From the literature evaluation of the contaminants found to be in the discharge, it is likely that certain compounds will be identified to be at or near toxic criteria, at least occasionally. The next step in the process is to develop a control strategy (possibly even alternate strategies) for each of these suspected

Table 4-2.   Priority Pollutant Toxicity Summary. (All measurements in μg / 1).

	Analytical Method	MDL	Chronic Instream Aquatic Criteria/TLs	Acute Instream Aquatic Criteria/TLs	Estimated (10-day average) Treated Effluent Concentration
Acenaphthene[a]	625-GC/MS	1.9	N/A	N/A	10.0
Acrolein[a]	624-GC/MS	N/A	21	68	≥100
Acrylonitrile[a]	624-GC/MS	N/A	2,600	7,550	≥100
Benzene	624-GC/MS	4.4	N/A	5,330	50
Benzidine[b]	625-GC/MS	44	N/A	2,500	25
Carbon tetrachloride (tetra-chloromethane)	624-GC/MS	2.8	N/A	35,200	50
Chlorobenzene	624-GC/MS	6.0	N/A	250	50
1,2,4-Trichlorobenzene	624-GC/MS	1.9	N/A	250	50
Hexachlorobenzene	625-GC/MS	1.5	N/A	250	10.0
1,2-Dichloroethane	624-GC/MS	2.8	20,000	118,000	1.0
1,1,1-Trichloroethane	624-GC/MS	3.8	N/A	18,000	≥100
Hexachloroethane	625-GC/MS	1.6	540	980	≥100
1,1-Dichloroethane	624-GC/MS	4.7	N/A	N/A	25
1,1,2-Trichloroethane	624-GC/MS	5.0	9,400	18,000	≥100
1,1,2,2-Tetrachloroethane	624-GC/MS	6.9	24,000	9,320	≥100
Chloroethane	624-GC/MS	N/A	N/A	N/A	50
Bis(2-chloroethyl)ether	625-GC/MS	5.7	N/A	238,000	≥100
2-Chloroethyl vinyl (mixed)	624-GC/MS	N/A	N/A	23,800	≥100
2-Chloronaphthalene	625-GC/MS	1.9	N/A	1,600	≥100
2,4,6-Trichlorophenol	625-GC/MS	2.7	970	N/A	10.0
p-Chloro-m-cresol	625-GC/MS	3.0	N/A	30	25
Chloroform (trichloromethane)	624-GC/MS	1.6	1,240	28,900	50
2-Chlorophenol	625-GC/MS	3.3	N/A	100	≥100
1,2-Dichlorobenzene	624-GC/MS	N/A	763	1,120	50

Table 4-2.   (Continued)

	Analytical Method	MDL	Chronic Instream Aquatic Criteria/TLs	Acute Instream Aquatic Criteria/TLs	Estimated (10-day average) Treated Effluent Concentration
1,3-Dichlorobenzene	624-GC/MS	N/A	763	1,128	50
1,4-Dichlorobenzene	624-GC/MS	N/A	763	1,120	20
3,3'-Dichlorobenzidine[b]	625-GC/MS	16.5	N/A	N/A	10
1,1-Dichloroethylene	624-GC/MS	2.8	N/A	11,600	≥100
1,2-trans-Dichloroethylene	624-GC/MS	1.6	N/A	11,600	≥100
Di-n-octyl phthalate	625-GC/MS	2.5	3	940	25
Diethyl phthalate	625-GC/MS	22.0	3	940	25
Dimethyl phthalate	625-GC/MS	1.6	3	940	25
1,2-Benzanthracene (benzo(a)anthracene)	624-GC/MS	7.8	N/A	N/A	1.0–10.0
Benzo(a)pyrene (3,4-benzopyrene)	625-GC/MS	2.5	N/A	N/A	1.0
3,4-Benzofluoranthene (benzo(b)fluoranthene)	625-GC/MS	4.8	N/A	N/A	1.0
11,12-Benzofluoranthene (benzo(k)fluoranthene)			N/A	N/A	
Chrysene	625-GC/MS	2.5	N/A	N/A	1.0
Acenaphthylene	625-GC/MS	2.6	N/A	N/A	1.0
Anthracene	625-GC/MS	3.5	N/A	N/A	10
1,12-Benzoperylene (benzo(ghi)perylene)	625-GC/MS	1.9	N/A	N/A	10
Fluorene	625-GC/MS	4.1	N/A	N/A	1.0
Phenanthrene	625-GC/MS	5.4	N/A	N/A	10
1,2,5,6-Dibenzanthracene (dibenzo(a,h)anthracene)	625-GC/MS	2.5	N/A	N/A	10

Compound	Method				
Indeno(1,2,3-cd)pyrene (2,3-o-phenylenepyrene)	625-GC/MS	3.7	N/A	N/A	1.0
Pyrene	625-GC/MS	1.9	N/A	N/A	1.0–10.0
Tetrachloroethylene	624-GC/MS	4.1	840	5,280	50
Toluene	624-GC/MS	6.0	N/A	17,500	50
Trichloroethylene	624-GC/MS	1.9	21,900	45,000	≥100
Vinyl chloride (chloroethylene)	601-GC/Hal	0.18	N/A	N/A	≥100
Aldrin	608-GC/ECD	0.004	N/A	3	1.0
Dieldrin	608-GC/ECD	0.002	0.0019	2.5	1.0
Chlordane (technical mixture and metabolites)	608-GC/ECD	0.014	0.0043	2.4	≥1.0
4,4'-DDT	608-GC/ECD	0.012	0.001	1.1	≥1.0
4,4'-DDE (P,P'-DDX)	608-GC/ECD	0.004	N/A	1,050	≥1.0
4,4'-DDD (P,P'-DDE)	608-GC/ECD	0.011	N/A	0.6	1.0
Alpha-endosulfan	608-GC/ECD	0.014	0.056	0.22	1.0
Beta-endosulfan[e]	608-GC/ECD	0.004	0.056	0.22	1.0
Endosulfan sulfate	608-GC/ECD	0.066	N/A	N/A	1.0
Endrin	608-GC/ECD	0.006	0.0023	0.18	1.0
Endrin aldehyde	608-GC/ECD	0.023	N/A	N/A	1.0
Heptachlor	608-GC/ECD	0.003	0.0038	0.52	1.0
Heptachlor epoxide (BHC-hexahlorocyclohexane)	608-GC/ECD	0.083	N/A	N/A	N/A
Alpha-BHC[e]	608-GC/ECD	0.003	N/A	100	N/A
Beta-BHC	608-GC/ECD	0.006	N/A	100	N/A
Gamma-BHC (Lindane)[e]	608-GC/ECD	0.004	0.08	2	N/A
PCB-1242 (Aroclor 1242)	608-GC/ECD	0.068	N/A	100	N/A
PCB-1254 (Aroclor 1254)	608-GC/ECD	N/A	0.014	2	1.0
PCB-1221 (Aroclor 1221)	608-GC/ECD	0.10	0.014	2	1.0
PCB-1232 (Aroclor 1232)	608-GC/ECD	0.10	0.014	2	1.0
PCB-1248 (Aroclor 1248)	608-GC/ECD	0.80	0.014	2	1.0
PCB-1260 (Aroclor 1260)	608-GC/ECD	0.15	0.014	2	1.0

Table 4-2. (*Continued*)

	Analytical Method	MDL	Chronic Instream Aquatic Criteria/TLs	Acute Instream Aquatic Criteria/TLs	Estimated (10-day average) Treated Effluent Concentration
PCB-1016 (Aroclor 1016)	608-GC/ECD	0.04	0.014	2	1.0
Toxaphene	608-GC/ECD	0.24	0.013	1.6	N/A
Metals					
Antimony	204.1 (AA, flame)	200	1,600	9,000	
Arsenic	206.2 (AA, furnace)	1.0	190 (As^{3+})	360 (As^{3+})	
Beryllium	210.1 (AA, flame)	5.0	5.3	130	
Cadmium	213.1 (AA, flame)	5.0	$Exp\ (0.7852(lnH) - 3.490)$	$Exp\ (1.128(lnH) - 3.828)$	
Chromium, total	218.1 (AA, flame)	5	$11 + Exp\ (0.819(lnH) + 1.561)$	$16 + Exp\ (0.8190(lnH) + 3.688)$	
Chromium, VI	307B*1[a] (Color.)	N/A	11	16	
Copper	220.1 (AA, flame)	20	0.1×96 hr LC_{50}		
Lead	239.2 (AA, furnace)	1.0	0.01 xd 96 hr LC_{50}		
Mercury	245.1 (Cold vapor, man.)	0.2	0.012	2.4	
Nickel	249.1 (AA, flame)	40	0.01×96 hr LC_{50}		
Selenium	270.2 (AA, furnace)	2.0	35 (SeO3)	260 (SeO3); 760 (SeO3)	
Silver	272.1 (AA, flame)	10	0.12	$Exp\ (1.72(lnH) - 6.52)$	
Thallium	279.1 (AA, flame)	100	40	1,400	

Cyanide, free	DER (not EPA app.)	1	5.0	22
Phenols, total	420.1 (4AAP, man.)	5	N/A	100

N/A = Not available.

Adapted from: Commonwealth of Pennsylvania (1986), and Pennsylvania Water Quality Standards (1989).

[a] If acrolein and/or acrylonitrile is expected, use Method 603 as screening method.

[b] EPA says "When Benzidine is *known* to be present, screen with EPA 605." However, because HPLC is a generally unavailable procedure at this time, GC-MS enhanced to achieve a detection level more sensitive than EPA's MDL can be used. Permit monitoring requirements for these two chemicals can also be set using EPA 625 as the acceptable analytical procedure.

[c] When hexachloropentadiene is known to be present, screen with EPA 612.

[d] When N-Nitrosodimethylamine and/or N-Nitrosodiphenylamine are known to be present, screen with EPA 607.

[e] When alpha-BHC, gamma-BHC (Lindane), alpha-Endosulfan (1), beta-Endosulfan, and/or Endrin are known to be present, screen with EPA 608.

contaminants. This amounts to identification of the suspected pollutant, its source, why it is present, and alternative means of elimination or reduction.

SOURCE TESTING

Once the sources of principal concern are identified, treatment characteristics for each source are evaluated with respect to the existing end-of-pipe treatment technology at the facility, typically biological treatment such as activated sludge. Each source discharge should be defined as to biological treatability so as to take maximum advantage of the existing effluent treatment system. Following this determination, methods for reducing or eliminating source contributions through source treatment should be specifically addressed. Gathering and utilizing this information for a large number of sources requires a well planned and organized program. This effort lends itself to presentation as a methodology flow diagram as presented in Fig. 4-1.

This methodology proceeds stepwise through the testing effort necessary to define the disposition of a given source. The alternatives are:

- Inclusion in the total wastestream without pretreatment
- Inclusion in the total wastestream after pretreatment
- Recovery, recycle, substitution, or elimination

In order to communicate the use of Fig. 4-1, the following examples will be described:

Source A: Biodegradable organics such as phenol
Source B: Biodegradable organics such as phenol with a toxic metal such as lead
Source C: Poorly biodegradable organics such as a nonylphenol surfactant
Source D: Generally untreatable source such as sulfonated organics

The analysis of source A, a biodegradable organic, would proceed as follows:

- Step 1 would result in no inhibition or toxicity; proceed to step 3
- Step 3 would indicate an adequate amount and rate of biodegradation; proceed to step 4
- Step 4 would show no toxicity of the biologically treated sample; discharge to effluent treatment system without pretreatment

The analysis of source B, a biodegradable organic combined with a toxic metal, would proceed as follows:

- Step 1 would show inhibition or toxicity; proceed to step 2
- Step 2 would result in removal of the metal through source treatment; return to step 1

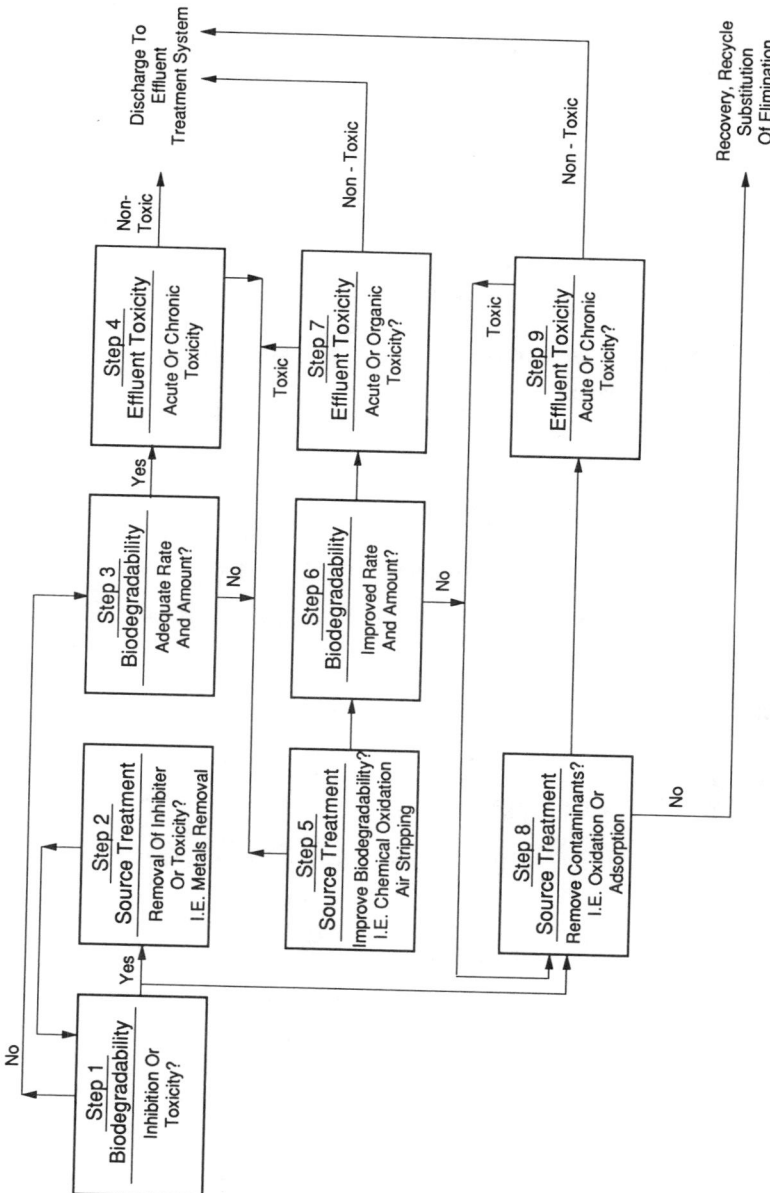

Fig. 4-1. Source testing and treatment flow diagram.

- Step 1 would show no inhibition after source treatment; proceed to step 3
- Step 3 would indicate an adequate amount and rate of biodegradation; proceed to step 4
- Step 4 may or may not show toxicity: if toxicity is indicated, proceed to step 5; if no toxicity is found, discharge to the effluent treatment system after source treatment for the inhibitor/toxicant

The analysis of source C, a poorly biodegradable organic such as nonylphenol surfactant would proceed as follows:

- Step 1 would not show inhibition or toxicity; proceed to step 3
- Step 3 would not indicate adequate amount or rate of biodegradation; proceed to step 5
- Step 5 would test source treatment options intended to improve biodegradability; proceed to step 6
- Step 6 would indicate that no improvement was achieved; proceed to step 8
- Step 8 would identify processes to remove the contaminants; proceed to step 9
- Step 9 would show no toxicity after treatment; discharge to the effluent treatment system after source treatment to remove the contaminants

The analysis of source D, a generally untreatable source, would proceed as follows:

- Step 1 would not indicate inhibition or toxicity; proceed to step 3
- Step 3 would not indicate an adequate amount or rate of biodegradability; proceed to step 5
- Step 5 would test source treatment options intended to improve biodegradability; proceed to step 6
- Step 6 would indicate that no improvement was achieved; proceed to step 8
- Step 8 would not identify a process to remove the contaminants; determine means of recovery, reuse, substitution, or elimination of the contaminants

There are, of course, many other possible pathways which could be followed depending on the character of the source. In some cases, biological reaction byproducts are toxic and end-of-pipe treatment such as carbon adsorption is required to achieve compliance. These technologies are discussed in detail in Chapters 9 and 10.

In order to be included in the overall wastestream to a biological treatment system, a source must be biologically treatable or at least noninhibitory to low contaminant levels and nontoxic in an acute or chronic test.

BIODEGRADABILITY

A biodegradable wastestream is defined as a stream which is noninhibitory and nontoxic to the microorganisms *and* treatable at a rate and to an amount adequate to meet effluent requirements. There are numerous optional biodegradability testing techniques, all with different end results and products, all with different advantages and disadvantages. It is usually recommended that no single test be relied upon, but that a second method be used to confirm results. The alternatives to be described herein are:

- Continuously Fed Batch Reactor (FBR)—provides rate and inhibition data
- Zahn Wellens—provides degradable amount data
- Glucose inhibition—provides inhibition data
- OECD Method 209 Inhibited OUR—provides inhibition data

There are numerous other methods of determining relative biodegradability which will not be described herein, such as BOD/COD ratios, respirometry, and biotreatability among others.

Inhibition

A variety of organic and inorganic chemicals will cause inhibition in the activated sludge process at some concentration level. Inhibition is indicated by either a sudden decrease in substrate utilization rate or by a decrease in oxygen utilization rate.

Volskay and Grady (1988) and Watkin and Eckenfelder (1988) have shown that inhibition can be:

- Competitive, where the inhibitor affects the base substrate utilization rate
- Noncompetitive, where only the inhibitor rate is influenced
- Mixed, in which both rates are influenced

The effects of substrate and inhibitor concentrations on the respiration rate of a microbial culture expressed as a fraction of the respiration rate in the absence of the inhibitor (control) have been shown by Volskay and Grady (1988).

While the relationships described above define the mechanism of inhibition, they are of limited use in evaluating most industrial wastewaters. In many cases, the inhibitor itself is not defined, variable sludge and substrate composition will influence inhibition, and interactions will frequently exist between inhibitors. For example, Watkin and Eckenfelder (1988) showed a variation in K_I (inhibition constant) of 6.5 mg/l to 40.4 mg/l of 2,4-dichlorophenol (DCP) for different sludges and operating conditions. Volskay and Grady (1988) showed that anywhere from 2.6 to 25 mg/l of pentachlorophenol could cause 50 percent inhibition of the oxygen utilization rate.

The inhibition constant is likely to be highly dependent on the specific enzyme system involved, which in turn is dependent on the history and population dynamics of the sludge. In some cases, the inhibition constant may be highly dependent on which metabolic pathways are present in any given microbial population. It is therefore apparent that each wastewater must be independently evaluated for inhibition. Depending on the wastewater in question, one or more protocols will be most applicable, either the FBR (Volskay and Grady 1988), the OECD Method 209 Inhibited OUR (Pennsylvania DER 1986), or the glucose inhibition test (Philbrook and Grady 1985).

Continuously Fed Batch Reactor (FBR) Procedure

Traditionally, rapid determination of inhibition thresholds and biological kinetic removal rates have been evaluated by batch studies. Such studies generally consisted of batch and/or Warburg-type respirometry tests. The FBR procedure was developed and refined to provide more accurate data in a shorter time. The FBR method (Larson and Schaeffer 1982) involves operating completely stirred, continuous-feed reactors without sludge recycle or effluent drawoff. The method allows for rapid determination of biological kinetic parameters in a relatively short time as compared to traditional batch and Warburg respirometry techniques. The method has two major advantages over other methods: (i) the test compound is continuously supplied to the reactor, so the reactor concentration of the test compound is gradually increased, thus avoiding an instantaneous shock effect as seen in other batch experiments; and (ii) actual compound concentrations are measured during the test as opposed to merely initial and final compound concentrations as in many respirometry techniques.

The objectives are to:

- Determine the maximum substrate removal rate of the wastestreams (q_{max})
- Determine if an inhibition level (S^*) exists of the wastewater toward activated sludge biomass

These values are useful in the determination of biodegradability and later in the design of the biological process. The use of these values will be further described in Chapter 6.

Procedure: A two-liter fed batch reactor with a volumetric input of 0.1 l/hr and an initial MLVSS concentration of approximately 2,000 mg/l is used in all testing. A schematic diagram of the continuously fed batch reactor configuration is shown in Fig. 4-2. During the 2.5 hour testing period, oxygen uptake rate, total organic carbon (TOC), and pH are determined at equal time intervals of 15 to 30 minutes. Depending on the wastewater, specific organic constituents can also be measured. Initial and final mixed liquor volatile solids

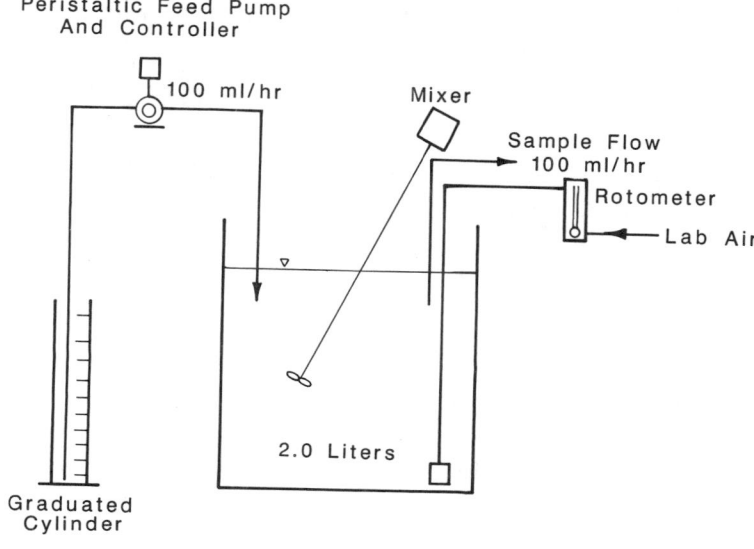

Fig. 4-2. Continuously fed batch reactor.

concentration (X_v), total nonvolatile (fixed) dissolved solids, reactor volume, and influent flow rates are also monitored. Adequate inorganic nutrients are added to the test sample as ammonium phosphate, $(NH_4)_2HPO_4$, with a TOC:N:P ratio of 100:10:1. Additionally, the feed is neutralized to a pH of 7.0 ± 0.2 and the reactor pH is maintained at 7.0 ± 0.2. The oxygen uptake rate is run in situ by periodically turning off the air and recording the decrease in dissolved oxygen.

The quantity of the feed to the FBR is established so as to (i) exceed the rate of utilization of substrate in the reactor, resulting in a net increase of substrate in the reactor, and (ii) insure the net increase in inorganic dissolved solids does not exceed 10,000 mg/l over the 2.5 hour test duration, to avoid salinity inhibition. The reactor is operated at an initial MLVSS concentration of approximately 2,000 mg/l. Prior to testing, the seed sludge is rinsed three times with equal volumes of tap water and sodium chloride. Sodium chloride is added to the rinse water at the same inorganic dissolved solids concentration as the unrinsed sludge. Rinsing the sludge should remove most of the background TOC. If possible, acclimated seed sludge should be used.

Data Analysis: Process wastestreams and individual compounds are tested for (i) maximum specific substrate utilization rate (q_{max}), in terms of mg TOC/g VSS-hour or a specific constituent; (ii) associated maximum specific oxygen uptake rate ($SOUR_{max}$), in terms of mg O_2/g VSS-hour; and (iii) existence of an inhibition threshold concentration (S^*), in terms of mg/l as TOC or a specific constituent.

A theoretical fed batch reactor response to an inhibitory compound at a mass input flow rate greater than q_{max} times X_v is shown in Fig. 4-3. Two important points should be recognized upon examination of this figure. First, the specific maximum substrate utilization rate is calculated as the difference in slopes between the projected linear concentration buildup, assuming no biological reaction (line A in Fig. 4-3), and the actual linear concentration buildup (line

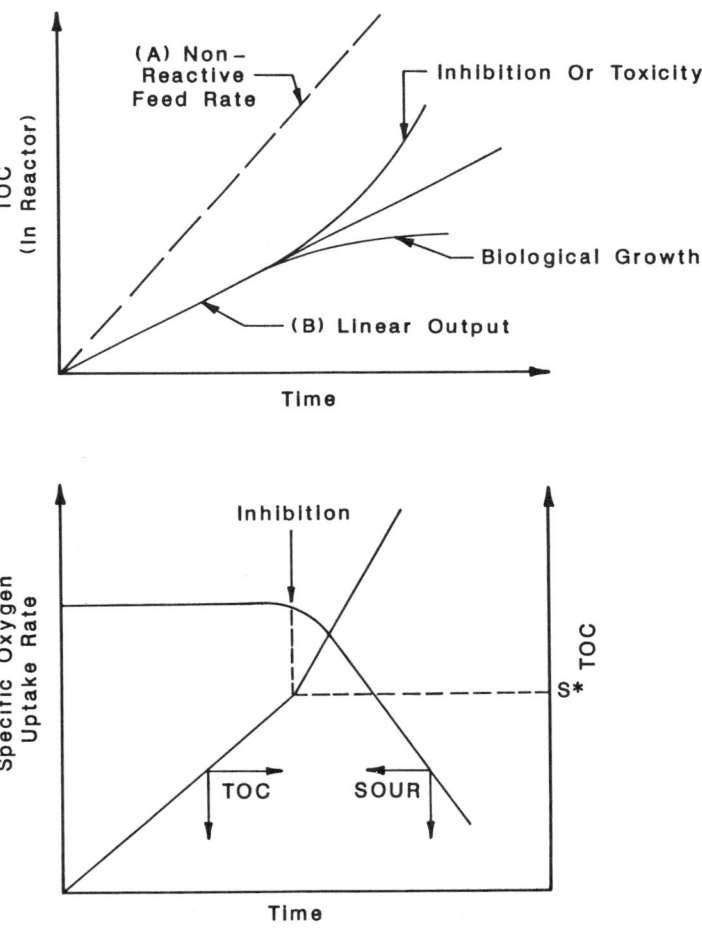

Fig. 4-3. Theoretical fed batch reactor output.

B in Fig. 4-3) divided by the biomass concentration. Second, if the test compound exhibits inhibition, then an upward deflection of the substrate concentration will be seen on the substrate concentration versus time curve. This upward deflection approximates the concentration at which inhibition first occurs and is usually accompanied by a rapid decrease in specific oxygen uptake rate.

Effect of Salinity on Reactor Performance: Preliminary tests have been conducted to determine the threshold salinity inhibition concentration for several sources of activated sludge. Fed batch reactor runs were made with sodium chloride solutions. A base substrate of mixed glucose-glutamic acid mixture was fed to the reactors at a 250 mg TOC/1-hour rate along with sufficient inorganic nutrients, such as ammonium phosphate. Sodium chloride was added at 8,900 and 13,500 mg NaCl/1-hour along with the glucose-glutamic acid solutions. The results of one of these test runs (8,900 mg NaCl/1 hour) are presented in Fig. 4-4. It can be observed that, regardless of the rate of salinity increase in the fed batch reactor, a salinity inhibition threshold of approximately 10,000 mg/l change in dissolved solids was obtained. Tests conducted at lower dissolved solids increases showed no inhibition.

Inhibition Threshold: The inhibition threshold is determined as follows. The point at which the SOUR begins to decline is determined by constructing two intersecting lines. The last SOUR point before the decline is chosen as the pivot point. A horizontal line is constructed by assuming a constant SOUR prior to inhibition and is taken as the average value of all SOUR values up to and

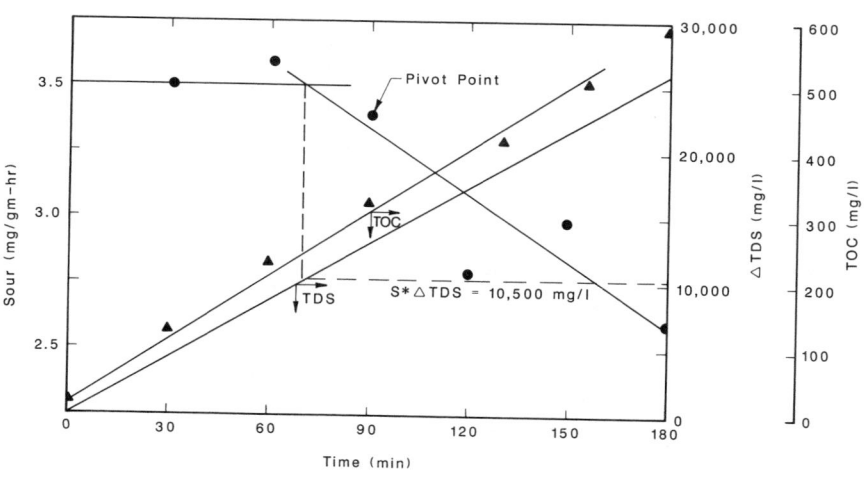

Fig. 4-4. Salinity inhibition in fed batch reactor at 8,900 mg NaCl/1-hr, 250 mg TOC/1-hr, 2,200 mg/l MLVSS.

including the pivot point. The negatively sloping line is constructed by linear regression of all points after the decline in SOUR including the pivot point. The intersection of these two lines is assumed to be the point at which inhibition is first seen. The threshold inhibition concentration is then read directly as the specific concentration in the reactor at the time corresponding to the intersection of the two lines.

The inhibitory response in the fed batch reactor was studied for a known inhibitory substance, namely, 2,4-dichlorophenol (DCP). Figs. 4-5, 4-6, 4-7, and 4-8 show the response at four different mass input rates: 20, 25, 40, and 50 mg DCP/1-hour, respectively. These demonstrate inhibition thresholds of 14, 15, 17, and 15 mg/l as DCP, respectively. The inhibition threshold was found to be unaffected by mass loading rate and an average inhibition threshold concentration of 15 mg/l DCP with a standard deviation of 1.4 mg/l was obtained.

Figure 4-9 shows a typical response of a nontoxic wastestream in the fed batch reactor. Here, maximum specific oxygen uptake rate ($SOUR_{max}$) of 11.7 mg O_2/g VSS-hour was rapidly attained and corresponded to a maximum substrate utilization rate q_{max} of 4.9 mg TOC/g VSS-hour. Since there was no decrease in $SOUR_{max}$ nor in q_{max} over the test duration, the tested wastestream was not found to be inhibitory up to a TOC concentration of 200 mg/l.

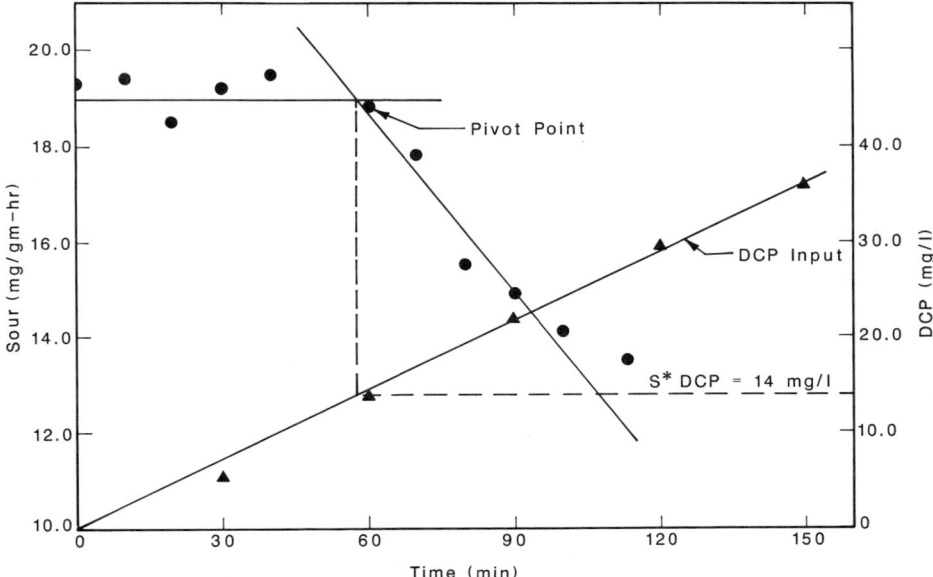

Fig. 4-5. DCP inhibition in fed batch reactor at 20 mg DCP/1-hr, at 1,600 mg/l MLVSS.

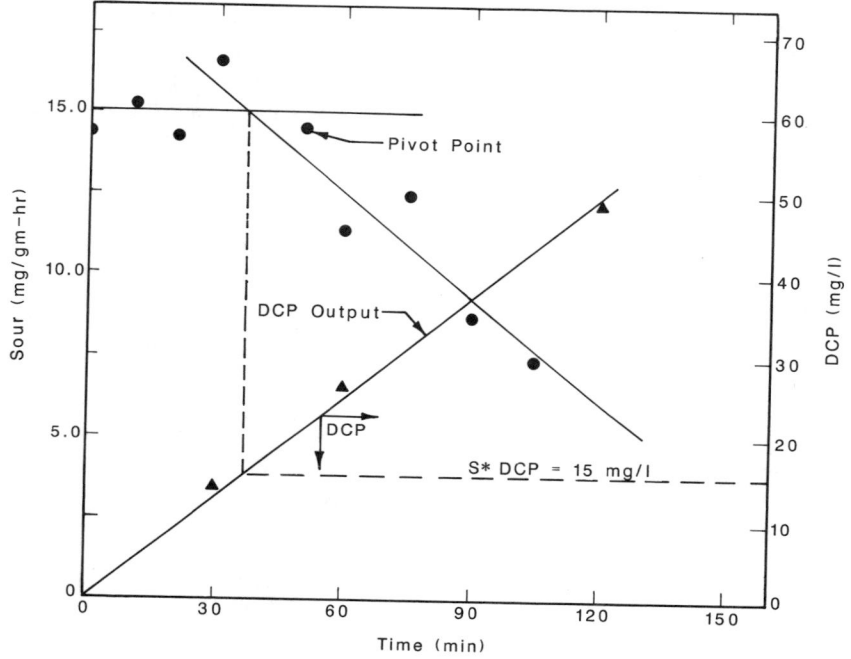

Fig. 4-6. DCP inhibition in fed batch reactor at 25 mg DCP / 1-hr at 560 mg / l MLVSS.

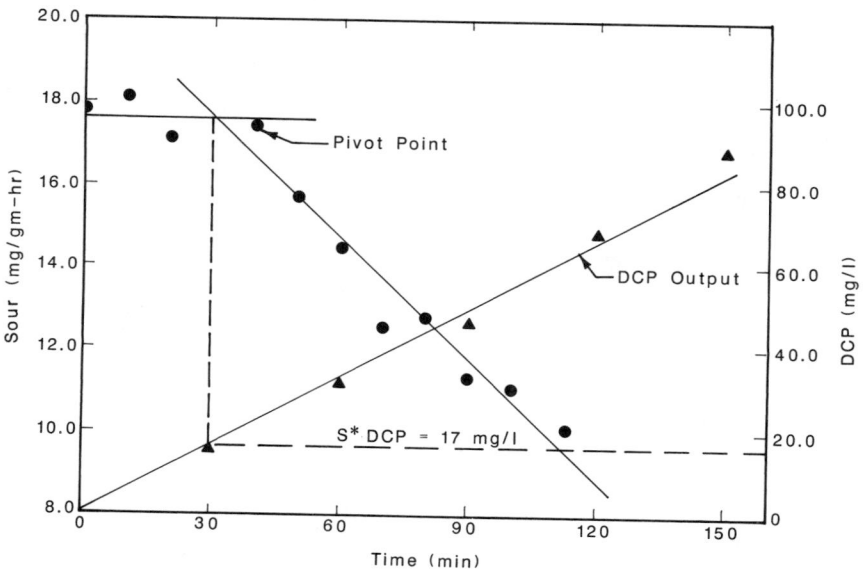

Fig. 4-7. DCP inhibition in fed batch reactor at 40 mg DCP / 1-hr, 1,600 mg / l MLVSS.

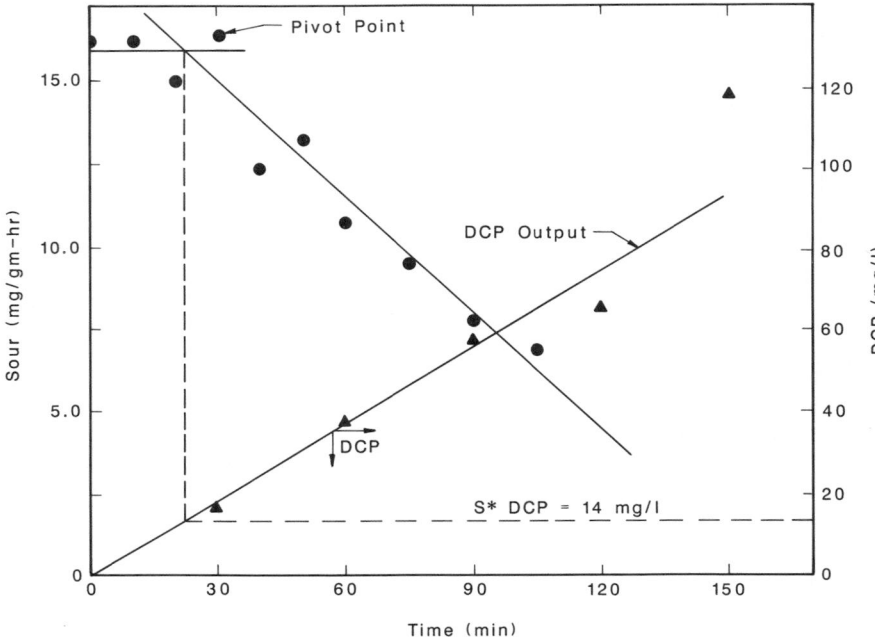

Fig. 4-8. DCP inhibition in fed batch reactor at 50 mg DCP / 1-hr, 560 mg / 1 MLVSS.

Although very few wastestreams demonstrate biological inhibition, Fig. 4-10 shows a marked inhibitory response. A poor TOC removal rate and a dropping oxygen uptake rate were seen early in the test. The inhibition threshold concentration can best be described as less than 12 mg/l TOC.

Uncoupling Effect: Several acidic aromatic and lipid soluble organic compounds have been demonstrated to uncouple oxidative phosphorylation. The result of uncoupling of oxidative phosphorylation is uncontrolled respiration and oxidation of primary substrates and intracellular metabolites. At relatively low concentrations, uncoupling is observed as highly elevated oxygen consumption rates and no effect on cell growth. At higher concentrations, inhibition and toxicity are seen by dramatic reductions in both oxygen utilization rates and cell growth. The effects for the oxidation of glucose in the presence of DCP are shown in Fig. 4-11.

Zahn Wellens Procedure

The Zahn Wellens procedure (ZWP) is designed to determine the ultimate amount of biodegradability of a compound or a mixture of compounds present

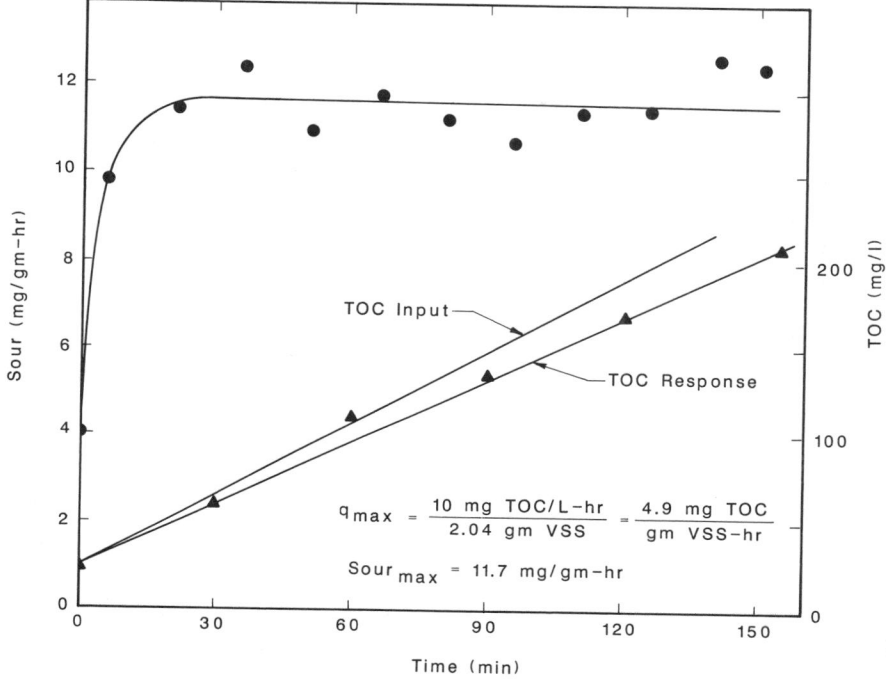

Fig. 4-9. Typical response of non-inhibitory wastestream in fed batch reactor, 2,150 mg/l MLVSS.

in wastewater streams (Zahn and Wellens 1974). The biodegradability of a compound is determined by the percentage of the TOC degraded after fourteen days. It is assumed that after fourteen days acclimation will have occurred, and all the biodegradable materials present in the tested wastewater stream have been decomposed. The FBR procedure determines the maximum biodegradation rate, while the ZWP determines the amount of organics that can be biodegraded. One must, therefore, use both techniques to determine the applicability of a biological system to treat a specific wastewater stream.

Procedure: The ZWP is designed to determine the decrease of the TOC in the test sample in the presence of activated sludge bacteria and nutrient salts during fourteen days. The activity of the activated sludge is monitored by simultaneous testing of a reference substrate (aniline).

Drinking water should be used as the dilution water for the activated sludge to which the samples will be added. However, if the chlorine content in the drinking water is too high, it will result in decreased degradation of the reference substance.

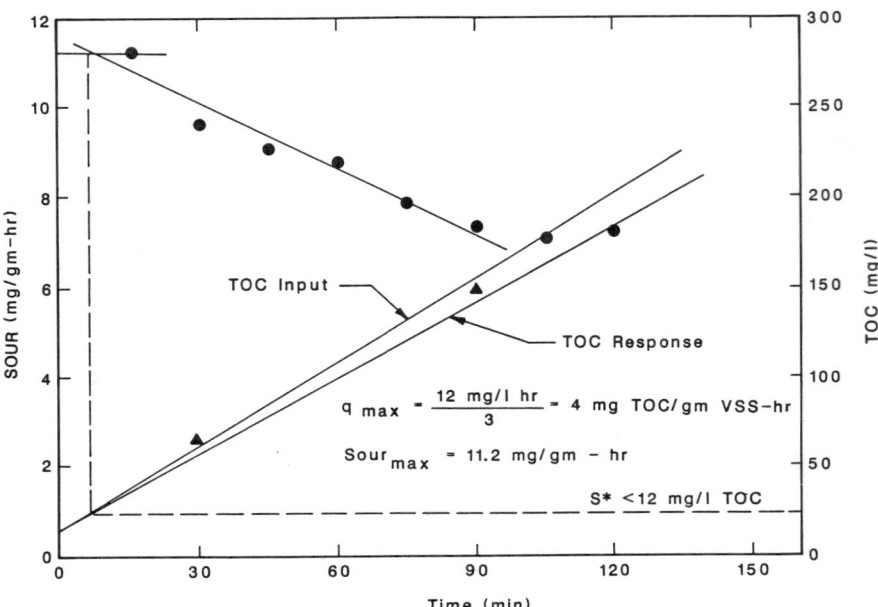

Fig. 4-10. Typical response of inhibitory wastestream in fed batch reactor, at 3,180 mg/l MLVSS.

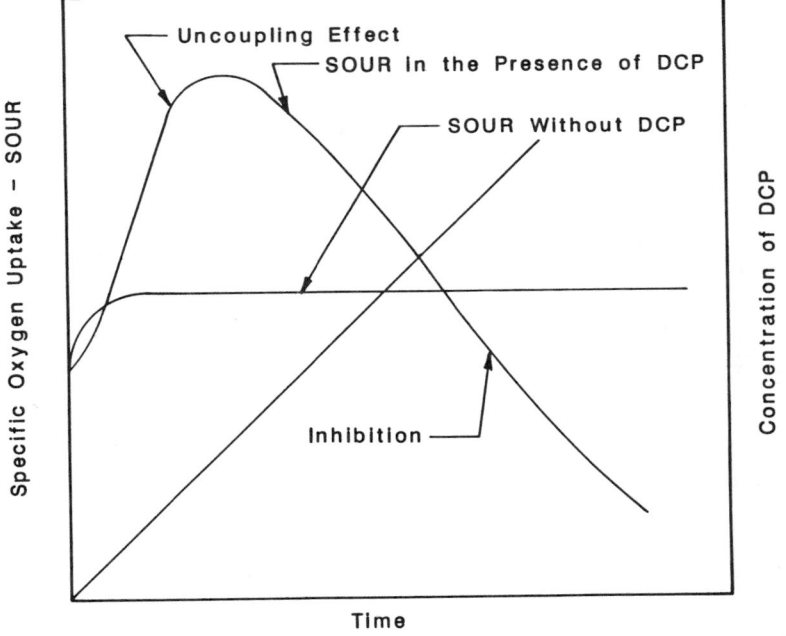

Fig. 4-11. Effect of uncoupling on the FBR response of glucose in the presence of DCP.

A nutrient salt solution is made by dissolving analytical grade nutrient salts in one liter of drinking water as follows:

- 38.5 g ammonium chloride
- 33.4 g sodium dihydrogen phosphate
- 8.5 g potassium dihydrogen phosphate
- 21.75 g dipotassium hydrogen phosphate

This solution is added to the sludge to be used at a dosage of 5 ml per liter.

The activated sludge to be used in the test should be fresh and washed with thirty volumes of drinking water to reduce the TOC to 20 to 40 mg/l or lower. Since washing the sludge can require six to seven hours, an automatic system is preferable. If the fresh sludge cannot be washed immediately (< one hour), it should be aerated until washed. At least 10,000 mg sludge is necessary for each test. In order to attain this, the sludge can be concentrated by sedimentation, but centrifugation is not recommended.

Each test is run in a mixed and aerated three liter beaker. The compressed air used should be cleaned with a cotton-wool strainer. A thermometer should be immersed into the test vessel to measure the temperature.

The test sample and analytical grade aniline are diluted with drinking water to a TOC concentration of approximately 400 mg/l. These will be added to two test chambers. A third test chamber will be a blank control containing water and activated sludge without sample. If the sample contains volatile compounds which may be air-stripped during the test, a control chamber with water and sample, but without activated sludge, is set up.

Test Conditions: The temperature in the test vessel should be 22 ± 3°C. The pH should be 6.0 to 7.5, and the dissolved oxygen level should be at least 2 mg/l. The testing duration will be 14 days and analysis will consist of TOC determination.

For the test sample add one liter activated sludge at a concentration of at least 10,000 mg/l into the three liter beaker. In a liter cylinder, add 10 ml nutrient salt solution to 500 ml drinking water, add enough sample (neutralized to pH 7) to correspond to 800 mg TOC/l when diluted to one liter, and fill the cylinder with drinking water to one liter. Add this solution to the three liter beaker with the activated sludge. The test chamber should now contain activated sludge at a concentration of 5,000 mg/l and sample at a concentration of 400 mg/l TOC.

For the aniline control, prepare the sludge mixture in the beaker just as before. Prepare the aniline solution as before except substituting aniline for the sample. Combine the aniline solution in the three liter beaker with the activated sludge. This test chamber should also contain activated sludge at a concentration of 5,000 mg/l and aniline at a concentration of 400 mg/l TOC.

For the blank control, prepare the sludge mixture in the beaker just as before. Prepare the water solution as before but without sample or aniline. Combine these solutions into the three liter beaker. This test chamber should contain activated sludge at a concentration of 5,000 mg/l and no sample or aniline.

For the stripping control, bypass the sludge mixture preparation and make up the one liter of sample solution just as before. Add this solution to the three liter beaker and add an additional liter of drinking water. This test chamber should contain sample at 400 mg/l TOC and no activated sludge.

Immediately remove a sample from each chamber prior to any aeration or agitation, and analyze it for TOC (TOC_A = initial value). Also record pH, dissolved oxygen, and temperature. Begin aeration and agitation and remove the second sample after three hours. Additional samples are taken three times a week for fourteen days. Prior to sampling, replenish the volume of each chamber to two (2) liters with drinking water.

Data Analysis: TOC elimination is determined by

$$E(\%) = 1 - \frac{(TOC_T - TOC_B)}{TOC_A} \times 100 \qquad (4\text{-}1)$$

where:

TOC_A = TOC value at the start of the test, mg/l
TOC_T = TOC value after fourteen days, mg/l
TOC_B = TOC value of blank after fourteen days, mg/l
E = TOC elimination, percent

The test can be discontinued prior to fourteen days if 70 percent elimination is obtained before that time.

The following descriptions are applied to the test results:

Readily degradable: ≥ 70 percent elimination
Moderately degradable: 20 to 70 percent elimination
Poorly degradable: ≤ 20 percent elimination

The test is valid only if the aniline test shows an elimination value of 70 percent or greater within fourteen days.

Comparison with FBR: Forty-eight samples which had been determined by the FBR procedure to be degradable were also tested by ZWP for verification. Of the forty-eight samples analyzed, seventeen were classified as readily degradable, twenty-three were classified as moderately degradable, and four were classified as poorly degradable.

Glucose Inhibition Test

Larson and Schaeffer (1982) developed a simple and rapid toxicity test based on the inhibition of glucose uptake by activated sludge in the presence of toxicants. The end result of the test is the determination of the concentration of the sample or toxicant which causes a 10 percent (EC_{10}) or 50 percent (EC_{50}) inhibition of glucose uptake as compared to a control. The test was subsequently modified for application to a variety of industrial wastewaters.

Procedure:

1. Place 10 ml of sample into a centrifuge tube
2. Add 1 ml of the stock glucose solution
3. Add 10 ml of activated sludge into the centrifuge tube and aerate at low rate
4. After exactly sixty minutes, add two drops of concentrated HCl and transfer tubes to the centrifuge
5. After centrifugation, measure glucose concentration of the clarified liquid according to the glucose test kit instructions
6. Sludge control—substitute 10 ml of deionized water for the sample in step 1 and perform steps 2 through 5 as before
7. Glucose control—place 30 ml of deionized water in centrifuge tube; add 1 ml of stock glucose solution and add two drops of concentrated HCl; do not add sludge or aerate

Data Analysis:

$$\% \text{ Inhibition} = \frac{C - C_B}{C_0 - C_B} \tag{4-2}$$

C = Final glucose concentration in sample solution
C_B = Final glucose concentration in sludge control sample
C_0 = Initial glucose concentration (glucose control)

The data can be plotted on log-probability paper as shown in Fig. 4-12 and the EC values determined. Under some conditions, the lowest amount showing inhibition (EC_{10}) may be of interest, while in source comparisons a more substantial inhibition (EC_{50}) may be of interest.

Modified OECD Method 209

The Organization for Economic Cooperation and Development (OECD) Method 209 (Volskay and Grady 1988) protocol involves measurement of oxygen uptake rate from a synthetic substrate with activated sludge to which a test compound

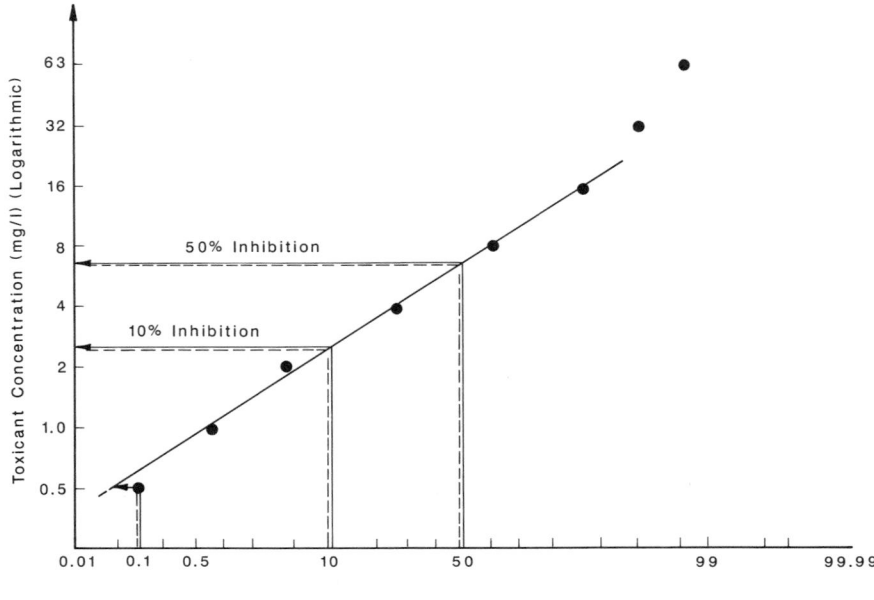

Fig. 4-12. Typical results of glucose inhibition test.

or wastestream has been added at various concentrations. The result of the test is determination of an EC_{50} in mg/l of a specific toxicant such as Dichlorophenol or in terms of COD or TOC. The EC_{50} is the concentration at which the specific oxygen uptake rate is inhibited to 50 percent of the specific oxygen uptake rate of the uninhibited control. The oxygen uptake rate is measured immediately after addition of the test compound and again after 30 minutes of aeration, with the latter being used for data analysis. The synthetic substrate used is the OECD stock solution containing the following quantities per liter: 16 g of peptone; 11 g of beef extract; 3 g of urea; 0.7 g of NaCl; 0.4 g of $CaCl_2 \cdot 2H_2O$; 0.2 g of $MgSO_4 \cdot 7H_2O$; and 28 g of K_2HPO_4. The OECD recommends the use of 3,5-dichlorophenol as a reference toxicant to ensure that the test is working properly and that the biomass has the appropriate sensitivity. The EC_{50} value for 3,5 DCP should lie between 5 and 30 mg/l for the test to be valid.

Volskay and Grady (1988) employed this protocol with modifications to determine the toxicity of selected priority organics. Since many of the compounds were volatile, the test was modified by using more dilute cell and substrate concentrations and by conducting the test in vessels that were sealed by the insertion of a polyfluoroethylene plug. This protocol is recommended for wastewaters containing high concentrations of volatile organics. The results obtained by Volskay and Grady (1988) for a variety of organics are summarized in Table 4-3.

Volskay-Grady Procedure: The Volskay-Grady procedure is intended for determining the inhibition of pure compounds when blended with a cosubstrate versus results on the cosubstrate alone:

1. Add microorganisms and OECD stock substrate to attain final concentrations of 75 mg/l and 375 mg/l, respectively, after addition of the test sample volume
2. Stir and aerate the mixture for 5 minutes to achieve oxygen saturation; add dilution water at the same volume as the test samples; measure DO after 5 minutes and 30 minutes; reaerate only if the DO drops below 40 percent of saturation
3. Add the test sample at varying concentrations bracketing the level anticipated to be experienced
4. Express the oxygen consumption for the test sample as a percentage of the oxygen consumption for the control; plot this percentage versus the log concentration of the test sample; manually determine the line of best

Table 4-3.   Selected EC_{50} Values Determined by the Modified OECD Procedure.

Compound	Observed EC_{50}[a] (mg/l)	EC_{50} Observed by Others[b] (mg/l)
Carbon tetrachloride	24	100
Chlorobenzene	140	—
Chloroform	500	—
1,2-Dichloropropane	520	—
2,4-Dimethylphenol	190	—
2,4-Dinitrophenol	110	—
Furfural	760	—
Nitrobenzene	320	100
4-Nitrophenol	72	59, 98, 126
Pentachlorophenol	2.6	2.5, 24, 25
Phenol	520	800
Tetrachloroethylene	170	—
1,1,1-Trichloroethane	360	—
1,1,2-Trichloroethane	440	—
Trichloroethylene	260	1,000

Source: Volskay and Grady (1988).
[a]After 30 minutes of contact.
[b]Data generated by OECD procedure only.

fit of the data points in the region of 50 percent oxygen consumption. The EC_{50} value is the 50 percent intersection of this line.

Modified Procedures: The OECD test can be modified for the evaluation of comingled industrial wastewaters:

1. Add wastewater to microorganisms at the present loading rate to the treatment plant
2. Determine the baseline oxygen uptake rate without inhibition as in step 2 above
3. Add progressively increasing concentrations of the suspect wastewater to the mixture defined by step 1 and repeat step 2
4. Plot the results as in step 4 above.

Results of use of the modified procedure on plastics derivatives wastewater suspected of inhibition is shown in Fig. 4-13.

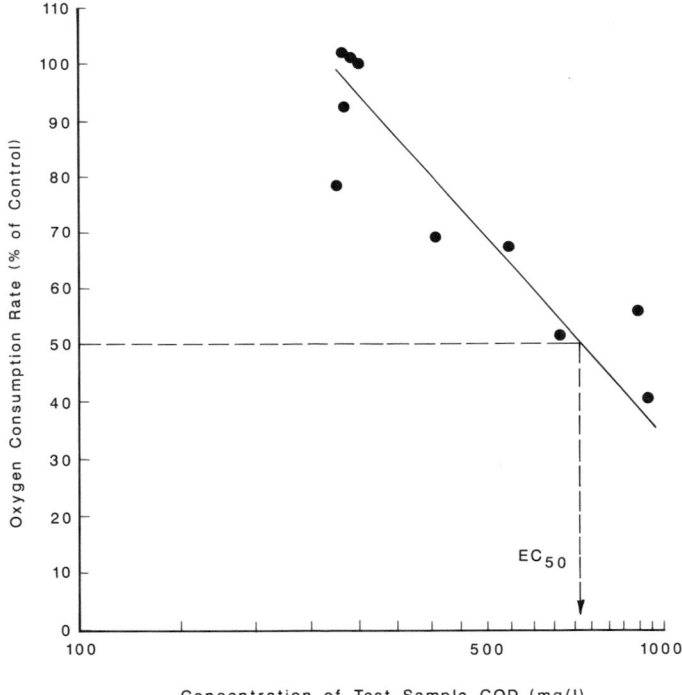

Concentration of Test Sample COD (mg/l)

Fig. 4-13. Determination of OECD EC_{50} based on COD for a plastics derivatives wastewater.

TOXIC UNITS

When dealing with conventional parameters, it is a fairly straightforward materials balance procedure to determine if biological treatment of a source will comply with effluent requirements after blending with other sources. The concentrations of these parameters can be converted to mass, summed, and converted back to a concentration. Effluent toxicity expressed as percent of whole effluent, however, cannot be similarly added.

Table 4-4 shows the results of biological treatment of seven sources at a dyes and plastics plant. Each source was diluted to simulate its contribution to the final effluent. Streams can be identified which contribute to the effluent TOC, BOD, ammonia, and metals. Each of these can then be addressed and these seven streams can safely be added to the treatment system. Conversion of toxicity to toxic units allows the same to be done with this parameter.

Definition of LC_{50}:   If the LC_{50} of compound y is x mg/l and a sample contains only y as a contaminant, the dilution determined to result in mortality to 50 percent of test organisms (an LC_{50}) has a concentration of compound y of x mg/l. Therefore, for a diluted sample of y to result in an LC_{50}, it must, by definition, contain x mg/l. If an unknown concentration of y must be diluted to 10 percent to achieve an LC_{50}, it had a 5 times higher concentration than a sample which only had to be diluted to 50 percent to achieve an LC_{50}. The final concentration in both diluted samples had to be x mg/l in order to result in an LC_{50}. If an LC_{50} of an unknown sample of y is determined to be exactly the

Table 4-4.   Treatment of Individual Source Wastestreams.

Source Area	Products or Operations	Treated 48-hr LC_{50}, %	BOD Removal, %	TOC Removal, %	Average OUR, g/g-day	NH_3-N	Cu	Ni
1	Wash down, spray drying products	15–25	0	2	0.043	<1	0.09	0.02
2	Epoxy resins	5–15	65	8	0.099	<1	0.09	0.32
3	Azo dyes	15–25	67	13	0.029	10	1.4	0.10
4	Dye standardization, spray drying, milling	5–25	83	34	0.047	54	0.15	0.07
5	Intermediates	5–15	91	50	0.140	40	0.08	0.17
6	Azo dyes	5–15	94	33	0.110	15	2.2	0.13
7	Epoxy resins, additives stabilizers, antioxidants	15–25	97	32	0.120	47	0.03	0.30

[a]All streams were introduced to activated sludge systems simulating the existing system operating at F/M of 0.01. Due to analytical error, source 2 ended up at higher F/M.

100 percent strength of the sample, then the concentration of y in the sample is exactly x mg/l.

Toxic Unit: A toxic unit (TU) is a unitless measurement of the concentration of a contaminant in a sample (C_s) compared to the concentration of that contaminant resulting in an LC_{50} (C_{LC50}).

$$TU = C_s/C_{LC50} \qquad (4\text{-}3)$$

Whole effluent toxicity (WET) is a unitless (percent being unitless) measurement of the concentration of a contaminant resulting in an LC_{50} (C_{LC50}) compared to the original concentration in the sample (C_s).

$$WET\ (\%) = (C_{LC50}/C_s) \times 100 \qquad (4\text{-}4)$$

Combining the above expressions:

$$TU = 100/WET\ (\%) \qquad (4\text{-}5)$$

Therefore, to calculate toxic units, either divide the sample concentration (C_s) by the LC_{50} concentration (C_{LC50}), or take the inverse of the whole effluent toxicity (WET) times 100.

Purpose: The purpose of toxic units is to simplify preparing material balances for toxicity, i.e., adding samples, checking for closure, etc. Assuming samples have common contamination, but at varying concentrations (C_{s1}, C_{s2}, C_{s3}, . . .), and using whole effluent toxicity, those would be summed as follows:

$$Sum = \frac{C_{LC50}}{C_{s1}} + \frac{C_{LC50}}{C_{s2}} + \frac{C_{LC50}}{C_{s3}} + \ldots \qquad (4\text{-}6)$$

Without common denominators, summing these expressions would be very cumbersome. However, the toxic unit expression is the inverse of WET and therefore has common denominators and can be summed, as follows:

$$Sum = \frac{C_{s1}}{C_{LC50}} + \frac{C_{s2}}{C_{LC50}} + \frac{C_{s3}}{C_{LC50}} + \ldots = \frac{C_{s1} + C_{s2} + C_{s3}}{C_{LC50}} \qquad (4\text{-}7)$$

However, we must adjust for volume of each component sample in the final sample:

$$Sum = \frac{C_{s1}}{C_{LC50}}(V_1/V) + \ldots = \frac{C_{s1}V_1 + C_{s2}V_2 + C_{s1}V_3}{C_{LC50}V} \qquad (4\text{-}8)$$

or

$$Total\ TU = TU_1(V_1/V) + TU_2(V_2/V) + TU_3(V_3/V) \qquad (4\text{-}9)$$

Table 4-5 shows an example of the summing of toxic units in a materials balance. The whole effluent toxicity of each stream was determined after a 25 : 1 dilution. The dilution was made to obtain an adequate range of results. The

Table 4-5.   Use of Toxic Units Expression to Prepare Materials Balance

Stream No.	Untreated Toxicity,[a] LC$_{50}$, %	Toxic Units[b]	Volume in Sample, ml	Dilution Factor	Actual Toxic Units in Composite, TU × 25/DF	Untreated Additive Toxicity,[d] (LC$_{50}$, %)
35	60	1.67	17.2	407	0.10	
38	37	2.70	12.0	583	0.12	
39	>87	1.15	5.6	1,250	>0.02	
41	<87	>1.15	28.8	243	<0.12	
43	24	<4.17	14.4	486	0.21	
48	60	1.67	32.6	215	0.19	
51	24	4.17	17.6	398	0.26	
52	63	1.59	28.0	250	0.16	
54	46	2.17	20.0	350	0.16	
61	35	2.86	6.4	1,094	0.07	
Dilution Water	———		6,817			
		23.30	7,000	———	1.41	71

[a]Sample diluted 25 : 1.
[b]100 ÷ LC$_{50}$.
[c]Toxic Units × (25 ÷ Dilution Factor).
[d]100 ÷ Actual Toxic Units.

toxic units were then computed. The volume of each full-strength component to make up the total was then used to compute a dilution factor for each sample into the total. From that, the toxic units to be added from each component could be computed and summed. The whole effluent toxicity of the composite could then be computed, in this example, 71 percent effluent LC$_{50}$. This composite sample was then made up and tested and, as can be seen in Table 4-6, it was determined to have an LC$_{50}$ of 75 percent, which agrees well with the calculated value of 71 percent.

We can then conclude that the toxicity of these samples is additive. There have occasionally been reports of synergism, where the toxicity of the composite is worse than the sum of the components, but this is rare. More commonly, the composite is less toxic than the sum due to components not sharing the same toxic mechanisms.

Table 4-6 also shows that after batch biodegradation, the composite toxicity

Table 4-6.   Confirmation Test Results.

Days After Startup	Treated LC$_{50}$, %
7	75
18	20
23	35
30	21

got worse. This relatively common occurrence has been attributed to the production of toxic biological end-products in the degradation process. This phenomenon is discussed in detail in Chapter 6.

SOURCE REDUCTION AND TREATMENT

As described previously in Fig. 4-1, source treatment requirements can include any of the following, depending on the nature of the source:

- Pretreatment to remove a bio-inhibitor or bio-toxicant
- Pretreatment to improve biodegradation rate or amount
- Removal of contaminants in order to discharge
- Treatment for recovery, recycle, substitution, or elimination

Where possible, process modification should be evaluated initially as the preferred solution. This might include substitution of alternative raw materials, modification of reaction conditions of tie up or precipitate the component, recovery procedures, or other possibilities. Due to its very nature, this treatment alternative, when successful, will normally be the most cost-effective solution.

One example of this type of solution is the discontinuation of sulfonation of insoluble organics in the chemical industry. Sulfonation is sometimes employed to solubilize organic material to avoid solid waste disposal problems; however, these sulfonated compounds are often quite toxic, relatively poorly biodegraded, and almost nonadsorbable on carbon. Discontinuing this practice will not affect the product but can dramatically improve the process discharge toxicity. Another example of process modifications is the chelation or precipitation of heavy metals discharged from a source. This can be accomplished through various means such as pH adjustment or chemical addition in the process or in a source treatment system.

After process modification is considered, the second effort should consist of conducting treatment tests on the identified sources. Depending on the nature of the contaminant, the processes typically found to be applicable are indicated in Table 4-7. The applications shown are not intended to eliminate consideration of any of the processes, but only to indicate applicability.

Chemical Oxidation

In some cases, adequate source treatment may be achieved by chemical oxidation. Common oxidants include permanganate, ozone, and hydrogen peroxide. The chemical degradation of refractory organics may take several forms dependent upon the nature of the compound under attack, physical parameters such as contact time, temperature, and pH as well as the oxidant used. These include primary degradation, in which a structural change occurs in the parent

Table 4-7.   Application of Source Treatment Technologies.

	Removal of Bioinhibitor or Biotoxicant	Improve Biodegradation Rate	Remove Contaminants for Discharges	Recovery, Recycle, Substitution, or Elimination
Metals precipitation	X		X	X
Chemical oxidation	X	X		
Wet air oxidation			X	
Macroreticular resin adsorption			X	
Carbon adsorption			X	
Membrane Processes			X	X
Stripping			X	
Filtration	X			

compound and results in improved biodegradability; intermediate degradation, resulting in reduced toxicity; and ultimate degradation, resulting in complete destruction to CO_2, H_2O, and other organics. The use of chemical oxidants to provide ultimate degradation of organic compounds may be extremely expensive, and will require the largest oxidant demand. However, a primary or intermediate degradation of compounds may be carried out with a much smaller oxidant demand and, therefore, when integrated with biological treatment, may represent a cost-effective means of reducing toxicity. These alternative treatment approaches are shown in Figure 4-14. This subject is thoroughly covered in Chapter 11.

The use of ozone to pretreat a mixed plastics additive and dyestuff wastewater following biological treatment is summarized in Table 4-8. While TOC reduction was not dramatic, the toxic units reduction was quite effective. It should be noted that conversion of refractory organics to biodegradable organics may result in a significant increase in effluent BOD.

Chemical oxidation using catalyzed hydrogen peroxide has been successful in some cases. Table 4-9 shows treatment results for two wastewaters. Although the organic removals were nominal, a substantial reduction in toxicity was observed for wastewater A. For wastewater B, the reverse was observed. It becomes obvious that the applicability of chemical oxidation for toxicity reduction must be considered on a case-by-case basis.

Wet Air Oxidation

Wet air oxidation (WAO) has been successfully demonstrated in a number of applications for source treatment. WAO is based on a liquid phase reaction between organic material in the wastewater and oxygen supplied by compressed air. The reaction takes place flamelessly in an enclosed vessel, which is pres-

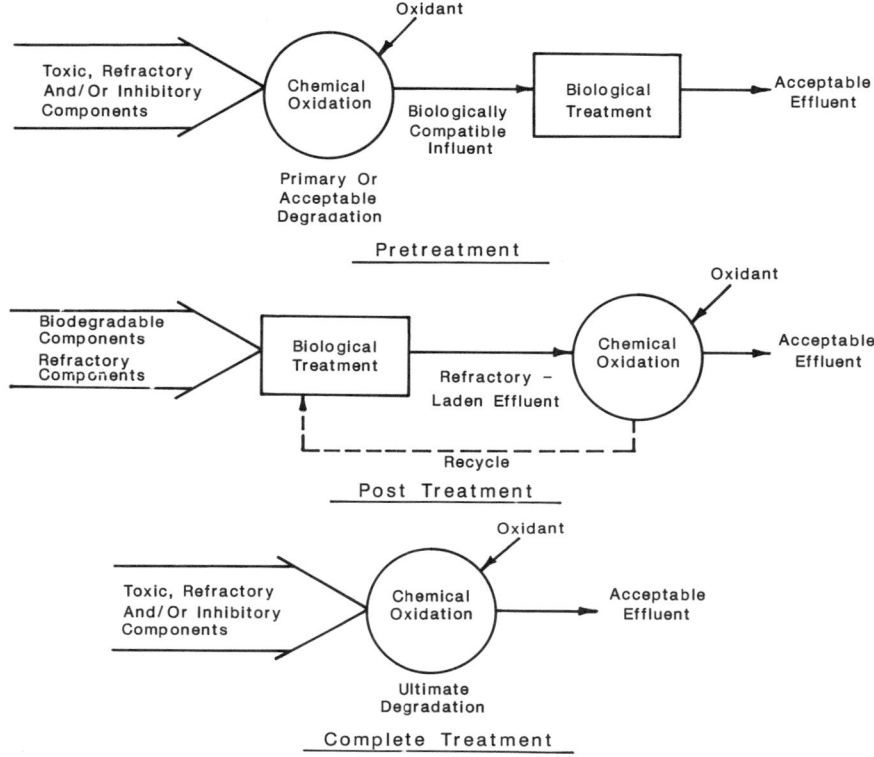

Fig. 4-14. Modes of chemical oxidation of toxic wastewaters.

surized and at high temperatures, typically 2,000 psig and 550°F. The system temperature, initiated by a startup boiler, is maintained through autothermal combustion of organics once the reaction starts. Typical performance data of WAO on concentrated organics are shown in Table 4-10 (Chudoba 1985). Changes in various parameters following WAO for two (2) wastewaters are

Table 4-8. Results of Ozone Oxidation Treatment

Ozone Dosage, mg O_3/l	Toxic Units[a]	TOC, mg/l
1	17.9	152
100	2.0	96
300	1.7	108
600	1.15	88

[a]*Mysidopsis bahia*

Table 4-9.   Catalyzed Hydrogen Peroxide H₂O₂ Oxidation.

	TOC, mg/l	COD, mg/l	BOD, mg/l	Toxic Units
Wastewater A				
Before H₂O₂	92	301	135	6.14
After H₂O₂	52	184	57	3.37
Removal, %	43.5	38.9	57.8	45.1
Wastewater B				
Before H₂O₂	2,150	2,040	300	41.7
After H₂O₂	95	850	42	33.3
Removal, %	95.6	58.3	86.0	—

Table 4-10.   Typical Performance Data for Wet Air Oxidation of Concentrated Organics.

	Influent, mg/l	Effluent, mg/l	Reduction, %
Total phenols	28,700	66	99.8
Organic sulfur	3,010	180	94.0
Sulfide sulfur	333	1	99.7
Cyanide	25,400	82	99.7
Pesticides (malathion, carbaryl, methoxychlor, dinoseb)	169	0.93	99.5
Total COD	76,000	2,500	96.7

Table 4-11.   Wet Air Oxidation of Concentrated Wastewater Streams[a]

	Wastewater	
	A	B
Influent COD, g/l	28.8	198.2
Reduction, percent	79.1	86.3
Influent BOD, mg/l	228	82,500
Reduction, percent	99.5	83.8
Influent color, APHA units	1 × 10[b]	—
Reduction, percent	98.5	—
Influent toxic units[b]	90.9	31.3
Reduction, percent	73.2	96.3

[a] Batch, bench-scale WAO treatment at 260°C for 60 minutes.
[b] 48 hr LC₅₀ mysid shrimp at 1:25 diln.

summarized in Table 4-11. As was the case for chemical oxidation, reduction in toxicity was not directly correlated to reduction in COD or BOD. Additional information on WAO is provided in Chapter 11.

Macroreticular Resin Adsorption

Although not yet widely applied, macroreticular resins may be excellent candidates for removal of specific nonpolar organic compounds, leaving other compounds for removal through more conventional means. Such resins are highly specific and thus can be formulated to remove one (1) compound, or a class of compounds. Evaluation and application of this technology requires a great deal more information about the specific compounds involved and significantly more sophisticated analytical capability than most other processes. If properly tested and applied, however, resin columns can be quite cost-effective with or without solvent recovery and reuse.

Macroreticular resins are effective for source reduction due to mechanisms similar to activated carbon, i.e., adsorption of wastewater constituents. Resins are applicable to wastestreams requiring removal of specific compounds which are known to be toxic or inhibitory to downstream processes. The systems are typically operated as a series of columns, with replacement of the lead column after all adsorption sites are exhausted. Greater than 99 percent removal of selected compounds is achievable using macroreticular resins as shown in Table 4-12. For further information on macroreticular resin adsorption, see Chapter 9.

Table 4-12. Macroreticular Resin Treatment of Selected Compounds.

Compound	Influent	Effluent	Percentage Reduction
Carbon tetrachloride	20,450	490	97.6
Hexachloroethane	104	0.1	99.9
2-Chloronaphthalene	18	3	83.3
Chloroform	1,430	35	97.6
Hexachlorobutudiene	266	<0.1	>99.9
Hexachlorocyclopentadiene	1,127	1.5	99.9
Napthalene	529	<3	>99.4
Tetrachloroethylene	34	0.3	99.1
Toluene	2,360	10	99.6
Aldrin	84	0.3	99.6
Dieldrin	28	0.2	99.3
Chlordane	217	<0.1	99.9
Endrin	123	1.2	99.0
Heptachlor	40	0.8	98.0
Heptachlor epoxide	11	<0.1	>99.1

Source: Patterson (1985).

Stripping

Air, steam, or inert gas stripping of volatile organics or ammonia is a simple and quantifiable process to evaluate on a laboratory scale. If stripping is warranted, it can be considered in a packed tower or tank at the source or preceding the end-of-pipe treatment system. Normally, air pollution considerations will dictate that stripping systems be covered and the off-gas collected for subsequent treatment. Removals greater than 99 percent are achievable through air stripping of selected compounds, as shown in Table 4-13 (U.S. EPA 1986). For reference and comparison, Henry's Law constants and compound solubilities in water are also provided. The application and design of the stripping process are described in Chapter 8.

Membrane Processes

Membrane technology has advanced to where it can realistically be considered for treatment of concentrated, low-volume wastestreams. Membrane processes currently available range from ultrafiltration to reverse osmosis. As with chemical oxidation and carbon adsorption, membrane processes are relatively non-

Table 4-13. Air Stripping of Selected Compounds.

	Henry's Law Constant,[a] dimensionless	Solubility, mg/l	Observed Removal, %
Benzene	0.23	1,780	90
Carbon tetrachloride (tetrachloromethane)	1.26	800	89
Chlorobenzene	0.16	448	97
1,1,1-Trichloroethane	0.21	4,400	99
Chloroform (trichloromethane)	0.14	7,840	99
1,2-Dichlorobenzene	0.081	100	93
1,3-Dichlorobenzene	0.11	123	95
1,4-Dichlorobenzene	0.11	79	97
1,2-trans-Dichloroethylene	0.22	6,300	84
Ethylbenzene	0.27	152	>99
Methylene chloride (dichloromethane)	0.13	16,700	>99
Bromoform (tribromomethane)	0.022	3,190	92
Dichlorobromomethane	0.088	—	98
Chlorodibromomethane	0.033	—	97
Naphthalene	0.017	30	91
Nitrobenzene	0.0001	1,900	28
Toluene	0.25	515	96
Trichloroethylene	0.49	1,000	98

Source: Patterson (1985)
[a]Experimentally determined values at 25°C.

specific in nature and will retain all molecules falling within certain size ranges according to the membranes selected. The main concerns for application of reverse osmosis are membrane type, membrane area required, and membrane life, each of which has an effect on cost. Membrane processes are capable of reducing salts (total dissolved solids) and various organics. The permeate can be reduced in organic and inorganic constituents from 50 to 90 percent, while the concentrate contains all of these removed contaminants in 10 to 25 percent of the original volume. In application to organics, it is likely that thermal oxidation or catalyzed chemical oxidation will be the only practical disposal options for the concentrate. Reverse osmosis followed by thermal oxidation of the concentrate, although quite expensive as a wastewater treatment process, will typically be substantially less expensive than direct thermal oxidation of the entire wastestream.

Filtration

Occasionally, direct filtration at the source is applicable for removal of insoluble components which have adsorbed substances such as organics, inorganics and heavy metals. Although possibly not toxic in their insoluble form, if not removed at the sources, these substances may desorb downstream due to effects of dilution water or pH adjustment.

Filtration following heavy metals precipitation may be advantageous for removal of metal-hydroxide particle carryover. The metals precipitation process is typically carried out at a pH above 9.0, depending on the metal of interest, while the typical biological system is operated at pH of 7.0. With this pH change, precipitated metals associated with the floc carryover could resolubilize, resulting in a potential toxicity problem.

Types of filtration devices range from monomedia sand beds to microscreens to ultrafiltration units. For gross solids removal, processes such as mono- or multimedia granular filtration units or microscreens are appropriate down to particle sizes of 1–5 microns. Ultrafiltration actually borders on being a membrane process which removes fine particles and specific molecular sizes. This process is applicable for removal of constituents of a known molecular size, with the concentrated sidestream receiving an alternate mode of final disposal such as incineration.

REFERENCES

Chudoba, J. 1985. Inhibitory Effect of Refractory Organic Compounds Produced by Activated Sludge Micro-organisms on Microbial Activity and Flocculation. *Water Research*, **19**(2): 197.

Larson, R. J., and S. L. Schaeffer. 1982. A Rapid Method for Determining the Toxicity of Chemicals to Activated Sludge. *Water Research*, **16**(675).

OECD. 1987. The Use of Biological Tests for Water Pollution Assessment and Control. Environment Monograph 11.

Patterson, J. W. 1985. *Industrial Wastewater Treatment Technology*, 2nd Edition. Boston, MA: Butterworth Publishers.

Pennyslvania Code Title 25. June 1989. Water Quality Toxics Management Strategy, pp. 122–133.

Pennsylvania Department of Environmental Resources, Bureau of Water Quality Management. 1986. Toxics Management Strategy Guidance Manual. Harrisburg, PA.

Philbrook, D. M., and C. P. L. Grady. 1985. Evaluation of Biodegradation Kinetics for Priority Pollutants. *Proc. 40th Ind. Waste Conf., Purdue University*.

U.S. EPA. January 1986. Project Summary: Evaluation of Emerging Technologies for the Destruction of Hazardous Wastes. EPA/600/52-85/069.

Volskay, V. T. and C. P. L. Grady. 1988. Toxicity of Selected RCRA Compounds to Activated Sludge Microorganisms. *WPCF*, **60**(10): 1850.

Watkin, A. 1986. Evaluation of Biological Rate Parameters and Inhibitory Effects in Activated Sludge. Ph.D. Thesis, Vanderbilt University.

Watkin, A. T. and W. W. Eckenfelder. 1988. A Technique to Determine Unsteady-State Inhibition Kinetics in the Activated Sludge Process. *Water Sci. Tech.*, **21**: 593–602.

Zahn, R. and Wellens, H. 1974. A Simple Procedure to Test the Biological Degradability of Products and Wastewater Components. *Chemiker-Zeitung*, **98**(228).

5

Removal of Metals to Nontoxic Levels

Perry W. Lankford

INTRODUCTION

Heavy metals are found in a variety of industrial wastewaters, including plating, chemical, petrochemical, paint, and pigment manufacture as well as most others to a lesser extent. It is sometimes quite important to reduce the levels of metals prior to biological treatment to avoid toxicity or accumulation of metals on the biological mass. It is also important to reduce metal levels to comply with BAT discharge permits and avoid aquatic toxicity of the effluent. There are still other conditions where drinking water maximum contaminant levels (MCLs) must be considered. Metals removal is most typically achieved through source treatment, chemical precipitation, or incidental removal in biological or adsorption processes.

Although conventional treatment technologies for most inorganic pollutants are well established, unusual treatment requirements or wastewater constituents may require unconventional applications. Advanced technologies such as ion exchange or reverse osmosis can be appropriate, but must consider the presence of fouling substances or high background salt levels.

TREATMENT REQUIREMENTS

Toxic Levels of Metals

There is a dramatic variation between the toxic levels of metals to various organisms. Many times those metals which one would think would be quite toxic are not. In other cases, a metal might be quite toxic to one life form, but not to another. In still other cases, the availability of the metal to the organism is affected by the hardness of the solution. Table 5-1 provides typical data on two

Table 5-1.   Acute Toxicity of Selected Metals (96-hr LC_{50}, all values in µg /l).

	Fathead Minnow	Daphnia	Rainbow Trout
Arsenic	15,600	5,278	13,340
Cadmium	38.2	0.29	0.04
Chromium, hexavalent	43,100	6,400	69,000
Copper	3.29	0.43	1.02
Lead	158	4	158
Mercury	—	5	249
Nickel	440	54	—
Selenium	1,460	710	9,000
Silver	0.012	0.002	0.023
Zinc	169	8.9	26.20

fish and a daphnid. From these data the variability can be observed; for example fathead minnows are affected by silver at 0.0121 µg/l, but by hexavalent chromium at 43,100 µg/l, six orders of magnitude difference. On the other hand, cadmium is toxic to rainbow trout at 0.04 µg/l, but fathead minnows at 38.2 µg/l, three orders of magnitude difference.

Water Quality Criteria

The U.S. EPA has published recommended Water Quality Criteria (U.S. EPA 1986) in compliance with the Clean Water Act. These criteria are instream levels at which there should be no effect on aquatic life. Other criteria are included for human health protection based upon ingestion of water and fish. The U.S. EPA updates these several times a year based upon continuing research in the field.

Criteria are provided for acute and chronic levels of metals for both freshwater and marine waters in Table 5-2. As can be observed, the freshwater and marine chronic criteria for the various metals vary from 0.012 to 1,000 µg/l and 0.025 to 560 µg/l, respectively, each four orders of magnitude. Since the chronic criteria are generally the most stringent and of greatest long-term concern, the subsequent discussion will focus on chronic and not acute levels.

The criterion for some of the metals is variable, depending upon the solution or receiving stream hardness. As stated previously, the relationship to hardness is based upon the availability of the metal ion to the organism. As shown in the footnotes to Table 5-2, these relationships are exponential and are more clearly seen in graphical form. Figures 5-1, 5-2 and 5-3 present this relationship for the six metals involved. Three figures were necessary because of the ranges for

Table 5-2. Water Quality Criteria for Metals (all values in µg/l).

	Freshwater Chronic	Marine Chronic
Aluminum	87	—
Arsenic, trivalent	190	36
Cadmium	0.66^a	9.3
Chromium trivalent	120^b	—
hexavalent	11	560
Copper	6.5^c	2.9^f
Iron	1,000	—
Lead	1.3^d	5.6
Mercury	0.012	0.025
Nickel	88^e	8.3
Selenium (selenite, SeO_3^{-2})	35	54
Silver	0.12	2.3^f
Zinc	59^g	86

Sources: U.S. EPA (1986); U.S. EPA (1988).
[a] At hardness of 50 mg/l; relationship is $e^{(0.7852 \ln H - 3.490)}$
[b] At hardness of 50 mg/l; relationship is $e^{(0.8190 \ln H + 1.561)}$
[c] At hardness of 50 mg/l; relationship is $e^{(0.8545 \ln H - 1.465)}$
[d] At hardness of 50 mg/l; relationship is $e^{(1.273 \ln H - 4.705)}$
[e] At hardness of 50 mg/l; relationship is $e^{(0.8460 \ln H + 1.1645)}$
[f] Acute criteria, no chronic available.
[g] At hardness of 50 mg/l; relationship is $e^{(0.8473 \ln H + 0.7614)}$

the different metals. Lead appears on both Fig. 5-1 and Fig. 5-2 in order to provide functional graphs.

Drinking Water MCLs

The U.S. EPA has established Maximum Contaminant Levels (MCLs) for numerous organic compounds and metals in drinking water. Although these are not directly applicable treatment standards, they do occasionally arise as a consideration in treatment programs. Table 5-3 provides these values. As can be observed through comparison with the previous criteria, MCLs are usually not as stringent as water quality criteria.

BAT Levels for Metals

Under the authority of the Clean Water Act, the U.S. EPA has developed Best Available Treatment Technology (BAT) effluent limitations guidelines for industrial point source dischargers. A range of concentration values for various industrial categories is observed for each pollutant regulated.

In 1981, U.S. EPA Region V developed a single BAT-equivalent value for each subject metal, based upon a technology performance evaluation (Patterson

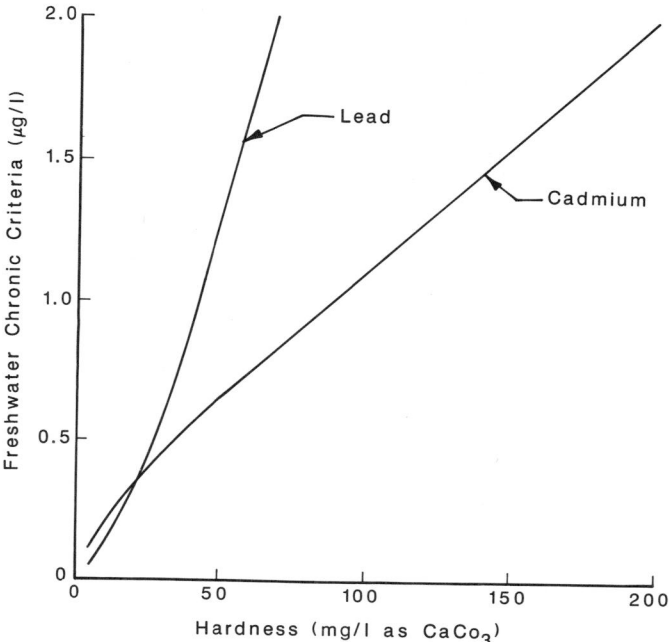

Fig. 5-1. Relationship of freshwater chronic criteria to hardness for lead and cadmium.

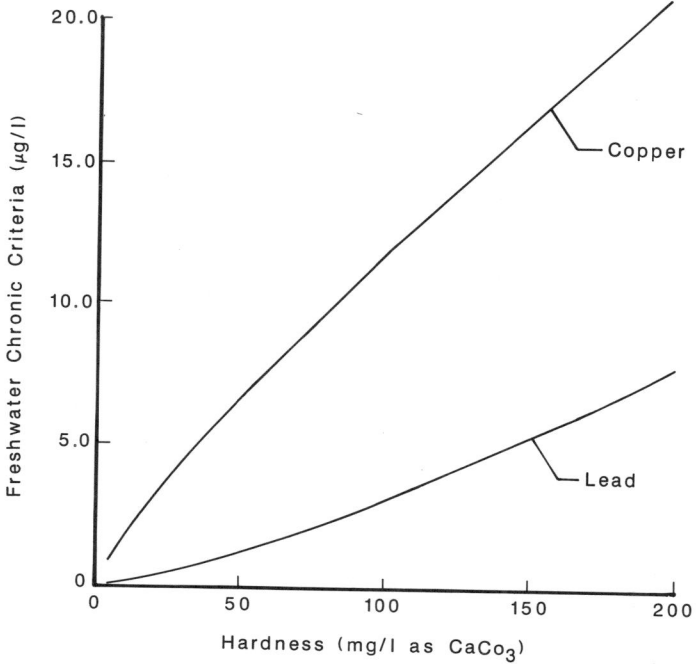

Fig. 5-2. Relationship of freshwater chronic criteria to hardness for copper and lead.

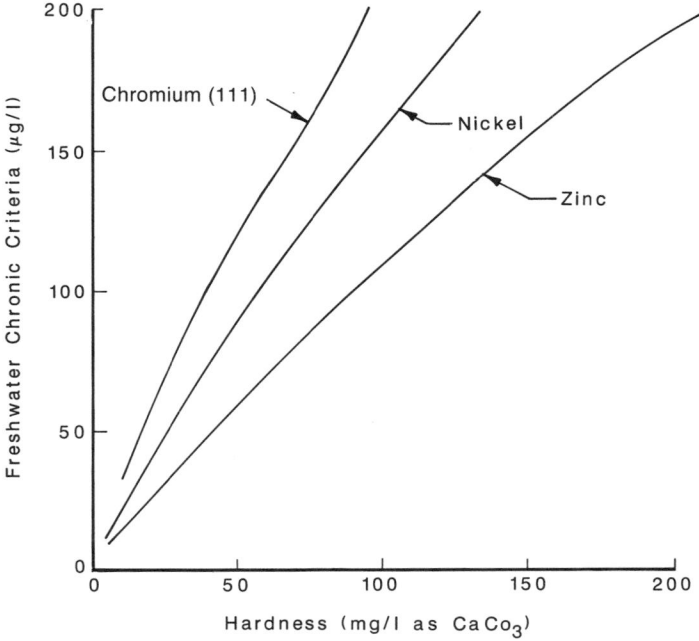

Fig. 5-3. Relationship of freshwater chronic criteria to hardness for chromium (III), nickel, and zinc.

Table 5-3.   Drinking Water MCLs for Metals.

	MCL, μg/l
Arsenic	50
Barium	1,000
Cadmium	10
Chromium	50
Lead	50[a]
Mercury	2
Selenium	10
Silver	50

Source: U.S. EPA (1980).
[a]Reduction of lead MCL to 5 μg/l has been temporarily deferred.

Table 5-4.   Summary of BAT-Equivalent Treatment Performance.

	BAT-Equivalent Concentration, $\mu g/l$ (30 day average)	Treatment Technology
Arsenic	200	1. Arsenite oxidation 2. Lime precipitation, or iron or alum coprecipitation 3. Gravity clarification
Barium	1,000	1. Sulfate precipitation 2. Coagulation 3. Gravity clarification
Cadmium	100	1. High pH precipitation 2. Gravity clarification for lime or filtration for caustic
Chromium, hexavalent	50	1. Acidic reduction to trivalent chromium or ion exchange at pH below 6.0
Chromium, total	500	1. Precipitation 2. Gravity clarification for lime or filtration for caustic
Copper	400	1. Precipitation 2. Gravity clarification
Fluoride	10,000	1. High pH lime precipitation 2. Gravity clarification
Iron	1,500	1. Oxidation at neutral pH of ferrous to ferric iron 2. Precipitation 3. Gravity clarification or filtration
Lead	150	1. High pH precipitation 2. Gravity clarification for lime or filtration for caustic
Mercury	3	1. Ion exchange or coagulation plus filtration
Nickel	750	1. High pH precipitation 2. Gravity clarification and/or filtration
Silver	100	1. Ion exchange or ferric chloride coprecipitation plus filtration
Zinc	500	1. Precipitation at optimized pH 2. Gravity clarification and/or filtration

Source: Patterson (1981).

Table 5-5.   Comparison of Toxic Levels, Freshwater Quality Criteria and BAT Levels for Metals (all values in μg/l).

	Acute[a] Toxicity, *Daphnia*	Freshwater[b] Chronic Criteria @ 50 mg/l Hardness	BAT[c] Equivalent Level
Cadmium	0.29	0.66	100
Chromium			
hexavalent	6,400	11	50
total	—	117	500
Copper	0.43	6.5	400
Fluoride	—	—	10,000
Iron	—	1,000	1,500
Lead	4.02	1.3	150
Mercury	5	0.012	3
Nickel	54	88	750
Selenium	710	35	—
Silver	0.002	0.12	100
Zinc	8.9	59	500

[a]*Source*: QSAR.
[b]*Source*: U.S. EPA (1986).
[c]*Source*: Patterson (1981).

1981). These BAT-equivalent effluent values are summarized in Table 5-4, for 30-day average performance. The document provides that based upon performance data of full-scale systems, a 24-hour maximum discharge value of 1.5 to 2.0 times the 30-day average would be expected.

Comparison of BAT and Toxic Levels

As presented previously, the toxic levels of the various metals vary by six orders of magnitude and the freshwater chronic criteria vary by five orders of magnitude. On top of that, BAT levels vary from three to 10,000 μg/l, four orders of magnitude. Table 5-5 provides a comparison of toxic levels, freshwater chronic criteria and BAT level for each metal. As can be observed BAT-equivalent technology is typically not adequate to treat effluents to below toxic levels without consideration of instream dilution. Only through treatment in more concentrated source streams followed by dilution in the total stream can the water quality levels be met using the described technology.

PROCESS DESCRIPTIONS AND DESIGN CRITERIA

Chemical Precipitation and Coprecipitation

Chemical precipitation is the standard procedure for removal of metals from waters and wastewaters. Removal is primarily dependent upon two factors:

1. The theoretical solubility of the most soluble solid species formed, which is a function of the solid species (solubility equilibrium constant), pH (for weak acid or base counter ions), and the concentration of the precipitating agent (common ion effect)
2. Separation of the solids from aqueous solution

These factors make precipitation ineffective at some extremely low concentrations due to the excess quantities of the precipitating counter ions required for equilibrium and the resulting stabilization of solids that occurs from the excess of potential determining ions. In order to alleviate these potential difficulties, a "coprecipitant" may be used, typically iron or aluminum hydroxides, to act as a coagulating material and to adsorb any stray metals in the complex matrix that were not precipitated. Coprecipitation is primarily an adsorption phenomenon, wherein a precipitated metal is adsorbed, coagulated, or enmeshed onto a bulk-precipitated solid such as ferric hydroxide.

Heavy metals are generally precipitated as the hydroxide through the addition of lime or caustic to attain the pH of minimum solubility. However, several metals are amphoteric and exhibit a specific pH zone of minimum solubility; this zone varies with the metal in question as shown in Fig. 5-4. The solubilities for chromium and zinc are theoretically minimum at pH 7.5 and 10.2, respectively, and show a significant increase in concentration above and below these pH values. The hydroxide precipitation reactions for all cationic metals (M^{++}) are very similar, regardless of whether lime or caustic is used:

$$M^{++} \begin{cases} CO_3 \\ SO_4 \\ Cl_2 \end{cases} + Ca(OH)_2 \rightarrow M(OH)_2\downarrow + Ca^{++} \begin{cases} CO_3\downarrow \\ SO_4\downarrow \\ Cl_2 \end{cases} \qquad (5\text{-}1)$$

$$M^{++} \begin{cases} CO_3 \\ SO_4 \\ Cl_2 \end{cases} + 2NaOH \rightarrow M(OH)_2\downarrow + Na_2 \begin{cases} CO_3 \\ SO_4 \\ Cl_2 \end{cases} \qquad (5\text{-}2)$$

There are, however, dramatic differences in generation of precipitates. In the case of lime, in addition to the target metal, calcium carbonate, sulfate, and excess hydroxide will precipitate, a distinct disadvantage in terms of sludge production. For example, assuming copper is the target metal and sulfate is the anion present, lime will stoichiometrically result in 3.68 lb sludge per lb of copper removed, while caustic will result in 1.54 lb per lb. However, this sludge mass can be an advantage if coprecipitation is a factor in overall performance.

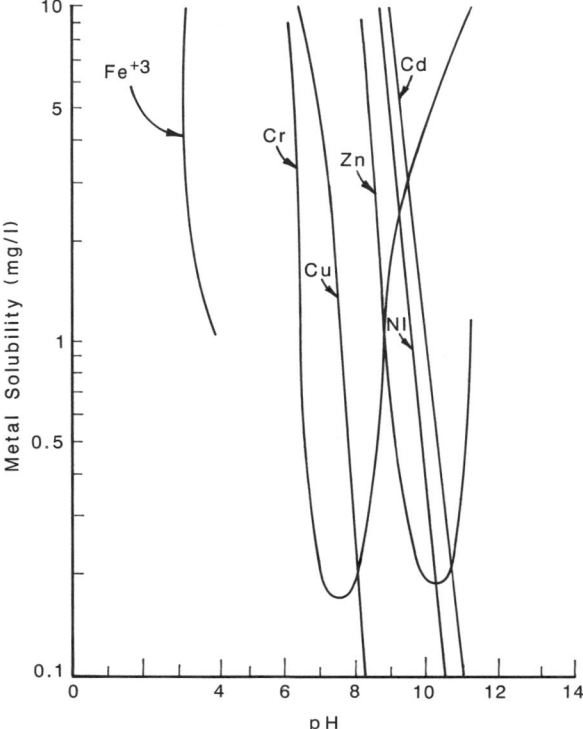

Fig. 5-4. Relationship of metals solubility to pH.

Heavy metals may also be precipitated as the sulfide and in some cases as the carbonate, as in the case of lead. The solubilities of sulfide and carbonate are typically much lower than hydroxide solubilities and therefore lower effluent levels can be reached. In practice solubility levels are, however, much higher than theory would predict. In order to meet low effluent requirements, it may be necessary in some cases to provide filtration to remove floc carried over from the precipitation/sedimentation processes.

There are a number of forms of sulfide which are effective precipitants, including those shown in the following general reaction:

$$M^{++} \begin{Bmatrix} CO_3 \\ SO_4 \\ Cl_2 \end{Bmatrix} + \begin{matrix} Na_2 \\ NaH \\ Mg \\ H_2 \\ Fe \end{matrix} \Biggr\} S \rightarrow MS\downarrow + \begin{matrix} Na_2 \\ NaH \\ Mg \\ H_2 \\ Fe \end{matrix} \Biggr\} \begin{Bmatrix} CO_3 \\ SO_4 \\ Cl_2 \end{Bmatrix} \qquad (5\text{-}3)$$

Due to the high solubility of the resulting compounds and the low solubility of metallic sulfides, this is a very efficient technique. For example, in the precipitation of copper, stoichiometrically only 1.50 lb of sludge is generated per pound of copper removed, even less than hydroxide precipitation.

Carbonate precipitation is especially effective for removal of lead and nickel. The reaction is as follows:

$$M^{++} \begin{Bmatrix} SO_4 \\ Cl_2 \end{Bmatrix} + Na_2CO_3 \rightarrow MCO_3\downarrow + Na_2 \begin{Bmatrix} SO_4 \\ Cl_2 \end{Bmatrix} \qquad (5-4)$$

Carbonate precipitates, where effective, are less soluble than hydroxides and produce less sludge overall due to the solubility of the more common carbonate salts. For example, for lead, carbonate precipitation will stoichiometrically produce 1.29 lb sludge per pound of lead removed, whereas lime with sulfate present will produce 1.82 lbs sludge per lb of lead.

When treating industrial wastewaters containing metals, it is frequently necessary to pretreat the wastewaters to remove substances which will interfere with the precipitation of the metals. Cyanide and ammonia form complexes with many metals that limit the removal that can be achieved by precipitation.

Cyanide can be removed by alkaline chlorination or other processes such as catalytic oxidation. However, cyanide wastewaters containing nickel or silver are difficult to treat by alkaline chlorination because of the slow reaction rate of these metal complexes.

Ammonia can be removed by stripping, break point chlorination, or other suitable methods prior to the removal of metals. The solubility of metals in a wastewater with and without ammonia removal is shown in Fig. 5-5.

For arsenic and iron precipitation, oxidation may be required utilizing chlorine or permanganate. For chromium waste treatment, hexavalent chromium must first be reduced to the trivalent state (Cr^{+3}) and then precipitated with lime. This is referred to as the process of reduction and precipitation. Chapter 11 provides more information on chemical oxidation.

Adsorption

Adsorption of anions and cations can take place on a variety of surfaces including activated carbon, aluminum oxides, iron oxides, silica, clays, and synthetic materials (zeolites, resins). These processes are highly pH dependent, with each metal having a distinct pH where the onset of adsorption occurs. The chemical characteristics are such that higher pH values favor adsorption of cations and lower pH values favor adsorption of anions such as arsenate and selenite. In addition, the presence of complexing agents can interfere with the adsorption of cationic metal species. A major problem in use of adsorption for metals is competition from major background ions, such as calcium, magnesium, or

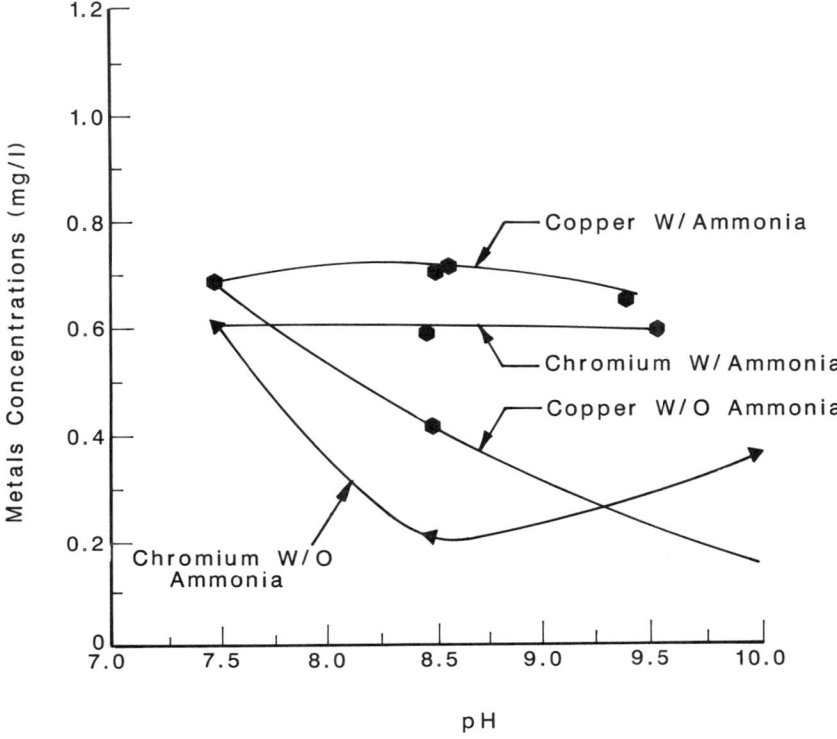

Fig. 5-5. Comparative optimum pH values for metal removal with and without ammonia.

sodium which do not need to be removed. At high concentrations, these exhibit little effect on adsorption of cationic heavy metals, but it is unclear what their effect might be at low levels.

The adsorption of cationic metal ions on activated carbon has not been completely explained. There are several schools of thought including adsorption of complexed organics and surface precipitation due to the sulfur content of the carbon. Regardless of the explanation, there is empirical evidence that carbon adsorption is quite effective on some metals under correct conditions.

Ion Exchange

Ion exchange is a process similar to adsorption except that an ion of like charge, for example Na^+ or H^+, is chemically exchanged for the heavier cationic metals and then later used in concentrated form as a regenerant.

Sodium exchange is the most common and the reaction is as follows, where

M^{++} represents the cationic metal and R represents the exchange resin:

$$Na_2R + \left.\begin{matrix} Ca \\ Mg \\ M^{++} \end{matrix}\right\}\left(\begin{matrix} (HCO_3)_2 \\ SO_4 \\ Cl_2 \end{matrix}\right) \rightarrow \left.\begin{matrix} Ca \\ Mg \\ M^{++} \end{matrix}\right\}R + Na_2\left(\begin{matrix} (HCO_3)_2 \\ SO_4 \\ Cl_2 \end{matrix}\right) \qquad (5\text{-}5)$$

As seen above, the hardness ions, calcium and magnesium, are also removed, requiring a larger mass of resin or causing the given mass of resin to prematurely exhaust. It would not be unusual to find two orders of magnitude more calcium and magnesium than the target metal.

Regeneration of sodium resin is performed with concentrated sodium chloride brine, as follows:

$$\left.\begin{matrix} Ca \\ Mg \\ M^{++} \end{matrix}\right\}R + 2NaCl \rightarrow Na_2R + \left.\begin{matrix} Ca \\ Mg \\ M^{++} \end{matrix}\right\}Cl_2 \qquad (5\text{-}6)$$

This then produces a concentrated brine containing all of the removed ions. Although the metals of concern are still in a soluble form in the brine, the volume is much less than the original.

Little significant work has been done on ion exchange for low levels of metals in a mixed background matrix, as required for wastewater toxicity reduction. Work done to date demonstrates a 50 percent removal of ppb level metals (Cd, Cu, Pb, and Zn) using a synthetic zeolite. It is not known whether more specific pH conditions or a particular background would improve these results. For low levels, a two stage anionic/cationic resin process may be required, or a mixed bed with possibly a two-stage regeneration process.

A new development in ion exchange resins has recently been reported by Rohm and Haas, called IRC-718 Resin (Etzel 1988). This promising resin has reportedly been able to remove specific metals to quite low levels without removing the background hardness. The mass of resin required can therefore be expected to be a fraction of that of conventional zeolite. These results are presented in Table 5-6.

Table 5-6. Performance of New Ion Exchange Resin.

Constituent	Influent mg/l	Conventional Resin Effluent mg/l	IRC-718 Resin Effluent mg/l
Hardness	30	2–5	30
Zinc	4	<0.05	<0.05
Lead	1.2	<0.01	<0.01

Source: Etzel (1988).

Reverse Osmosis

The removal of metals by reverse osmosis has been demonstrated in the laboratory and on a relatively small scale in the field. Reported evidence suggests that the rejection ratio for metal ions is independent of the initial concentrations, a great advantage when low influent levels are treated. However, the same reports suggest that the rejection ratios rapidly deteriorate from 80 to 50 percent. Reverse osmosis is not a currently applicable technology due to the lack of operational data. Reverse osmosis is rather expensive, with high capital costs and short membrane life. Operational costs are quite high due to the high pressure membranes (around 200 psi) required for low level concentrations.

DEVELOPMENT OF DESIGN CRITERIA

Jar Testing (Eckenfelder et al. 1980)

Chemical treatment of metals is a two step process as summarized previously, precipitation followed by separation. In order to attain the latter, coagulation or coprecipitation is sometimes necessary. Laboratory jar testing procedures can be conducted to confirm not only the optimum precipitation conditions and performance, but also coagulation and separation requirements. Coagulation is the chemical process of destabilization of colloidal or suspended particles so that solids separation occurs. This can be studied through conventional jar testing or zeta potential procedures.

Zeta Potential (Eckenfelder et al. 1980)

The zeta potential is a measurement of the stability of a particle and indicates the potential which would be required to penetrate the layer of ions surrounding the particle for destabilization. Therefore, the higher the zeta potential, the more stable the particle. The purpose of coagulation is to reduce the zeta potential by adding specific ions and then to induce motion for the destabilized particles to agglomerate.

The measurement of the zeta potential can be accomplished by the determination of the electrophoretic mobility (EM) of the charged particles under an applied voltage:

$$\zeta = \frac{4\pi\mu}{DV}\,(EM) \tag{5-7}$$

where:

ζ = zeta potential
D = dielectric constant of the medium

V = applied potential
EM = electrophoretic mobility
μ = viscosity

For practical usage in the determination of zeta potential, the above can be reduced to:

$$\zeta = \frac{113,000\mu}{D}(EM) \qquad (5\text{-}8)$$

in which zeta potential ζ is in mv, viscosity μ in poise, and electrophoretic mobility EM in μm-cm/volt-sec.

$$EM = \frac{(G)(L)}{(V)(T)} \qquad (5\text{-}9)$$

where:

G = length of grid divisions (μm)
L = length of electrophoresis cell (cm)
V = voltage (volts)
T = time (seconds)

Zeta potential measurement involves tracing the path of colloidal particles under an applied voltage over a measured distance. A test cell containing the liquid and equipped with a platinum or molybdenum electrode is viewed by the use of a stereoscopic microscope and the velocity of the colloidal particles is calculated.

Laboratory Procedures (Eckenfelder et al. 1980)

Because of the complex reactions involved, laboratory experimentation is essential to establish the optimum pH and coagulant dosage for coagulation of a wastewater. Two procedures can be followed for this purpose: the jar test, in which pH and coagulant dosage are varied to attain the optimum operating conditions; and zeta potential control, in which coagulant is added to attain zero zeta potential. The procedures to determine the optimum coagulant dosage using these two tests are outlined below:

Zeta Potential Procedure   (Eckenfelder et al. 1980):

(a) Place 1,000 ml of sample in a beaker.
(b) Add the coagulant in known increments. (The optimum pH should be established either by zeta potential or by a jar-test procedure.)
(c) Rapid-mix the sample for 3 minutes after each addition of coagulant; follow with a slow mix.

(d) Determine the zeta potential after each reagent addition and plot the results as shown in Fig. 5-6.
(e) If a polyelectrolyte is to be used as a coagulant aid, it should be added last.

Jar-Test Procedure   (Eckenfelder et al. 1980):

(a) Using 200 ml of sample on a magnetic stirrer, add coagulant in small increments at a pH of 6.0. After each addition, provide a 1 minute rapid mix followed by a 3 minute slow mix. Continue addition until a visible floc is formed.
(b) Using this dosage, place 1,000 ml of sample in each of six beakers in a standard jar test apparatus such as Fig. 5-7.
(c) Adjust the sample pH to pH 4.0, 5.0, 6.0, 7.0, 8.0, and 9.0 with standard acid or alkali.

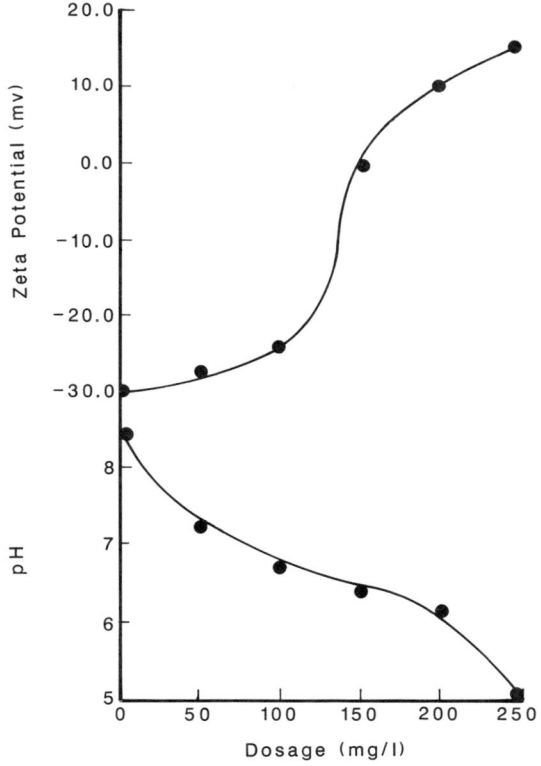

Fig. 5-6. Correlation of zeta potential and pH to coagulant dosage. (*Source:* Eckenfelder 1980.)

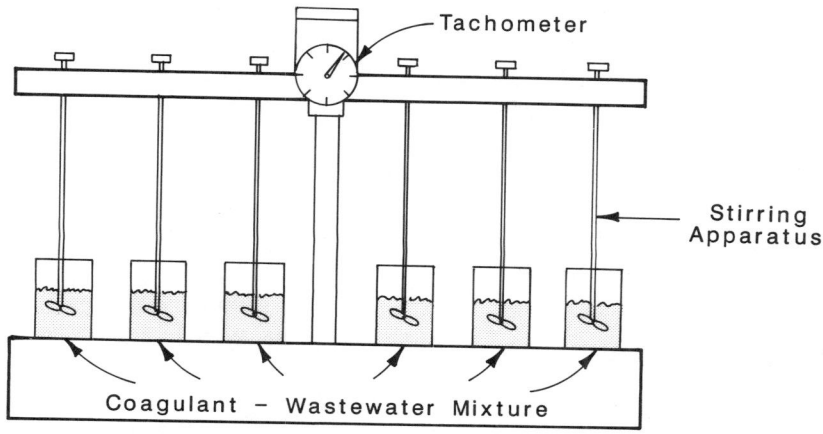

Fig. 5-7. Laboratory scale jar test assembly for coagulant studies. (*Source:* Eckenfelder 1980.)

(d) Rapid-mix each sample for 3 minutes; follow this with 12 minute floc-culation at slow speed.
(e) Measure the effluent residual parameter of interest, e.g., filtered concentration of each settled sample.
(f) Plot the percent removal of the significant parameters versus pH and select the optimum pH (Fig. 5-8).
(g) Using this pH, repeat steps (b), (d), and (e), varying the coagulant dosage.

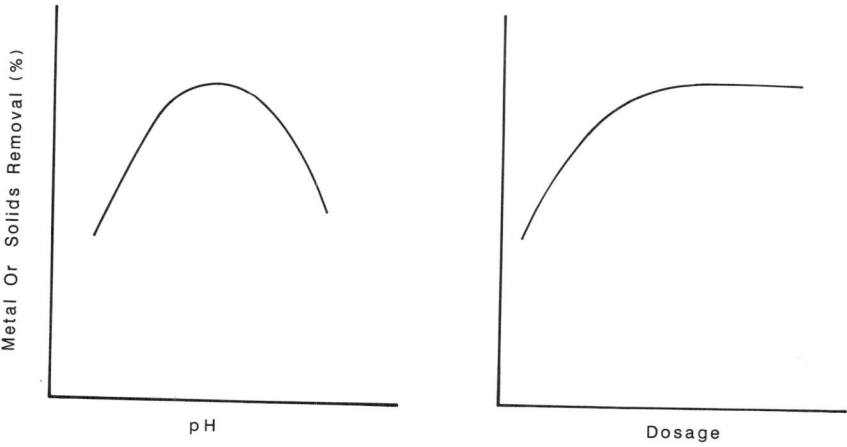

Fig. 5-8. Typical plots of jar-test analysis. (*Source:* Eckenfelder 1989.)

(h) Plot the percent removal versus the coagulant dosage and select the optimum dosage (Fig. 5-8).
(i) If a polyelectrolyte is used, repeat the procedure, adding polyelectrolyte toward the end of the rapid mix period.

Isotherm

Adsorption and ion exchange can be tested simply in the laboratory by the isotherm procedure outlined in Chapter 9.

PERFORMANCE OF METALS REMOVAL PROCESSES

This section provides a brief summary of the most effective reported treatment technologies for each of the subject pollutants and the performance thereof.

Arsenic

In aqueous systems, arsenic exists as either the arsenite ion (AsO_2^-; As^{+3}) or arsenate ion (AsO_4^{-3}; As^{+5}). Treatment methods for arsenic include lime or sulfide precipitation, or coprecipitation (sometimes described as precipitation/coagulation) with iron or aluminum hydroxide. The oxidation state of the arsenic influences the efficiency of each of these treatment processes. Sulfide precipitation is effective for arsenate but ineffective for arsenite. Lime precipitation is preferred over sulfide precipitation due to higher treatment efficiency, but requires higher pH (pH 12 +). Caustic precipitation is less effective than lime. Iron or aluminum coprecipitation is more effective than lime precipitation, but yields more sludge, which is often more difficult to dewater than is the lime precipitate sludge.

Both aluminum and iron coprecipitation are strongly influenced by treatment pH, with aluminum treatment efficiency declining at pH above 7, and iron treatment efficiency declining above pH 9. Both processes perform better on arsenate than arsenite.

Thus, for all precipitation treatment processes for arsenic, enhanced performance is observed when the arsenic is present as arsenate. Most effective treatment may require a chemical oxidation step, to convert any arsenite to arsenate. Chlorination has been used to achieve this oxidation.

Each of the precipitation processes is, in addition, influenced by the efficiency of suspended solids removal. Clarification normally provides adequate solids removal, due to the quantities of lime, iron, or aluminum salts required for precipitation.

Table 5-7 presents a summary of reported achievable effluent levels of arsenic for different processes.

Table 5-7.   Achievable Effluent Levels of Arsenic.

Technology	Achievable Effluent Concentration, $\mu g/l$
Ferric sulfate coprecipitation with carbon adsorption, from high level	60
Alumina adsorption	10–60
Sulfide (sodium or hydrogen) precipitation with filtration	10–50
Carbon adsorption, from low level	4–60
Ferric hydroxide coprecipitation	5
Ferric sulfate/alum coprecipitation with filtration	3–6

Source: Patterson (1985); Eckenfelder (1989).

Barium

Barium is removed from solution by precipitation as barium sulfate. Barium sulfate has a maximum theoretical solubility at 25°C of approximately 1.4 mg/l as barium at stoichiometric concentrations of barium and sulfate. But the solubility level of barium can be reduced in the presence of excess sulfate, to effluent levels as low as 30 to 300 $\mu g/l$. Adequate reaction time must be provided and the use of coagulant, such as iron or alum for clarifier supernatant suspended solids control may be needed.

Barium has also been removed through ion exchange and electrodialysis, although these processes would be far more expensive than chemical precipitation.

Cadmium

Cadmium is typically removed from wastewaters by precipitation or ion exchange. In some cases, electrolytic and evaporative recovery processes can be effectively employed providing the wastewater is in a concentrated form. Cadmium forms an insoluble and highly stable hydroxide at an alkaline pH, with a soluble cadmium level of approximately 1,000 $\mu g/l$ at pH 8 and 50 $\mu g/l$ at pH 10 to 11. Lime provides a good settling precipitate, while caustic requires filtration. Coprecipitation with iron hydroxide has reduced cadmium to 8 $\mu g/l$ at pH 6.5; and 50 $\mu g/l$ at pH 8.5.

Cadmium is not precipitated in the presence of complexing ions, such as cyanide. In these cases, it is necessary to pretreat the wastewater to destroy the complexing agent, for example, cyanide destruction is necessary prior to cadmium precipitation. A hydrogen peroxide oxidation precipitation system has been developed which simultaneously oxidizes cyanides and forms the oxide of cadmium, thereby yielding cadmium where recovery of the cadmium is feasible.

Table 5-8 summarizes reported achievable effluent levels of cadmium.

Table 5-8.   Achievable Effluent Levels of Cadmium.

Technology	Achievable Effluent Level, $\mu g/l$
Hydroxide precipitation pH 8 to 10.5	100–1,000
Ferric hydroxide coprecipitation, pH 6 to 10	40–50
Hydroxide precipitation with filtration, pH 10 to 11.5	0.7–80
Sulfide and lime precipitation with filtration, pH 8.5 to 10	2–29
Sulfide precipitation	5–10
Ion exchange	1–50

Sources: Patterson (1985); Eckenfelder (1989).

Chromium

Reduction of hexavalent chromium from a valence state of +6 to +3 and subsequent hydroxide precipitation of the trivalent chromium ion is the most common method of hexavalent chromium control. The standard reduction treatment technique is to lower the waste stream pH to 2.0 to 3.0 with sulfuric acid, and convert the hexavalent chromium to trivalent chromium with a chemical reducing agent such as sulfur dioxide, sodium bisulfite, or ferrous sulfate. One common source of the latter is spent pickle liquor and, in the subsequent precipitation step, the iron will function as a coagulant at the expense of about fourfold greater sludge production. The efficiency of conversion of the hexavalent to trivalent chromium is dependent upon the reaction time, treatment pH, and type and concentration of reducing agent used. Close process control is necessary to achieve effective chromium reduction.

Ion exchange has been employed successfully for direct hexavalent chromium removal. In the ion exchange system, wastewater pH is a critical factor in successful treatment. At a pH below 4, the chromic acid attacks the resin, while at a pH above 6, ion exchange efficiency decreases due to the increase in the ratio of dichromate to chromate and the poorer removal of dichromate.

Total chromium is the sum of the hexavalent plus trivalent chromium. In chromium control by reduction-precipitation, effective treatment requires the successful accomplishment of three sequential steps: hexavalent chromium reduction to trivalent chromium; precipitation of the trivalent chromium; and removal of the precipitated chromium. Treatment performance will deteriorate due to incomplete achievement at any one of these three stages.

However, because of the need to precipitate other metals as well, a typical pH of about 8 is often reported for mixed chromium-metals wastes. Caustic is more common than lime in newer or upgraded treatment plants due to its ease of handling. In the gravity clarification stage, lime treated wastes settle better than caustic treated wastes. Filtration of caustic treated wastes may be necessary to achieve the required performance.

Table 5-9 summarizes reported achievable effluent levels of chromium.

Table 5-9.   Achievable Effluent Levels of Chromium.

Technology	Achievable Effluent Level, $\mu g/l$
Hexavalent chromium	
Ferrous sulfate	10–1,000
Metabisulfite	1–500
Sulfur dioxide reduction	10–100
Bisulfite reduction	5–100
Bisulfite with hydrazine	100
Ion exchange	10–1,000
Trivalent chromium	
Precipitation, pH 10	20–170
Precipitation with filtration, pH 10	50
Lime and sulfide precipitation with filtration	10

Adapted from: Patterson (1985); Eckenfelder (1989).

Copper

The standard treatment method for copper is precipitation. Although most authorities agree that optimum copper precipitation occurs between pH 9.0 and 10.3, effective treatment has been observed at much lower operating pH values. Poor performance in copper treatments seems more often to result from insufficient solids removal than from inadequate pH control.

Both lime and caustic are widely used to precipitate copper. However, in copper sulfate wastewaters, the addition of lime may result in massive calcium sulfate sludge formation. In this instance, filtration should be avoided, since the slowly forming calcium sulfate will tend to cement the filter. With both lime and caustic treatment, gravity clarification can provide effective treatment. The caustic sludge yield can be significantly less than from lime treatment, although the caustic sludge dewatering properties are poorer.

Cupric oxide has a minimum solubility between pH 9.0 and 10.3 with a reported solubility of 10 $\mu g/l$. Field practice has indicated that the maximum technically feasible treatment level for copper by chemical precipitation is 20 to 70 $\mu g/l$ as soluble copper. Precipitation with sulfide at pH 8.5 will result in effluent copper concentrations of 10 to 20 $\mu g/l$.

Low residual concentrations of copper are difficult to achieve by precipitation in the presence of complexing agents such as cyanide and ammonia. Removal of the complexing agent by pretreatment is essential. Copper cyanide complexes are, however, effectively removed by adsorption on activated carbon.

Table 5-10 provides a summary of achievable effluent copper levels.

Fluoride

Treatment options for fluoride are limited to two alternative precipitation processes, with significant differences in performance and associated sludge yield.

Table 5-10.   Achievable Effluent Levels of Copper.

Technology	Achievable Effluent Level, $\mu g/l$
Hydroxide precipitation with filtration, pH 8.5	10–400
Hydroxide plus sulfide precipitation with filtration, pH 8.5	10–200
Ion exchange	10–50
Sulfide precipitation	10
Lime and sulfide precipitation with filtration	5

Adapted from: Patterson (1985); Eckenfelder (1989).

Lime precipitation to form calcium fluoride has been the dominant technology for fluoride control. At the high treatment pH required (pH 12+), and associated high lime dosage, filtration is risky, due to the tendency of the gravity clarified effluent to cement the filter. Thus, within the solubility constraints associated with calcium fluoride formation, most effective treatment with lime depends upon highly efficient solids removal by gravity clarification. The precipitated solids are reported to have poor settling characteristics.

The second treatment alternative is fluoride removal by alum addition, essentially a coprecipitation process. Much lower effluent fluoride levels are achieved with alum coprecipitation than with lime precipitation and best removal appears to result at pH 6 to 7. Treatment efficiency in this pH range depends upon the alum dosage (mg/mg fluoride), with enhanced treatment at higher (200+ mg/mg) alum dosages. A major disadvantage of the process is the voluminous sludge produced, which even at moderate alum dosages can amount to as much as 40 percent of the original wastewater volume.

Table 5-11 provides a summary of achievable effluent levels of fluoride.

Table 5-11.   Achievable Effluent Levels of Fluoride.

Technology	Achievable Effluent Level, $\mu g/l$
Lime precipitation	6,000–10,000
Lime and alum coprecipitation	1,000–3,000
Alum coprecipitation	500–2,000
Alumina contact bed	100–3,000

Adapted from: Patterson (1985); Eckenfelder (1989).

Iron

In aqueous systems, iron exists in the ferric (Fe^{+3}) or ferrous (Fe^{+2}) form, depending upon conditions of pH and dissolved oxygen concentration. In precipitation treatment, ferrous iron is more soluble than ferric iron. Therefore, most effective iron precipitation treatment incorporates conversion of any ferrous to ferric iron before precipitation. At neutral pH and in the presence of oxygen, ferrous iron rapidly oxidizes to the ferric form, which readily hydrolyzes to the insoluble precipitate, ferric hydroxide.

The principal treatment process for iron is oxidation-precipitation-clarification. The iron wastewater is first neutralized to pH near 7, where ferrous iron oxidation rapidly occurs and where the solubility of ferric hydroxide is at its minimum. Following pH adjustment, the waste is aerated to provide oxygen for the iron oxidation process. Where iron complexing agents are present in the waste, the rate of oxidation is slowed. Sufficient aeration time is therefore essential in order to achieve a high degree of oxidation. Freshly precipitated ferric hydroxide has a low specific gravity, which makes settling difficult without extremely large clarifiers or additional treatment such as filtration. Lime is frequently used for pH adjustment and, where sulfuric acid is present as in a pickling waste, large quantities of calcium sulfate are also precipitated.

In some industrial wastes, ferrous and ferric iron may exist in the presence of cyanide. Extremely stable iron cyanide complexes result. Such species present considerable difficulties for both iron and cyanide treatment. No truly effective treatment method has been reported for such wastes.

Table 5-12 provides a summary of achievable effluent levels for iron.

Table 5-12. Achievable Effluent Levels of Iron.

Technology	Achievable Effluent Level, $\mu g/l$
Oxidation/precipitation, pH 7 to 10	400–600
Oxidation/precipitation with filtration, pH 7 to 10	50–250
Two stage hydroxide precipitation	10
Sulfide precipitation	10

Adapted from: Patterson (1985); Eckenfelder (1989).

Lead

Precipitation of lead is extremely effective, except in instances such as the tetraethyl lead industry where significant concentrations of organic lead occur. Lead in the organic form is not amenable to precipitation, and the organic component must be destroyed by chemical means such as chlorination, before the lead can be precipitated.

Although the literature contains conflicting values of the optimum pH for lead precipitation, ranging from pH 6 to above pH 10, there is strong evidence that best precipitation occurs in the pH range of 9 to 10. However, at this high pH, other metals in the wastewater with a lower optimal pH may not be effectively precipitated.

Lead wastewaters precipitated with lime have good settling properties, while caustic treated wastes may require filtration to achieve an equivalent effluent lead level. Lead is effectively precipitated as carbonate by the addition of soda ash, resulting in effluent dissolved lead concentrations of 10 to 30 μg/l at a pH of 9.0 to 9.5. Precipitation with lime at pH 11.5 results in effluent concentrations of 19 to 200 μg/l. Precipitation as the sulfide can be accomplished with sodium sulfide at a pH of 7.5 to 8.5.

Table 5-13 provides a summary of reported achievable effluent levels of lead.

Table 5-13. Achievable Effluent Levels of Lead.

Technology	Achievable Effluent Level, μg/l
Hydroxide precipitation, pH 11.5	10–200
Carbonate precipitation, pH 9 to 9.5	10–30
Sulfide precipitation	10
Lime and sulfide precipitation with filtration	5
Ion exchange	50

Adapted from: Patterson (1985); Eckenfelder (1989).

Manganese

Similar to iron treatment, technology for the removal of manganese involves conversion of the soluble manganous ion to an insoluble precipitate. Removal is accomplished by oxidation of the manganous ion and separation of the resulting insoluble oxides and hydroxides. However, compared to iron, manganous ion has low reactivity with oxygen and simple aeration is not an effective technique below pH 9. It has been reported that even at higher pH levels, organic matter in solution can combine with manganese and prevent its oxidation by simple aeration. A reaction pH above 9.4 is required to achieve significant conversion of manganous ion to insoluble manganese dioxide.

The presence of copper ion enhances air oxidation of manganese, and chlorine dioxide rapidly oxidizes manganese to the insoluble form. Permanganate has successfully been employed, and ozone in conjunction with lime has been successful for the oxidation and removal of manganese.

Mercury

Many types of mercury treatment technology have been described in the technical literature. Among the most effective are the ion exchange and coagulation treatment processes. Typically, ion exchange treatment of mercury involves the formation of a negatively charged mercuric chloride complex by addition of chlorine or hypochlorite (to oxidize any metallic mercury present), or chloride salts, and removal of the mercuric chloride complex on an anion exchange resin.

Most experience with ion exchange treatment has been with chlor-alkali wastes which contain high background chloride levels.

Removal of mercury by precipitation has been reported for a variety of mercury-containing wastewaters. The process has been applied with success to both organic and inorganic mercury. Iron and alum coagulants are reported to produce equivalent mercury removal results, although the alum coagulant may display poorer settling properties than the iron coagulant. In both instances, effluent filtration is necessary to achieve best treatment results.

Activated carbon adsorption is an alternative to precipitation, although less effective at higher influent mercury levels.

Table 5-14 provides a summary of achievable effluent levels of mercury.

Table 5-14. Achievable Effluent Levels of Mercury.

Technology	Achievable Effluent Level, $\mu g/l$
Sulfide precipitation	10–20
Sulfide precipitation, hydroxide coprecipitation	1
Alum coprecipitation	1–10
Iron coprecipitation	0.5–5
Ion exchange	1–5
Carbon adsorption, high influent	20
Carbon adsorption, low influent	0.25

Adapted from: Patterson (1985); Eckenfelder (1989).

Nickel

Precipitation treatment is the standard practice for control of nickel in industrial wastes, although in specialized circumstances such as for recovery in plating plants, reverse osmosis has been utilized. This later technology, although effective, is principally utilized for alkaline nickel wastes.

Best precipitation treatment is achieved at a pH above 9.5, which may cause deterioration of the precipitation treatment of other metals in the wastewater. Even when lime is employed as the treatment chemical, nickel hydroxide precipitates have rather poor settling characteristics. Poor performance in nickel treatment more often results through inadequate solids separation than through incomplete precipitation of the soluble nickel. Unless extended clarifier detention time is provided, filtration appears necessary to achieve good control of total nickel.

Nickel forms insoluble nickel hydroxide upon the addition of lime, resulting in a minimum solubility of 120 $\mu g/l$ at pH 10 to 11. Nickel can also be precipitated as the carbonate or the sulfate associated with recovery systems. In practice, lime addition (pH 11.5) may be expected to yield residual nickel concentrations in the order of 150 $\mu g/l$ after sedimentation and filtration.

In the presence of complexing agents such as ammonia, EDTA, or cyanide, nickel can form extremely stable soluble complexed species which interfere with conventional precipitation treatment.

Recovery of nickel can be accomplished by ion exchange or evaporative recovery, providing the nickel concentrations in the wastewaters are at a sufficiently high level.

Table 5-15 provides a summary of reported achievable effluent levels of nickel.

Table 5-15.   Achievable Effluent Levels of Nickel.

Technology	Achievable Effluent Level, $\mu g/l$
Lime precipitation, pH 9 to 12	500
Lime precipitation with filtration, pH 9 to 12	50–150
Lime plus sulfide precipitation with filtration	40

Adapted from: Patterson (1985); Eckenfelder (1989).

Selenium

Selenite ion (SeO_3^{-2}) is the most common form of selenium in wastewaters, except for the solid forms of selenide resulting from mining, ore milling and pigment and dye manufacturing. Other forms decompose in water to form the selenite ion. Particulate selenide is effectively removed by sedimentation and filtration. As expected for an anion, lime precipitation of selenite is relatively ineffective although lime coprecipitation is somewhat effective. Activated carbon adsorption is also ineffective, although there have been reported cases of excellent removal using carbon after lime precipitation. Iron coprecipitation has achieved 80 percent removal, improving at lower pH. Weak-base anion exchange is an effective technique for removing selenite, but after chemical oxidation to selenate, the anion exchange step was more effective.

Table 5-16 provides a summary of reported achievable effluent levels of selenium.

Table 5-16.   Achievable Effluent Levels of Selenium.

Technology	Achievable Effluent Level, $\mu g/l$
Lime precipitation with filtration, pH 11.5	200–300
Lime precipitation with filtration and carbon adsorption, pH 11.5	20–50
Ferric hydroxide coprecipitation, pH 6.2	10–50
Alumina adsorption	5–20
Oxidation and anion exchange	1–10

Adapted from: Patterson (1985); Eckenfelder (1989).

Silver

Treatment technology for silver is influenced more by the value of the recovered metal than by the limitations of discharge permits. Many treatment/recovery techniques are employed, although the final polishing step is often ion exchange. Effective removal processes include hydroxide precipitation at pH 11, coprecipitation with ferric chloride or alum, and sulfide precipitation. Cyanide in plating wastes interferes with silver removal, therefore oxidation pretreatment is required.

Table 5-17 provides a summary of reported achievable effluent levels for silver.

Table 5-17.   Achievable Effluent Levels of Silver.

Technology	Achievable Effluent Level, $\mu g/l$
Ferric hydroxide coprecipitation, pH 6.2 to 6.4	25
Alum coprecipitation, pH 6.2 to 6.4	25
Hydroxide precipitation, pH 11	20
Hydroxide precipitation with filtration, pH 11	15
Ion exchange	1–5

Adapted from: Patterson (1985); Eckenfelder (1989).

Zinc

Hydroxide precipitation is the standard practice in treatment of zinc wastewaters. There is a great deal of confusion in the technical literature regarding the optimum pH for zinc precipitation. Optimum performance has been cited at pH values as low as 9.0 to 9.5, and as high as pH 11 and above. Zinc is an amphoteric metal, with increasing solubility at both higher and lower pH. It is possible that constituents (such as complexing agents) other than zinc in the various wastewaters may influence zinc precipitation efficiency as a function of pH.

As with other metals, both lime and caustic are used as the precipitating chemical. The effluent quality of full-scale zinc treatment systems appears to be influenced most by the efficiency of suspended solids removal. Many systems employing only gravity clarification for solids control exhibit high effluent solids and associated high effluent zinc. Therefore, zinc removal requires treatment at the best pH value for that specific wastewater, plus efficient suspended solids removal by gravity clarification and/or filtration.

Table 5-18 provides a summary of reported achievable effluent levels for zinc.

Table 5-17. Achievable Effluent Levels of Zinc.

Technology	Achievable Effluent Level, $\mu g/l$
Lime precipitation, pH 8 to 10	400–1,000
Electrolytic	300–700
Lime precipitation with filtration, pH 8 to 10	10–300
Ion exchange	50
Reverse osmosis	5–50
Two-stage lime precipitation with filtration	5–15
Sulfide precipitation	10

Adapted from: Patterson (1985); Eckenfelder (1989).

REFERENCES

Eckenfelder, Jr., W. W. 1989. *Industrial Water Pollution Control*, 2nd Edition. New York: McGraw-Hill.

Eckenfelder, Jr., W. W., C. E. Adams, Jr., and D. L. Ford, 1980. *Development of Design and Operational Criteria for Wastewater Treatment*, Boston, MA: CBI Publishing.

Etzel, J. R. 1988. *Industrial Pretreatment Technologies for Heavy Metal Removal and Treatment of Heavy Metal Sludges to Render Them Non-Hazardous*. Virginia WPCA.

Patterson, J. W. 1981. *Guidance for BAT-Equivalent Control of Selected Toxic Pollutants*. U.S. EPA Region V Enforcement.

Patterson, J. W. 1985. *Industrial Wastewater Treatment Technology*, 2nd Edition. Stoneham, MA: Butterworth Publishers.

QSAR, Quantitative Structural Activity System

U.S. EPA. 1988. Methods for Aquatic Toxicity Identification Evaluations Phase I Toxicity Characterization Procedures. EPA 60013-881036. Duluth, MN

U.S. EPA. December 31, 1980. *Deviation of Site-Specific Water Quality Criteria for the Protection of Freshwater Aquatic Life and its Uses*, U.S. EPA Environmental Research Laboratories (Draft).

U.S. EPA. 1986. *Quality Criteria for Water*, EPA 440/5-86-001.

Volskay, V. T. and C. P. L. Grady. 1988. Toxicity of Selected RCRA Compounds to Activated Sludge Microorganisms. *J. WPCF*, **60**(10): 1850.

6

Aerobic Biological Treatment

W. Wesley Eckenfelder, Jr.

INTRODUCTION

A majority of the priority or toxic organics can be removed using biological treatment. Their actual removal, however, may occur through one or more mechanisms—sorption, stripping, or biodegradation. The removal of priority pollutants in a biological process is summarized in Table 6-1.

REMOVAL MECHANISMS

Sorption

Limited sorption on biological solids occurs for a variety of organics, but this phenomenon is not a primary mechanism of organic removal in the majority of cases. An exception is lindane, as reported by Weber and Jones (1983), who showed that while no biodegradation occurred, there was significant sorption. It is probable that other pesticides will respond in a similar manner in biological wastewater treatment processes.

Stripping

Volatile organic compounds (VOCs) will air-strip in such biological treatment processes as trickling filters, activated sludge, and aerated lagoons. Depending on the VOC in question, both air-stripping and biodegradation may occur. The stripping of VOCs in biological treatment processes is currently receiving considerable attention in the United States since current legislation severely limits permissible emissions of VOCs. Stripping of VOCs in biological treatment processes is discussed in Chapter 8.

Table 6-1. Comparison of Biodegradability of Toxic Organic Pollutants as Predicted by EPA Study to the Five-Plant Study Results.

	EPA Percent Degradation[a]	Five-Point Study Percent Removal[b]
Volatile Compound		
Acrylonitrile	100	99
Benzene	100	100
Bromomethane	48	100
Bromodichloromethane	67	89
Carbon tetrachloride	100	100
Chlorobenzene	100	56[c]
Chloroethane	NA	91
Chloroform	100	99
Dibromochloromethane	55	100
1,1-Dichloroethane	100	79[c]
1,2-Dichloroethane	100	99
1,1-Dichloroethene	100	100
t-1,2-Dichloroethene	100	98
1,2-Dichloropropane	92	97
1,3-Dichloropropane	100	98
Ethylbenzene	100	99
1,1,2,2-Tetrachloroethane	36	93
Tetrachloroethene	100	27[c]
1,1,1-Trichloroethane	100	38[c]
1,1,2-Trichloroethane	59	72[c]
Trichloroethene	100	40[c]
Toluene	100	100
Vinyl chloride	NA	100
Base/Neutral Compound		
Acenaphthene	100	95
Acenaphthylene	98	93
Anthracene	92	97
Benzo(a)anthracene	35	56[c]
Benzo(b)fluoranthene	100	—[c]
Benzo(a)pyrene	NA	—[c]
Bis (2-ethylhexyl) phthalate	95	64
Butylbenzylphthalate	100	87[c]
Dibenzo(a,h)anthracene	NA	—[c]
Di-*n*-butylphthalate	100	99
1,3-Dichlorobenzene	35	100[c]
1,2-Dichlorobenzene	29	94
Diethylphthalate	100	99
Dimethylphtalate	100	99
Dioctylphthalate	94	90
Fluoranthene	100	83[c]
Fluorene	77	93
Isophorone	100	100
Naphthalene	100	100

Table 6-1. (*Continued*)

	EPA Percent Degradation[a]	Five-Point Study Percent Removal[b]
Nitrobenzene	100	98
Pyrene	100	84[c]
1,2,4-Trichlorobenzene	24	82
Acid Compound		
2-Chlorophenol	100	41
2,4-Dichlorophenol	100	91
2,4-Dimethylphenol	100	100
2,4-Dinitrophenol	100	84
2-Nitrophenol	100	77
Pentachlorophenol	100	36
Phenol	100	98
2,4,6-Trichlorophenol	100	45

[a]Includes volatilization losses at 25°C and initial concentration of 5 mg/l.
[b]Based on arithmetic means.
[c]Influent concentration less than 40 ppb.
NA = Not Available.

Biodegradation

Most of the priority or toxic organics will biodegrade, even though some degrade very slowly. In addition, the biomass must be acclimated to the specific organic, particularly when a dissimilar biomass such as a municipal biological sludge is initially used to treat the wastewater. As shown in Fig. 6-1, depending on the organic, total acclimation may take a considerable period of time. For example, the maximum biodegradation of ethylbenzene did not occur until after more than four weeks of acclimation. In another investigation, Tabak and Barth (1978) treated benzidine starting with municipal activated sludge and required six weeks to attain complete acclimation as shown in Fig. 6-2.

As stated above, initial acclimation may take as much as several weeks, however, Watkin (1986) found that acclimated biomass possesses a "genetic memory." In this work, the addition of the test substrate, dichlorophenol (DCP), was discontinued to an activated sludge reactor for ten sludge ages. After resuming the DCP feed to the reactor, DCP biodegradation resumed immediately. Initial acclimation of the reactor had required three weeks.

In some cases, a readily degradable cosubstrate, such as methanol or glucose, must be added to effect rapid biodegradation of a specific toxic organic. However, this is not usually a requirement for most wastewater treatment plants, since a wide variety of degradable organics are usually present.

In summary, when adequate acclimation and a suitable cosubstrate are provided, there are conditions under which extremely low levels of priority organics

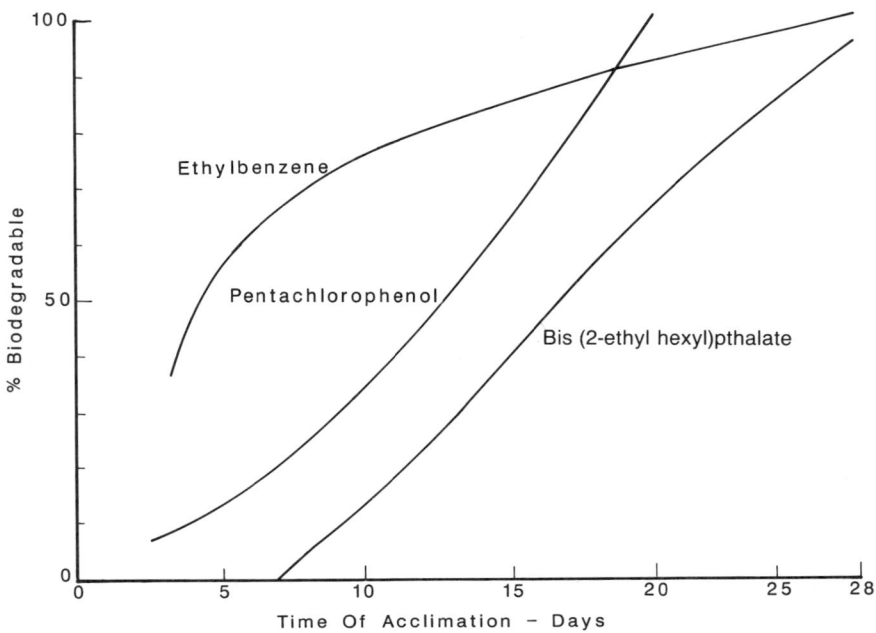

Fig. 6-1. Effect of time of acclimation on biodegradability.

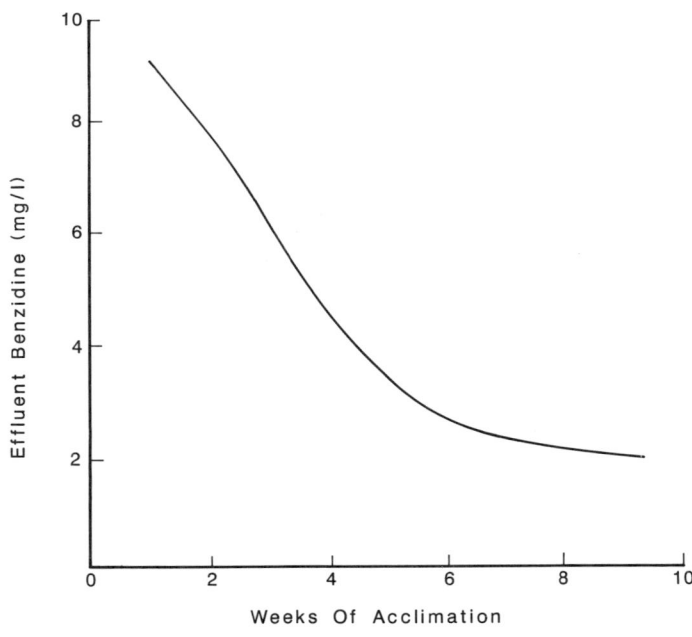

Fig. 6-2. Acclimation for the degradation of benzidine.

128

can be attained biologically. For example, phenolics have been treated in the activated sludge process to effluent levels of less than 10 $\mu g/l$.

REMOVAL KINETICS

Specific Organic Compounds

The kinetic removal mechanism for specific organics in an aerobic biological process has been defined by Monod:

$$\mu = \frac{\mu_m S}{K_s + S} \quad \text{and} \quad q = \frac{q_m S}{K_s + S} \tag{6-1}$$

in which

μ = specific growth rate, 1/day
μ_m = maximum specific growth rate, 1/day
S = substrate concentration, mg/l
K_s = substrate concentration when the rate is one-half the maximum rate, mg/l
q = specific substrate removal rate, 1/day
q_m = maximum substrate removal rate, 1/day

In complete mix activated sludge (CMAS), the Monod equation can be expressed:

$$S_0 - S = \frac{q_m S}{K_s + S} \cdot X_v t \tag{6-2}$$

in which

S_0 = initial substrate concentration, mg/l
X_v = volatile suspended solids under aeration, mg/l
t = liquid retention time, days

Solving for S yields:

$$S = \frac{-B + (B^2 + 4 S_0 K_s)^{1/2}}{2} \tag{6-2a}$$

in which

$$B = q_m X_v t + K_s - S_0$$

The SRT in the activated sludge process can be defined as follows:

$$\theta_c = \frac{X_v t}{a(S_0 - S) - b x_d X_v t} \tag{6-3}$$

in which

θ_c = SRT, days
a = yield coefficient, g/g
b = endogenous coefficient, 1/day
X_d = degradable fraction of the VSS

In complete mix activated sludge (CMAS), the effluent substrate concentration is directly related to the SRT (θ_c). Combining Eqs. (6-2) and (6-3) yields:

$$S = \frac{K_s(1 + bX_d\theta_c)}{\theta_c(q_m a - bX_d) - 1} \qquad (6\text{-}4)$$

This relationship for dichlorophenol (DCP) is shown in Fig. 6-3.

For the case of a plug flow activated sludge (PFAS) reactor, the performance equation derived from the Monod relationship is:

$$\frac{1}{\theta_c} = \frac{\mu_m(S_0 - S)}{(S_0 - S) + CK_s} - bX_d \qquad (6\text{-}5)$$

where

$C = (1 + \alpha) \ln [(\alpha S + S_0)/(1 + \alpha)S]$
$\alpha = R/Q$
S_0 = influent substrate concentration prior to mixing with the recycle

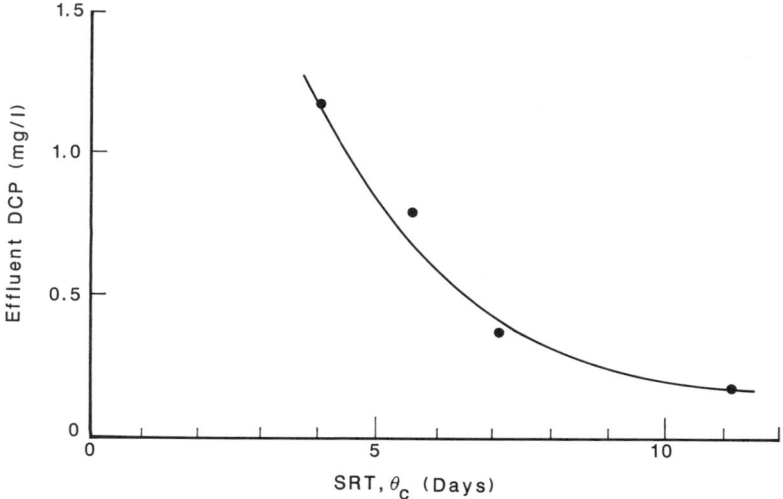

Fig. 6-3. Effect of SRT on DCP removal.

Effluent levels of specific priority organics; e.g., phenol, can be computed from Eqs. (6-4) or (6-5) depending on the configuration of the reactor, complete mix or plug flow.

EXAMPLE 6-1

Compute the SRT required in a complete mix activated sludge plant to reduce phenol from 10 mg/l to 15 μg/l, where:

$$q_m = 1.8 \text{ g/g VSS} \cdot \text{day @ } 20°C$$
$$\theta = 1.1$$
$$K_s = 100 \ \mu\text{g/l}$$
$$a = 0.6$$
$$bX_d = 0.05 \text{ day}^{-1} \text{ @ } 20°C$$
$$= 0.033 \text{ @ } 10°C$$

Equation (6-4) can be rearranged to yield

$$\theta_c = \frac{K_s + S}{aq_m S - bX_d(K_s + S)}$$

$$= \frac{0.1 + 0.015}{0.6 \cdot 1.8 \cdot 0.015 - 0.05(0.115)}$$

$$= 11.0 \text{ days}$$

What is the increase in SRT if the temperature is reduced to 10°C

$$q_{m\,10°} = q_{m\,20°} \cdot 1.1^{-10}$$

$$= 1.8/2.6$$

$$= 0.69$$

$$\theta_c = \frac{0.1 + 0.015}{0.6 \cdot 0.69 \cdot 0.015 - 0.033(0.115)}$$

$$= 47.6 \text{ days}$$

Calculate the SRT at 20°C for a plug flow system with a 50 percent sludge recycle rate

$$\frac{1}{\theta_c} = \frac{aq_m(S_0 - S)}{(S_0 - S) + CK_s} - bX_d$$

$$C = (1 + \alpha) \ln\left[(\alpha S + S_0)/(1 + \alpha)S\right] \quad \text{when} \quad \alpha = R/Q$$

for $\alpha < 1$,

$$\frac{1}{\theta_c} \cong \frac{aq_m(S_0 - S)}{(S_0 - S) + K_s \ln(S_0/S)} - bX_d$$

$$\frac{1}{\theta_c} = \frac{0.6 \cdot 1.8(10 - 0.015)}{(10 - 0.015) + 0.1 \ln(10/0.015)} - 0.05$$

$$= 0.964 \text{ day}^{-1}$$

$$\theta_c = \frac{1}{0.964}$$

$$= 1.04 \text{ days}$$

Total Organics

In most cases, K_s is very small and removal of an organic to low residual concentrations can be considered as a zero order reaction. It has been shown that for multicomponent substrates simultaneous removal of organics at different zero order rates can be approximated by the relationship (Adams et al. 1975; Daigger and Grady 1971):

$$\frac{S_0 - S}{X_v t} = KS/S_0 \tag{6-6}$$

in which S_0 and S represent the influent and effluent total organic content expressed as TOC, COD, or BOD. The coefficient K in Eq. (6-6) is a proportionality constant indicating that, as the fraction of organic remaining (S/S_0) decreases, the removal rate $(S_0 - S/X_v t)$ decreases since the more readily degradable substrates are removed first. Eq. (6-6) should not be used at effluent levels below 10 mg/l since most of the residual organics are oxidation byproducts and not the organics originally constituting the wastewater. It should also be noted that as the composition of the wastewater changes, K will also change depending on the zero order reaction rates of the respective constituents. Eq. (6-6) is employed to estimate process operating conditions to achieve defined effluent TOC, COD, or BOD levels.

EXAMPLE 6-2

A wastewater from a chemical plant has the following characteristics:

BOD = 425 mg/l
K = 5.0/day @ 20°C
DCP = 10 mg/l
K_s = 0.2 mg/l (for DCP)

$q_m = 0.8$ g/g $\cdot$ d (for DCP)
$a = 0.6$ (for DCP and BOD)
$b = 0.1$ @ 20°C (for DCP and BOD)

Permit requirements call for an effluent soluble BOD of 15 mg/l and DCP of 20 μg/l. Develop a process design to meet these requirements.

- *Requirements for BOD removal:*

$$\frac{S_0 - S}{X_v t} = K \frac{S}{S_0}$$

$$\frac{425 - 15}{X_v t} = 5.0 \frac{15}{425}$$

$$X_v t = 2,342 \text{ day} \cdot \text{mg/l}$$

$$F/M = \frac{425}{2,342}$$

$$= 0.18 \text{ day}^{-1}$$

Selecting:

$$X_v = 3,000 \text{ mg/l}$$

$$t = 0.78 \text{ days}$$

$$\Delta X_v = aS_R - bX_d X_v t$$

Assume $X_d = 0.58$. Then

$$\Delta X_v = 0.6(410) - 0.058(2,342)$$

$$= 110 \text{ mg/l}$$

$$\theta_c = \frac{X_v t}{\Delta X_v}$$

$$= \frac{2,342}{110}$$

$$= 21 \text{ days}$$

$$X_d = \frac{0.8}{1 + 0.2b\theta_c}$$

$$= \frac{0.8}{1 + 0.2(0.1)(21)}$$

$$= 0.56 \text{ (compared to assumed 0.58)}$$

The calculated X_d of 0.56 agrees well with the assumed value of 0.58 and, therefore, the BOD requirement will be met with an SRT of twenty-one days.

- *Requirements for DCP removal:*

$$\theta_c = \frac{K_s + S}{aq_m S - bX_d(K_s + S)}$$

Assume $x_d = 0.30$

$$\theta_c = \frac{0.2 + 0.02}{0.6 \cdot 0.8 \cdot 0.02 - 0.1 \cdot 0.030 \cdot 0.22}$$

$$= 73.3 \text{ days}$$

Check X_d

$$X_d = \frac{0.8}{1 + 0.2(0.1)(73.3)}$$

$$= 0.32 \text{ (compared to assumed } 0.30)$$

The calculated X_d of 0.32 agrees well with the assumed value of 0.30 and, therefore, the DCP requirement will be met with an SRT of seventy-three days, which exceeds the SRT required for BOD removal, therefore DCP removal will control the design.

In practice in a majority of cases, a high degree of back mixing takes place so that conditions in a plug flow tank may approach that of complete mixing. However, in order to take advantage of the kinetics of priority pollutant removal, multiple basins in series will offer distinct advantages over a single complete mix basin.

Equation (6-2) can be applied to predict the performance of multiple reactors in series as shown by the following example.

EXAMPLE 6-3

Design a single stage and a three-stage activated sludge reactor to reduce phenol. Compute the effluent phenol level in each case.

$S_0 = 10 \text{ mg/l}$
$bX_d = 0.05$
$\theta_c = 10 \text{ days}$
$q_m = 0.6$
$K_s = 0.2$
$a = 0.6$

For the CMAS (one-stage) $X_v t$ is computed from Eqs. (6-2) and (6-3) as 39.44.

$$B = 0.6(39.44) + 0.2 - 10$$

$$= 13.86$$

$$S = \frac{-13.86 + \left[(13.86)^2 + 4 \cdot 10 \cdot 0.2\right]^{1/2}}{2}$$

$$= 0.14 \text{ mg}/\text{l}$$

Hence, S for a single complete mix basin is 0.14 mg/l. We can now compute the performance of three basins in series. Each basin is considered as a complete mix basin in which the retention time is based on $(Q + R)$. The effluent from the third basin is assumed negligible. The concentration entering the first basin with a 50 percent recycle is

$$S_0 = \frac{10}{1.5}$$

$$= 6.67$$

For the same total basin volume as in the complete mix case $X_v t$ is reduced by a factor of 1.5

$$X_v t = 39.44/3 \cdot 1.5$$

$$= 8.76$$

$$B = 0.6 \cdot 8.76 + 0.2 - 6.67$$

$$= -1.21$$

$$S = \frac{+1.21 + \left[1.21^2 + 4 \cdot 0.2 \cdot 6.67\right]^{1/2}}{2}$$

$$= 1.9 \text{ mg}/\text{l}$$

In like manner, the effluent (S) from basin 2 is computed as 0.105 mg/l and from basin 3 as 0.005 mg/l.

INHIBITION

Most effluent permit levels are in the $\mu g/l$ range so that it is improbable that inhibition will be a factor. The Haldane equation (6-7) may be used to define maximum concentrations of specific pollutants to avoid inhibition:

$$\mu = \frac{\mu_m S}{S + K_s + S^2/K_i} \tag{6-7}$$

in which K_i is the inhibition constant determined through the continuously fed batch reactor (FBR) test described in Chapter 4.

Development of Design Criteria

In order to employ Eqs. (6-4) or (6-5), it is necessary to determine values for q_m, K_s, a, b, and X_d. The latter three parameters a, b, and X_d are readily determinable by conventional techniques (Eckenfelder 1989). K_s and q_m, however, present a greater challenge.

Templeton and Grady (1988) have recently shown that the history of the biomass will influence the value of the coefficients in Eq. (6-5). When bacterial cells are grown at a constant specific growth rate in continuous culture, physiological adaptation occurs. As a result, the bacterial cells achieve fixed levels of ribonucleic acid (RNA) protein and other macromolecules unique to their historical growth conditions. The values of q_m and K_s determined using a given culture will, therefore, reflect these growth conditions. Philbrook and Grady (1985) have shown that because of microbial competition each CMAS reactor will have associated with it a unique microbial community. Reactors which have been operated at high specific growth rates will contain communities which are characterized by high μ_m and low K_s values. In addition, μ_m may be influenced by the operating SRT due to the change in active mass. As the SRT is increased, nondegradable mass accumulates, resulting in a progressive decrease in active mass and presumably a decrease in μ_m as related to the total volatile suspended solids. It is therefore important to determine the coefficients with a biomass cultured at the appropriate sludge age with the appropriate substrate. In order to maintain the integrity of the results, a short term test such as the FBR should be used in order not to change the population dynamics.

The modified FBR test as described in Chapter 4 would appear to be most applicable to the determination of the kinetic coefficients q_m and K_s under appropriate operating conditions. In the test, biomass produced at the desired SRT is placed in a 2 liter reactor. A solution wastewater containing the desired compound(s) is added at a constant rate of 100 ml/hr. In order to determine q_m, the addition rate must exceed the degradation rate as shown in Fig. 6-4. When testing wastewater samples, the priority organic levels may be too low, requiring that the concentration be supplemented to provide a sufficient concentration to meet the conditions of the test. The data are only relevant so long as the concentration levels achieved in the test reactor are below the inhibition threshold.

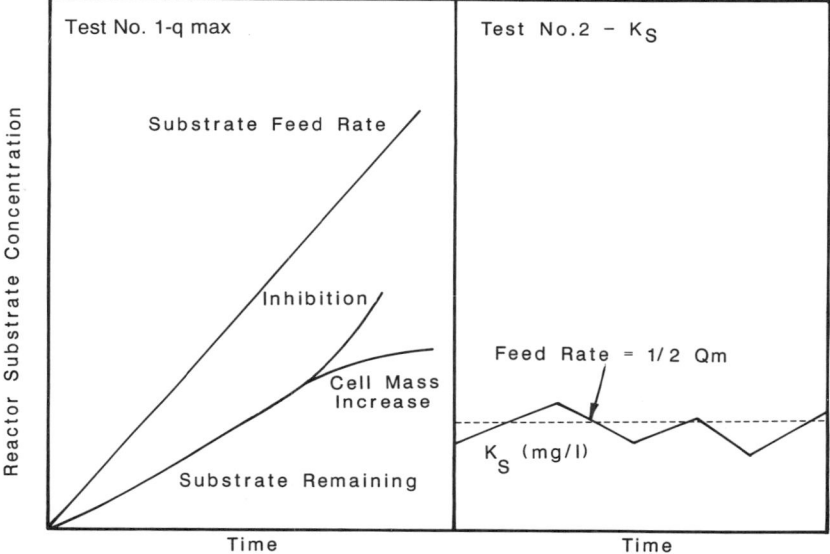

Fig. 6-4. Correlation of FBR results.

This can be determined from the shape of the concentration-time curve. The maximum degradation rate q_m is computed as the difference in slope of the addition rate and the residual accumulation.

A second FBR test is then conducted with the addition rate of the compound equal to one-half the maximum degradation rate (q_m) determined in the first test. As shown in Fig. 6-4, the steady state concentration observed in the reactor during this test will be K_s in mg/l. Since these coefficients have been observed to vary with SRT, it may be desirable to repeat the test with sludges acclimated to a range of SRT's if available or relevant.

Temperature is also an important variable in the FBR determination. The effect of temperature on q_m for phenol is shown in Fig. 6-5. These data are obtained from an oxygen activated sludge plant treating a wastewater containing acetone and phenol. The temperature coefficient θ is computed as 1.1.

EFFECT OF SHOCK LOADS ON EFFLUENT QUALITY

The effect of a sudden increased influent concentration of a specific priority pollutant can be estimated from Eqs. (6-2) and (6-3). The premise is made that the biomass is in equilibrium with a given influent concentration of pollutant. An increase in influent concentration causes an increase in effluent concentra-

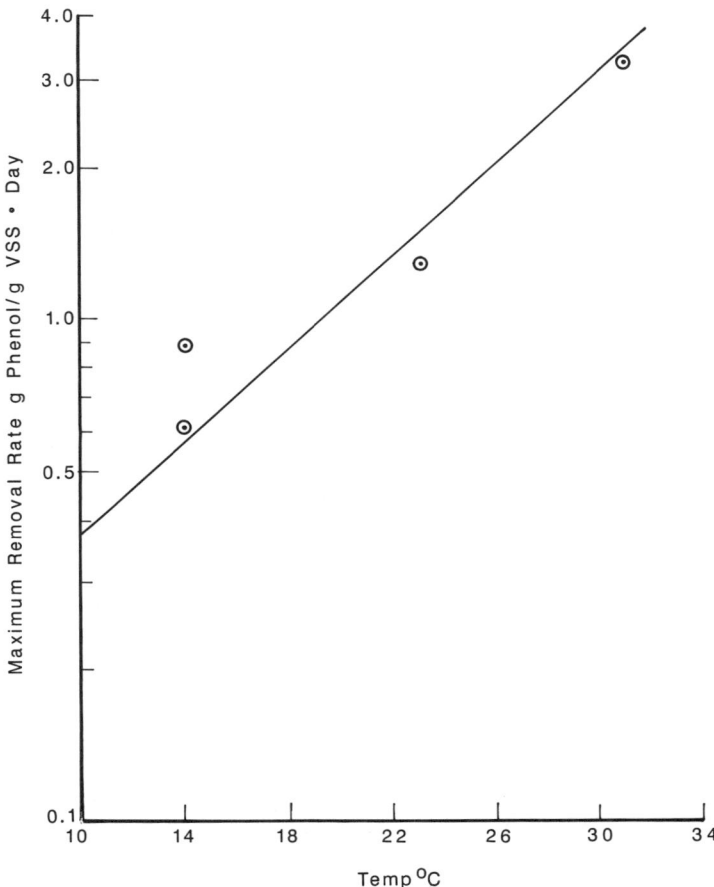

Fig. 6-5. Effect of temperature on phenol removal rate.

tion with a corresponding increase in removal rate. If the increased influent is maintained, the biomass concentration will increase causing the effluent to return to its original value. The significance of this is to regulate the influent concentration such that the maximum permissible effluent is not exceeded. Substrate removal kinetics at the increased influent concentration is described in Eq. (6-8) with the $X_v t$ corresponding to the pre-shock conditions taken from Eq. (6-7). The resulting expression is:

$$S_{02} - S_2 = \frac{q_m S_2}{K_s + S_2} \cdot \frac{\theta_c a (S_{01} - S_1)}{(1 + b\theta_c)} \tag{6-8}$$

where:

S_{01}, S_2 = the steady state (pre-shock) values of the influent and the effluent concentrations

S_{02}, S_2 = the shock load influent and effluent concentrations

Equation (6-8) assumes that there is no inhibition at the highest influent concentration (or the resulting concentration in a complete mix reactor).

Rearrangement of Eq. (6-8) yields

$$S_2 = \frac{-B + [B^2 + 4S_{02}K_s]^{1/2}}{2} \tag{6-9}$$

in which

$$B = K_s - S_{02} + \frac{q_m \theta_c a (S_{01} - S_1)}{1 + b\theta_c}$$

EXAMPLE 6-4

For the following conditions, compute the effluent priority pollutant level for a range of increased influent concentrations of 20 to 100 mg/l

$q_m = 1.0$ g/g · day
$K_s = 0.1$ mg/l
$bX_d = 0.05$
$a = 0.5$
$S_{01} = 10$ mg/l
$S_1 = 0.043$ mg/l

$X_v t$ and θ_c for the steady-state conditions are calculated from Eqs. (6-2) and (6-8), respectively. The values obtained are $X_v t = 33.12$ mg/l · day and $\theta_c = 10$ days. Substituting in Eq. (6-9) yields:

S_{02}, mg/l	S_2, mg/l
20	0.149
30	0.74
40	7.26
50	17.0
70	36.9
100	66.9

Multiple basins in series are an advantage in dampening a shock load effect, providing inhibition does not occur. For example, an increase in pollutant from a steady state value of 10 mg/l to 20 mg/l will result in an increase in effluent concentration to 0.85 mg/l in a single stage reactor while a three stage reactor of the same volume will result in an increase in effluent concentration to 0.255 mg/l.

SOLUBLE MICROBIAL PRODUCT FORMATION

Biodegradation in the activated sludge process can be described by the following relationships:

$$\text{Organics} + O_2 + N + P + \text{cells} \rightarrow \text{new cells} + CO_2 + H_2O + SMP$$

and

$$\text{cells} + O_2 \rightarrow CO_2 + H_2O + N + P + \text{cellular SMP}$$

As can be seen in the above equations, soluble microbial products (SMP) are generated both through the degradation of organics and through the endogenous degradation of the biomass. The SMP are oxidation byproducts which are largely nondegradable in the activated sludge process. Data for a peptone-glucose mixture and a synthetic fiber wastewater are shown in Fig. 6-6.

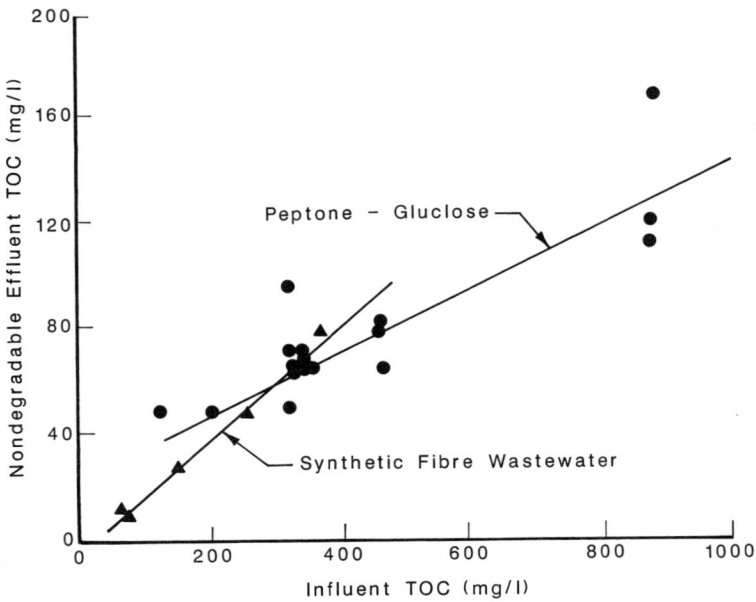

Fig. 6-6. Nonbiodegradable residual TOC from biooxidation.

Table 6-2. Molecular Weight Distribution of Biological Effluents.

Mol. Wt.	PA and Dyestuffs, % TOC		Glucose, % COD[a]
	Influent	Bioeffluent	
>10,000	—	11.5	45
500–10,000	—	14.5	16
<500	100	74.0	39

[a]Chuboda (1985).

Many of these byproducts are of high molecular weight, as shown in Table 6-2. It has been further found that some of these high molecular weight fractions have a high toxicity to some aquatic species. Results showing the toxicity of several wastewaters from the plastics industry before and after biological oxidation is shown in Fig. 6-7. While most of the wastewaters showed reduced toxicity following biooxidation, two of the wastewaters exhibited increased toxicity after biooxidation, making the oxidation byproducts suspect.

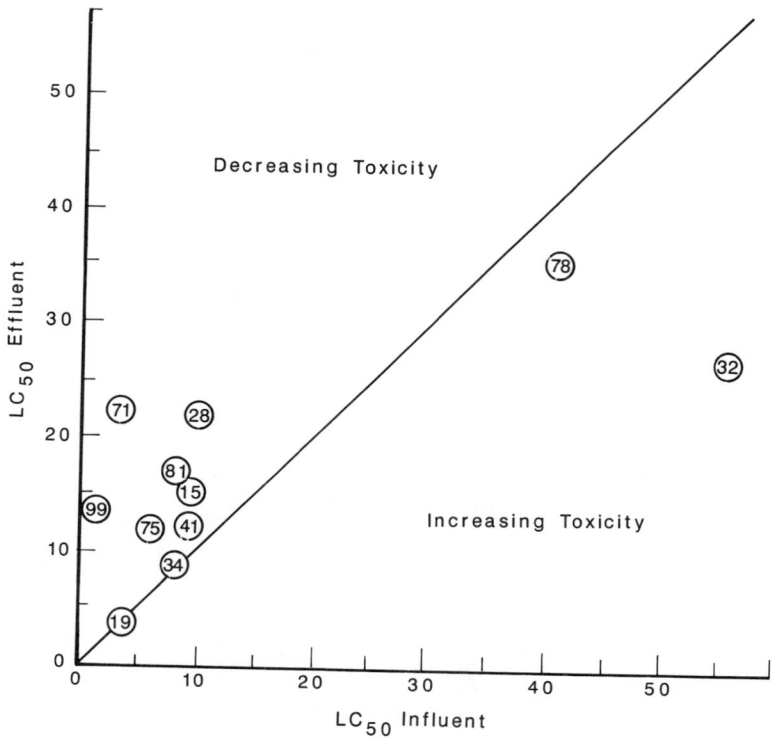

Fig. 6-7. Change in toxicity through biological treatment. Numbers in circles indicate TOC reduction.

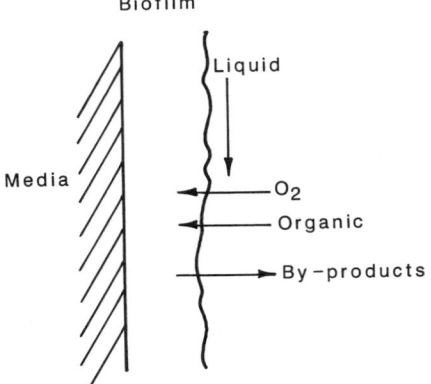

Fig. 6-8. Mechanism of organic removal in a fixed film process.

AEROBIC FIXED FILM REACTORS

Aerobic biological processes such as the activated sludge process are not applicable to wastewaters containing BOD levels below 50 mg/l since microbial growth is insufficient to prevent washout. Low concentration organic wastewaters are, therefore, treated in a fixed film process. Depending on the specific conditions and effluent requirements, such processes include trickling filters, rotating biological contactors (RBC), upflow or downflow aerated packed beds, or biologically activated carbon (BAC). The mechanism of organic removal in each of these processes is shown in Fig. 6-8.

At BOD concentrations above about 10 mg/l, supplementary oxygen must be supplied by aeration to the reactor as shown in Fig. 6-9. At higher substrate

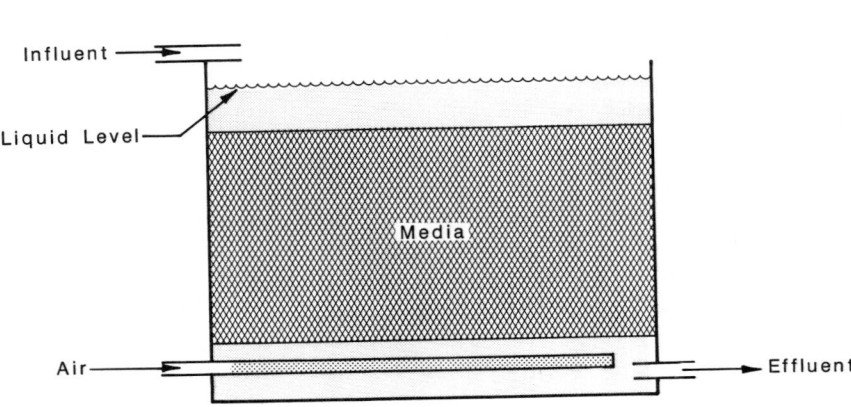

Fig. 6-9. Fixed film reactor with air addition.

levels, the organisms have a high growth rate and a high value of K_s as defined by Eq. (6-1). Rittmann and McCarty (1980) developed a biofilm model which considered mass transfer of the substrate through the bulk liquid to the biofilm, diffusion of the substrate through the biofilm, biological utilization of the substrate, growth of the biofilm and decay of the biofilm. They predicted that the concentration of substrate needed to keep the biofilm in a steady state (no net gain or loss) is:

$$S_{min} = K_s \frac{b}{aK - b} \qquad (6\text{-}10)$$

in which

S_{min} = minimum substrate concentration, mg/l
K_s = Monod constant, mg/l
b = endogenous coefficient, 1/days
a = cell yield coefficient
K = maximum specific utilization rate

Van der Kooij et al. (1980) found that the constant K_s for acetic acid was 4 μg/l. Values for other compounds are shown in Table 6-3. Data for acetate reduction through a column at different influent concentrations are shown in Fig. 6-10.

Bio-oxidation of low levels of specific organics can best be achieved in a biologically active carbon column in which the carbon serves as the media for biofilm growth: 2 to 3 mg/l of organic compounds can be treated without supplemental aeration when the influent wastewater is near saturation. Results from a BAC column study are shown in Table 6-4 (Shibata et al. 1987).

Table 6-3. Kinetic Coefficients for Oligiotropic Growth.

Substrate	μ_m (hr^{-1})	K_s, μg/l
Acetic acid	0.2	2.2
Methanol	0.081	1.3

Table 6-4. BAC Continuous Column Results.

	Influent	Effluent
D.O., mg/l	9.2	7.0
TOC, mg/l	3.61	0.44
Methanol, μg/l	1,260	0.2
IPA, μg/l	2,520	2.1
Acetone, μg/l	159	1.3

Detention time based on empty bed.
Volume = 10 m.

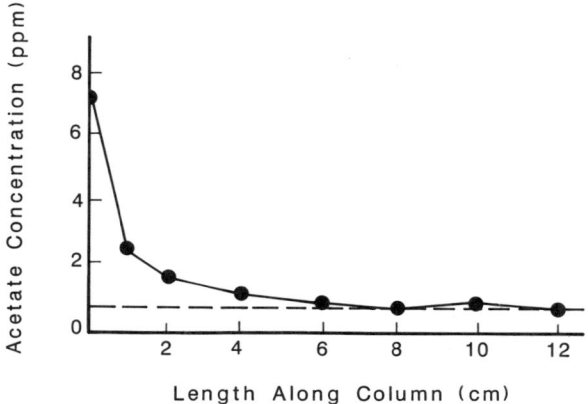

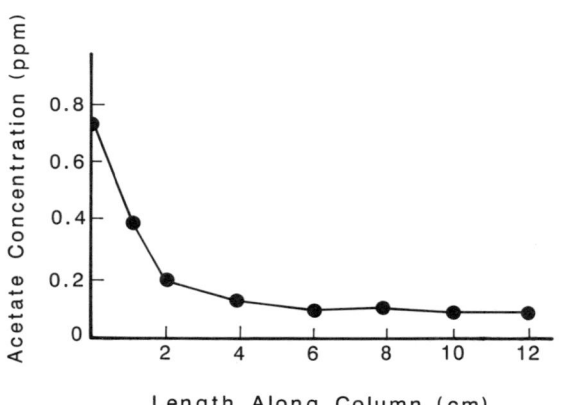

Fig. 6-10. Acetate removal in a column at two influent concentrations. (*Source:* Rittmann and McCarty 1980).

REFERENCES

Adams, C., W. W. Eckenfelder, and J. A. Hovious. 1975. Kinetic Model for Design of Completely Mixed Activated Sludge Treating Variable Strength Industrial Waste. *Water Research* **9**(37).

Chudoba, J. 1985. Inhibitory Effect of Refractory Organic Compounds Produced by Activated Sludge Microorganisms on Microbial Activity and Flocculation. *Water Research* **19**(2): 197.

Daigger, G. T. , and C. P. L. Grady. December 1971. Factors Affecting Effluent Quality from Fill and Draw Activated Sludge Reactors. *J. WPCF.* **43**(12): 1845

Eckenfelder, W. W. 1989. *Industrial Water Pollution Control*, 2nd Edition. New York: McGraw-Hill Book Co.

Philbrook, D. M., and C. P. L. Grady. 1985. Evaluation of Biodegradation Kinetics for Priority Pollutants. *Proc. 40th Ind. Waste Conf., Purdue University.*

Rittmann, B. E., and P. L. McCarty. 1980. Substrate Flux into Biofilms of any Thickness. *Biotech. and Bioeng.*, **22**(11): 2359–2374.

Shibata, K. et al. 1987. Biological Decomposition of Trace Organic Compounds, *Water Sci. Tech.*, **19**: 417–427.

Tabak, H. H., and E. F. Barth. 1978. Biodegradability of Benzidine in Aerobic Suspended Growth Reactors. *J. WPCF*, **50**: 552.

Templeton, L. L., and C. P. Grady. 1988. Effect of Culture History on the Determination of Biodegradation Kinetics by Batch and Fed Batch Techniques. *J. WPCF*, **60**(5): 651.

Van der Kooji et al. 1980. Growth of *Aeromonas hydrophila* at Low Concentrations of Substrates Added to Tap Water. *Appl. Environ. Microbiol.*, **39**: 1198–1204.

Watkin, A. 1986. Evaluation of Biological Rate Parameters and Inhibitory Effects in Activated Sludge. Ph.D. dissertation, Vanderbilt University.

Weber, W. J., and B. Jones. 1983. Toxic Substance Removal in Activated Sludge and PAC Treatment Systems. U.S. EPA NTIS No. PB-86-18242J/AS.

7

Anaerobic Biological Treatment

Richard E. Speece

DEMONSTRATED TREATMENT PERFORMANCE

Research studies such as those performed by the Pulp and Paper Institute of Canada have succeeded in producing nontoxic effluent using only anaerobic treatment on industrial wastewater. Chlorophenols, chloroguaiacols, chloroveratroles, and chlorocatechols have been successfully treated in a continuous upflow anaerobic sludge blanket reactor (Woods et al. 1988). Chlorinated methoxybenzenes were converted to chlorinated hydroxybenzenes in the first step of their anaerobic biodegradation. Chlorophenols and chlorocatechols were then dechlorinated and the dechlorination was strongly dependent on chlorine position with ortho most readily removed. Pentachlorophenol has been reduced by 99.9 percent in anaerobic digestion (Guthrie et al. 1984).

The Enso-Fenox process (Hakulinen and Salkinoja-Salonen 1982; Salkinoja-Salonen et al. 1979) has been very successfully used to remove chlorinated phenolic compounds from pulp bleaching effluents. It is a two-stage anaerobic/aerobic process consisting of a nonmethanogenic anaerobic fluidized bed followed by a trickling filter. Overall reduction of chlorophenols is 65 to 100 percent. It has been found that compounds containing only *meta* and *para* chlorines were not degraded significantly during the first month (Woods et al. 1988). Later, further acclimation resulted in *meta* and *para* dechlorination. The rate of dechlorination under anaerobic conditions increased with the degree of chlorination. Throughout seven months of continuous treatment, there was no evidence of dehalogenation of monochlorophenols or removal of chlorines from the *para* position to the hydroxyl. However, dehalogenation of *para*-chlorophenol has been demonstrated by others (Krumme and Boyd 1988). There were fundamental differences between the degradation of chlorinated methoxyphenols and chlorophenols which lead to significantly better conversion of the former.

Studies have been conducted (Krumme and Boyd 1988) on reductive dechlorination of chlorinated aromatic compounds under anaerobic conditions with chlorinated phenols as the sole carbon and energy source. Approximately 40 percent of the added chlorophenols was converted to CH_4 and CO_2. Substrate loading rates were 20 mg/l-d at hydraulic detention times of two to four days with 90 percent substrate conversion efficiency. Reductive dechlorination of mono-, di-, tri-, and pentachlorophenols has been demonstrated in anaerobic sewage sludge (Boyd and Shelton 1984).

Reductive dehalogenation has been demonstrated to be quite feasible (Bouwer 1988) for hexachloroethane, ethylene dibromide, tetrachloroethylene, chlorobenzene, 1,2-dichlorobenzene, and pentachlorophenol. All of these reductions appear to be energetically favorable under sulfate reducing conditions, with reduction of ethylene dibromide and hexachloroethane appearing to be possible under aerobic respiration and denitrification. Hydrogen sulfide is a strong reductant and nucleophile that can participate in substitution and addition reactions with several classes of organic functional groups.

The following constituents were tested in the laboratory (Blum et al. 1986) at their approximate concentrations in coal conversion wastewater (CCWW) and were anaerobically degraded in serum bottles: 1,000 mg/l phenol; 500 mg/l resorcinol; 1,000 mg/l cathechol; 500 mg/l p-cresol; 200 mg/l pyridine; 2,000 mg/l benzoic acid; 250 mg/l 4-methylcatechol; 500 mg/l, 4-ethylpyridine; and 2,000 mg/l hexanoic acid. The findings from this study confirm those of other researchers: phenol, resorcinol, cathechol, p-cresol, benzoic acid, and hexanoic acid were degraded anaerobically with gas production.

The degradation of p-cresol, o-methoxyphenol, and m-methoxyphenol has been demonstrated (Boyd et al. 1983) in serum bottle studies at 50 mg/l with wastewater sludge. Methane production was achieved. Catechol and resorcinol were found to be intermediates in the degradation of o-methoxyphenol and m-methoxyphenol, respectively. Again, o-cresol was resistent to degradation in these experiments.

Recent research has demonstrated the anaerobic biodegradability of several individual CCWW constituents, including phenol, resorcinol, catechol, p-cresol, benzoic acid, and hexanoic acid. Constituents shown to be more resistant to anaerobic biodegradation include o-cresol, 3,4-xylenol, 3,5-xylenol, and 2,3,5-triethylphenol. Attached anaerobic growth on granular activated carbon has been able to treat full strength CCWW (Suidan et al. 1987), which contains considerable inherent toxicity, some of which is nonbiodegradable. Adsorption followed by anaerobic biodegradation is the theory for the effectiveness of this configuration. Continuous operation requires that carbon be wasted regularly.

A serum bottle technique has been demonstrated (Shelton and Tiedje 1984) to determine whether an organic chemical is susceptible to anaerobic degradation to the end products of methane and carbon dioxide. Table 7-1 provides a summary of the organic compounds mineralized in this investigation.

It was observed in this work (Shelton and Tiedje 1984) that "some compounds which are relatively slowly degraded in whole sludge may not be appreciably degraded in 10 percent sludge. Laboratory studies have compared the degradation of six phthalic acid esters [dimethyl-, diethyl-, di-*n*-butyl-, butyl-benzyl-, di-(2-ethylhexyl)-, and di-*n*-octylphthalate] and six monosubstituted phenols (*o*-, *m*-, and *p*-cresol, and *o*-, *m*-, and *p*-chlorophenol) undiluted in 10 percent sludge. Of the six phthalic acid esters, four were mineralized in both undiluted and 10 percent sludge, whereas two were persistent in both. Of the six substituted phenols, five were mineralized in undiluted sludge (*o*-cresol was persistent), whereas two were persistent in both. Of the six substituted phenols, five were mineralized in 10 percent sludge; *m*- and *p*-chlorophenol were not mineralized in 10 percent sludge. For compounds which are slowly mineralized or are toxic at higher concentrations or both, e.g., *m*- and *p*-chlorophenol, the 10 percent sludge assay may yield false negatives; however, for many compounds it is believed that the 10 percent sludge assay should successfully reflect the activity of undiluted sludge." Tables 7-2 and 7-3 show the effect of different sludges and the reproducibility of a given sludge on anaerobic biodegradation.

A survey was conducted (Battersby and Wilson 1989) on the anaerobic biodegradation potential of seventy-seven organic chemicals by digestor sludge. Fig. 7-1 presents the typical net gas (CH_4 plus CO_2) production patterns for organic chemicals incubated anaerobically with diluted primary digestor sludge. The biodegradation potentials are summarized in Table 7-4, with the pattern of gas production referenced to the code in Fig. 7-1.

ACCLIMATION

A petrochemical may initially exhibit toxicity to an unacclimated population of methane-fermenting bacteria, but with acclimation the toxicity may be greatly reduced or disappear. In addition, the microorganisms may develop the capacity to actually degrade compounds which showed initial toxicity. Inducible enzymes may be synthesized and/or a shift in the predominant population may result allowing metabolism to commence.

Fifty-three petrochemicals were assayed for toxicity and biodegradation in anaerobic treatment (Chou et al. 1978a). The petrochemicals which could be anaerobically biodegraded in serum bottles are shown in Table 7-5 along with the lag times. Table 7-6 provides the utilization efficiencies of petrochemicals which were fed to anaerobic filters at a feed concentration of 2,000 mg/l.

Table 7-1. Summary of Organic Compounds Mineralized Under Anaerobic Conditions in 10% Sludge (8 weeks).

Conditions	Organic Compounds
Mineralized in 10% sludge from a secondary digester[a]	Acetylsalicylic acid
	Acrylic acid
	p-Anisic acid
	o-Anthranilic acid
	Benzoic acid
	Benzyl alcohol
	2,3-Butanediol
	Catechol
	m-Cresol
	p-Cresol
	Di-n-butylphthalate
	Dimethylphthalate
	Ethyl acetate
	2-Hexanone
	o-Hydroxybenzoic acid
	p-Hydroxybenzoic acid
	3-Hydroxybutanone
	3-Methylbutanol
	1-Octanol
	Phenol
	Phloroglucinol
	Phthalic acid
	Polyethylene glycol
	Protocatechuic acid
	Pyrogallol
Partially mineralized in 10% sludge from a secondary digester[b]	p-Aminobenzoic acid
	Butylbenzylphthalate
	4-Chloroacetanilide
	m-Chlorobenzoic acid
	Diethylphthalate
	Geraniol
	4-Hydroxyacetinilide
	p-Hydroxybenzyl alcohol
	2-Octanol
	Propionanilide
Mineralized in 10% sludge from a primary digester[a]	Butylbenzylphthalate
	m-Chlorobenzoic acid
	o-Chlorophenol
	Diethylphthalate
	o-Methoxyphenol
	m-Methoxyphenol
	o-Nitrophenol
	p-Nitrophenol

Source: Shelton and Tiedje (1984).
[a]Greater than 75% of theoretical methane production.
[b]Between 30 and 75% of theoretical methane production.

Table 7-2.   Percent Theoretical Methane Production from Nine Substrates by Sludges from Nine Municipalities[a].

	Source of Seed Sludge								
	Adrian	Jackson	Ann Arbor	St. Johns	Iona	Holt	Mason	Chelsea	Portland
Ethanol	62	78	86	78	71	33	94	54	52
Polyethylene glycol	61	43	ID[b]	78	51	53	82	38	98
p-Cresol	91	88	101	79	80	88	94	62	77
Phthalic acid	88	80	132	113	73	86	96	60	96
m-Cresol	82	103	85	0	0	77	91	ID	0
Di-n-butyl-phthalate	24	49	91	37	36	ID	0	0	0
2-Octanol	0	22	58	0	0	0	87	0	0
m-Chlorobenzoic acid	85	0	0	0	0	85	0	0	0
Propionanilide	0	0	36	33	0	23	0	0	0

Source: Shelton and Tiedje (1984).
[a]Fresh sludge (10 percent) from primary digesters incubated for 8 weeks, after which methane was measured by gas chromatography.
[b]ID, Insufficient Data: Generally due to leaky bottles or only one bottle showing evidence for degradation.

Table 7-3.   Reproducibility of Results with Different Batches of Sludge for Several Compounds and Sludges.

| Sludge | No. of Times Degradation Observed[a] | | No. of Times Degradation Tested | |
	p-Cresol	Phthalic acid	m-Chlorobenzoic acid	Di-n-butyl-phtalate
Jackson[b]	5/5	5/5	0/3	3/4
Adrian	2/3	3/3	1/2	2/3
Holt	3/3	3/3	2/2	0/2
Mason	2/2	2/2	0/2	1/2
Iona	2/2	1/2	0/2	1/2
Chelsea	1/2	1/2	—	0/2

Source: Shelton and Tiedje (1984).
[a]Degradation is defined as >75% of theoretical gas or methane production.
[b]Phenol and benzoic acid were both degraded in Jackson sludge two out of two times.

Table 7-4.   Anaerobic Biodegradation Potential of Phenol and Substituted Phenols.

Chemical	Net Gas Production, % Theoretical	Pattern of Gas Production[a]	Lag,[b] days	Biogradation Potential
Phenol	99 ± 8.0	2	6	+
2-Aminophenol	88 ± 15.4	2	15	+
3-Aminophenol	9 ± 17.7	3	>75	−
4-Aminophenol	20 ± 12.6	3	57	−
2-Chlorophenol	−16 ± 11.2	3	>65	−
3-Chlorophenol	−42 ± 13.5	4	>100	Inhibitory
4-Chlorophenol	−40 ± 11.3	4	>65	Inhibitory
2,4-Dichlorophenol	−90 ± 10.0	4	>80	Inhibitory
2,6-Dichlorophenol	−57 ± 28.1	4	>80	Inhibitory
3,5-Dichlorophenol	−381 ± 9.5	5	>80	Inhibitory
2,4,6-Trichlorophenol	−207 ± 14.1	4	>80	Inhibitory
Pentachlorophenol	EC_{50}, 9.2 mg/l[d]	5	>80	Inhibitory
2-Cresol	−16 ± 5.2	3	>65	−
3-Cresol	75 ± 15.0	2	40	±
4-Cresol	96 ± 4.3	2	7	+
2-Nitrophenol	−11 ± 13.8	3	>90	−
3-Nitrophenol	−62 ± 17.7	4	>90	Inhibitory
4-Nitrophenol	−126 ± 23.8	4	>90	Inhibitory
2,4-Dinitrophenol	−320 ± 14.4	5	>90	Inhibitory
2,5-Dinitrophenol	−328 ± 14.4	5	>90	Inhibitory
4-Nonylphenol	−51 ± 14.5	5	>65	Inhibitory
2-Phenylphenol	−93 ± 11.8	5	>80	Inhibitory
Catechol	90 ± 7.3	2	>17	+

Table 7-4   (*Continued*)

Chemical	Net Gas Production, % Theoretical	Pattern of Gas Production[a]	Lag,[b] days	Biogradation Potential
Benzoate, Substituted Benzoates, Phthalic Acid, and PAE				
Sodium benzoate	93 ± 7.5	2	2	+
3-Aminobenzoic acid	12 ± 20.4	2	64	−
4-Aminobenzoic acid	89 ± 12.3	2	26	+
2-Chlorobenzoic acid	16 ± 6.4	3	14	−
3-Chlorobenzoic acid	105 ± 13.9	2	29	+
4-Chlorobenzoic acid	44 ± 6.5	3	9	±
2-Nitrobenzoic acid	−2 ± 7.8	4	>80	Inhibitory
3-Nitrobenzoic acid	−25 ± 6.5	4	>80	Inhibitory
4-Nitrobenzoic acid	−19 ± 11.0	4	>80	Inhibitory
Phthalic acid	135 ± 7.5	2	9	+
Dimethyl phthalate	41 ± 8.3	2	16	±
Di-*n*-butyl phthalate	24 ± 9.6	2	23	−
bis(2-Ethylhexyl) phthalate	5 ± 9.5	3	>77	−
Glycols, Alkanes, and Sodium Stearate				
Ethylene glycol	106 ± 8.8	1	>1	+
Diethylene glycol	97 ± 5.8	1	1	+
Triethylene glycol	98 ± 2.4	1	1	+
Hexylene glycol	−8 ± 8.6	3	>60	−
Neopentyl glycol	1 ± 78	3	>60	−
n-Undecane	−4 ± 14.5	3	>65	−
n-Hexadecane	8 ± 15.1	3	>65	−
Sodium stearate	94 ± 10.9	1	1	+
Surfactants, Pesticides, and Organotins				
CTAB	−86 ± 6.8	4	>60	Inhibitory
Sodium dodecylbenzene sulfonate	−89 ± 7.2	5	>60	Inhibitory
Sodium 4-octylbenzene sulfonate	−117 ± 7.0	5	>60	Inhibitory
2,4-D	−30 ± 9.4	3	>75	−
2,4,5-T	−50 ± 6.4	5	>75	Inhibitory
MCPA	−29 ± 10.0	4	>75	−
MCPP	−30 ± 6.7	4	>75	−
Lindane (1,2,3,4,5,6-hexachlorocyclohexane, δ-isomer)	−97 ± 37.9	4	>75	Inhibitory
Dieldrin	−26 ± 9.9	3	>75	−
cis-Permethrin	−21 ± 17.3	3	>65	−
trans-Permethrin	−3 ± 17.6	3	>65	−

Table 7-4 (*Continued*)

Chemical	Net Gas Production, % Theoretical	Pattern of Gas Production[a]	Lag,[b] days	Biogradation Potential
Surfactants, Pesticides, and Organotins				
Butyltin tricholoride	−60 ± 6.7	5	>60	Inhibitory
Dibutyltin dichloride	−98 ± 7.5	5	>60	Inhibitory
Tributyltin chloride	−36 ± 7.3	5	>60	Inhibitory
Effect of Ring Structure on Biodegradation Potential				
Tetrahydrofuran	−9 ± 5.9	3	>60	−
Furan	−22 ± 6.0	3	>70	−
Pyrrole	−5 ± 3.4	3	>70	−
N-Methylpyrrole	−7 ± 16.5	3	>70	−
Thiophene	−21 ± 4.5	3	>70	−
Benzene	−10 ± 8.9	3	>80	−
Pyridine	−58 ± 12.6	2	30	±
Naphthalene	−32 ± 4.0	4	>70	−
1-Naphthol	−79 ± 8.1	5	>75	Inhibitory
2-Naphthol	−83 ± 9.9	5	>75	Inhibitory
1-Naphthoic acid	−16 ± 7.3	3	>75	−
Quinoline	97 ± 7.8	2	15	+
Anthraquinone	−150 ± 22.9	4	>75	Inhibitory
Effect of Substituent on the Biodegradation Potential of Monosubstituted Benzenes[a]				
Benzene	10 ± 8.9	3	80	−
Chlorobenzene	7 ± 8.8	3	80	−
Phenol	99 ± 8.0	2	6	+
Sodium benzoate	93 ± 7.5	2	2	+
Cumene	2 ± 2.8	3	60	−
Aniline	6 ± 20.3	3	60	−
N-Methylaniline	14 ± 16.5	3	42	−
Sodium Benzenesulfonate	10 ± 3.9	3	60	−

Source: Battersby and Wilson (1989).
[a]See Fig. 7-1.
[b]Time required to achieve 10% ThGP.
[c]+, Complete degradation (≥ 80% ThGP); ±, partial degradation; −, not degraded, (<30% ThGP).
[d]EC_{50}, concentration causing a 50% reduction in gas production.

The structural characteristics of compounds appeared to influence the length of lag period of cross acclimation. Data suggested that:

1. Nontoxic aliphatic compounds containing carboxyl, ester, or hydroxyl groups were all readily acclimated by acetate cultures. Butyric acid, valeric acid, propylene glycol, ethyl acetate, and acetate were all degraded within four days.

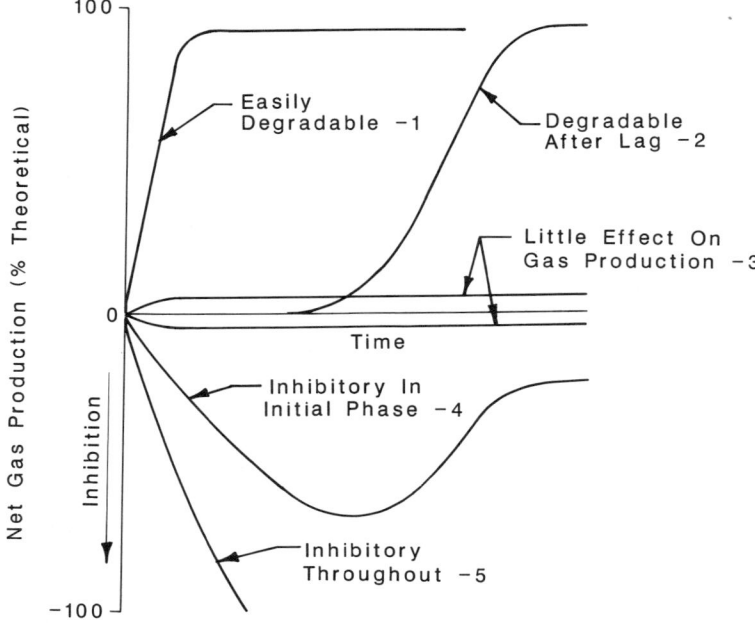

Fig. 7-1. Typical patterns of anaerobic net gas production. (*Source:* Battersby and Wilson 1989.)

2. Toxic compounds with carbonyl functional groups or double bonds all acclimated within a short period of time, although these compounds were usually toxic to unacclimated acetate cultures. For examples, butyraldehyde and sorbic acid required seven and ten days for acclimation, respectively.

3 Compounds with amino functional groups showed difficulty in acclimation. Aniline, 4-amino-butyric acid, and *sec*-butylamine were not acclimated in this study. 1-amino-2 propanol was the only compound acclimated, but it degraded at an extremely slow rate.

4. Compounds with dicarboxylic groups required a longer length of time to be acclimated as compared to compounds with one carboxylic group and the same carbon chain length. For example, succinic acid, adipic acid, and phthalic acid all had longer lag periods as compared to butyric acid, hexanoic acid, and benzoic acid, respectively.

5. The position of functional group also affected the length of lag period for acclimation. In the case of the hydroxyl group, primary butanol required 4 days for acclimation, and secondary butyl alcohol required 14 days, while tertiary butyl alcohol was not acclimated in the serum bottles.

Table 7-5. Petrochemical Removal Rates Observed in Serum Bottle Anaerobic Tests.

Compounds	Lag Time, days	Percent Substrate Removed	Removal Rate, mg/l day
1-Amino-2-propanol	9	65	22
Sec-Butyl alcohol	14	100	42
Phenol	7	98	42
Phthalic acid	14	90	50
Adipic acid	10	82	50
Valeric acid	3	100	72
Methyl ethyl ketone	8	100	72
Maleic acid	10	95	74
Glutaric acid	10	100	90
Propionic acid	2	100	90
Butanol	4	100	100
Propanol	4	100	110
Succinic acid	10	93	110
Hexanoic acid	5	100	112
Propylene glycol	4	100	125
Acetone	5	100	125
Fumaric acid	13	86	125
Vinyl acetate	3	100	125
Sorbic acid	10	76	142
Butyraldehyde	7	99	142
Methanol	6	66	182
2-Propanol	4	100	200
Glycerol	8	90	200
Benzoic acid	11	82	200
Crotonic acid	2	100	200
Isobutyric	3	100	250
Methyl acetate	3	100	250
Ethyl acetate	3	100	250
Butyric acid	3	100	284
Formic	4	89	286

Source: Chou et al. (1978a).

Figure 7-2 shows that even nitrobenzene was initially toxic and prevented acetate conversion to methane for 60 days. Methanogenesis subsequently acclimated to the nitrobenzene, even though it was not biodegraded.

At the end of 90 days acclimation, acetate, glycerol, butyric, and isobutyric acids, methyl acetate, ethyl acetate, acetaldehyde, methyl ethyl ketone, and adipic acid acclimated very well with high utilization efficiency (above 66 percent). propanol, 2-propanol, propylene glycol, propionate, acrolein, acetone, hexanoic, and benzoic acids had degradation ranging from 33 percent to 66 percent. Propanol, acrylic acid, 3-chloro-1,2-propandiol, and 1-amino-2-propanol acclimated poorly (below 33 percent of utilization).

Table 7-6. Utilization Efficiency of Petrochemicals in Anaerobic
Filters .

Petrochemicals	Time, days	Effluent Concentration, mg/l	Percent Utilization
Butanol	52	80	95
sec-Butanol	52	110	93
tert-Butanol	52	400	73
Butyraldehyde	52	270	82
Valeric acid	52	110	93
Vinyl acetate	52	140	91
4-Aminobutyrate	52	200	87
sec-Butylamine	52	510	66
2-Chloropropionate	110	1,560	14
Glutaric acid	110	70	96
Maleic acid	110	80	96
Sorbic acid	110	110	94
Hydroquinone	110	280	85
Crotonaldehyde	110	100	95
Crotonic acid	110	70	96
Acrylonitrile	110	1,500	17
Ethyl acrylate	110	100	95
Phenol	110	90	95
Succinic acid	110	90	95
Fumaric acid	110	100	95
Catechol	110	120	93
Nitrobenzene	110	350	81
Resorcinol	110	90	95

Source: Chou et al. (1978a).

The anaerobic filters proved to be greatly superior to the completely mixed reactors as acclimation systems. It was found that some petrochemicals could not be acclimated and metabolized in the completely mixed reactors because the 20-day solid retention time (SRT) tended to allow washout of the biomass. However, the exceptionally long SRT inherent to the anaerobic filters allowed retention of the biomass over the prolonged acclimated periods so that very high utilization efficiencies were eventually possible. The filters acclimated to 2,000 mg/l of 2-chloropropionic acid and acrylonitrile with only 14 percent and 17 percent utilization, respectively, at the end of 110 days. The remaining tests all acclimated well to their respective substrates with at least 80 percent utilization. At the end of 52 days of acclimation, butanol acclimated to 95 percent utilization efficiency, and tert-butanol and sec-butylamine also acclimated with 73 percent and 66 percent utilization. Sec-butanol,, valeric acid, vinyl acetate, and 4-amino-butyrate had good acclimation with utilization efficiencies of at least 87 percent at the end of 52 days.

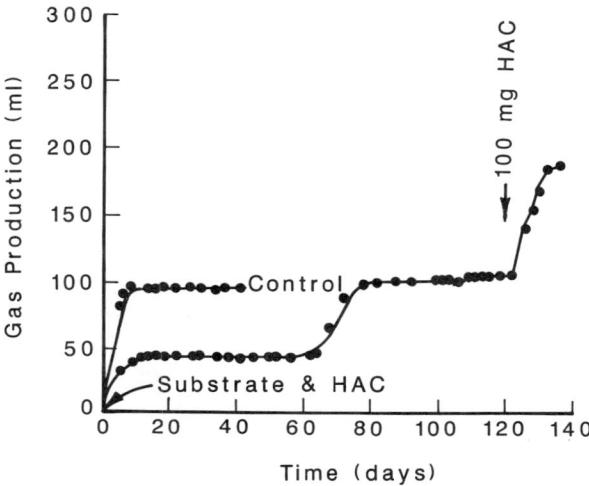

Fig. 7-2. Cross acclimation of acetate culture to nitrobenzene.

It should be pointed out that many compounds which were not acclimated in serum bottles were metabolized in anaerobic filters or completely mixed reactors. Therefore, a compound not showing acclimation does not necessarily mean that it cannot be acclimated; rather it is a matter of the time required and techniques of acclimation. For instance, propanol which was not acclimated by acetate cultures in serum bottles was acclimated in a Warburg flask after a 10 day lag period. It was also acclimated in a completely mixed reactor and had 54 percent removal during the steady state period when the reactor was operated at a 20 day hydraulic retention time. Acrylic acid was reported by an independent investigator to be metabolized.

Acetaldehyde was a toxic compound, but it had a high utilization efficiency from the start (85 percent). It did not show adverse effects as the substrate concentration increased sharply in the second 30 day period, and had 97 percent utilization efficiency at the end of 90 days acclimation. Acrolein, as stated earlier, was found to be an extremely toxic compound to unacclimated acetate cultures, but with acclimation showed 42 percent utilization and there were indications that over 90 percent utilization is possible with proper acclimation. Propanol, 1-amino-2-propanol, and 3-chloro-1,2-propandiol acclimated poorly by comparison with acrolein and acetaldehyde.

The mildly toxic compounds such as glutaric, maleic, and fumaric acids, hydroquinone, catechol, resorcinol, ethyl acrylate, and phenol all acclimated very well in the anaerobic filters. The utilization efficiency also improved with the increasing length acclimation time. Filters receiving crotonaldehyde and nitrobenzene, considered to be extremely toxic compounds with unacclimated acetate cultures, both acclimated very well with 95 percent and 81 percent utilization, respectively. The toxic compound vinyl acetate acclimated well with a utilization efficiency of 82 percent at the end of 52 days acclimation.

TOXIC INHIBITION

A common misconception about anaerobic degradation is that the process cannot tolerate toxic substances and that the biota "die" when exposed to toxicants. Cessation of methane production was presumed to indicate that the biomass had "died." This concept arose from municipal sludge digestion experiences when methane bacteria were inhibited. The result was that volatile acid production continued, exceeded the buffer capacity of the system, depressed the pH, and led to a "sour" digester. This is a rather complex phenomenon in which the original toxicant may perhaps play only a minor "triggering" role.

It is true that methanogens, like other organisms, can be inhibited by toxicants. Due to the relatively low fraction of substrate synthesized into cells by the methanogens and the associated prolonged generation times, the recovery period can be considerably extended if the toxicant is indeed bacteriocidal. However, studies on toxicity recovery of only the methanogens indicate that most toxicants exhibit a bacteriostatic or reversible effect on the methanogens at the lower concentration ranges normally encountered in industrial wastewaters (Speece 1983).

Effect of Molecular Structure

The effect of molecular structure of 52 petrochemicals was assayed with unacclimated acetate-enriched methanogens (Chou et al. 1978b). The following structural characteristics influenced the selection of the petrochemicals that were assayed: double bonds, position of functional group, length of carbon chain, branching, oxidation state, type of functional groups, number of identical functional groups, and configuration. The results showed a definite correlation between molecular structure and toxicity to unacclimated methanogens. Chlorine substitution, aldehydes, double bonds, and benzene rings exhibited toxicity. The addition of hydroxyl groups decreased toxicity. Table 7-7 shows the relative toxicity to the acetate utilizing methanogens of the petrochemicals assayed. Those not manifesting toxicity at 4,000 mg/l are not listed.

Figure 7-3 shows the cross acclimation characteristics of an acetate-enriched methane-producing culture to phenol. The control received only acetate. After approximately 30 days, the culture started to metabolize the phenol. Fig. 7-4 shows the eventual response of a methanogenic culture to 3-chloro-1,2-propandiol. It initially inhibited acetate conversion to methane for over 40 days. Even-

Table 7-7. Relative Toxicity of Petrochemicals Assayed.

Compound	Conc. Resulting in 50% Activity, mM	Compound	Conc. Resulting in 50% Activity, mM
1-Chloropropene	0.1	2-Chloropropionic acid	8
Nitrobenzene	0.1	Vinyl acetate	8
Acrolein	0.2	Acetaldehyde	10
1-Chloropropane	1.9	Ethyl acetate	11
Formaldehyde	2.4	Acrylic acid	12
Lauric acid	2.6	Catechol	24
Ethyl benzene	3.2	Phenol	26
Acrylonitrile	4	Aniline	26
3-Chloro-1,2-propandiol	6	Resorcinol	29
Crotonaldehyde	6.5	Propanal	90

tually, even though the chloropropandiol was not metabolized, the culture recovered its normal function of acetate conversion to methane.

pH Inhibition

A low pH is generally associated with a sour digester. The problem that causes a low pH may be complex. For instance, temporary exposure to a toxicant or to a reduced temperature may "trigger" an increase in volatile acids which

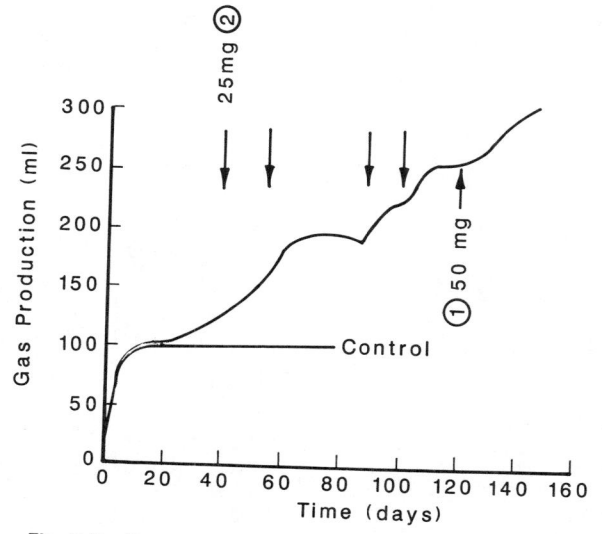

Fig. 7-3. Cross acclimation of acetate culture to phenol .

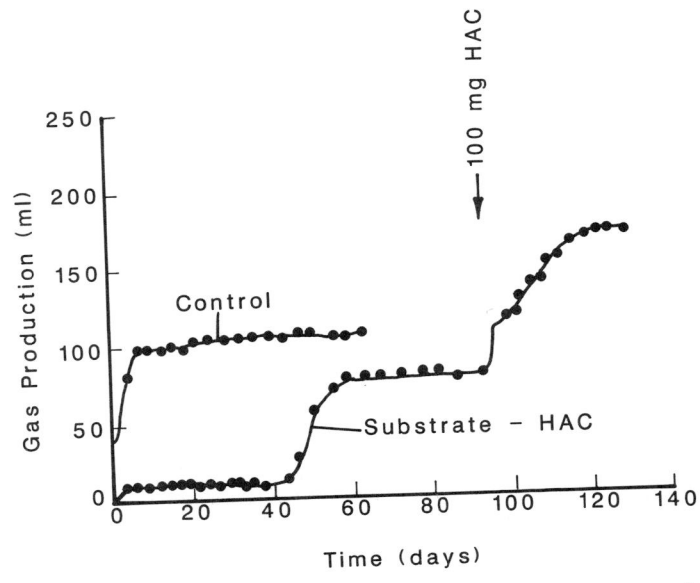

Fig. 7-4. Cross acclimation of acetate culture to 3-chloro-1,2,-propandiol.

exceeds the buffer capacity of the system, resulting in a low pH. The initial, adverse environmental factor may exist only temporarily, but the resulting low pH may persist for an extended period.

Response to adverse pH levels was studied (Clark and Speece 1970) in anaerobic filters. The pH of the feed solution was depressed by addition of hydrochloric and acetic acids and the hydraulic retention time was kept low so that the entire depth of the anaerobic filter was subject to essentially the same low level. Acetate conversion to methane was monitored and a "pH inhibition factor" was calculated as the fractional rate of conversion of acetate to methane in reference to the control rate prior to reduction in the pH of the feed.

For a pH drop to 5.5 for only 12 hours, the pH inhibition factor dropped to about 0.5, but recovered to normal within 1 day after the pH was restored. When the pH was dropped to 5 for 15 days, the pH inhibition factor dropped to about 0.25 within 3 days and held reasonably steady for the 15 day period of pH depression. On day 15, the pH was abruptly restored to normal, but the gas production did not recover abruptly. Instead, it gradually returned to the normal background rate in about 3 weeks. The rate of recovery of the gas production was a function of the duration of the pH depression, i.e., short exposure to low pH gave short gas production recovery times and longer exposure to low pH resulted in longer gas production recovery times. Repeated subjection of the filter to a feed solution of pH 5 resulted in repeated drops to the same pH inhibition factor of about 0.25. This occurred in spite of the fact that gas production never fully recovered between successive pH drops.

An increase in pH to the range of 9.0 to 9.5 resulted in a decline in the pH inhibition factor to less than 0.1 of the background rate. Restoration of the pH to normal was mirrored by a rapid increase in the gas production rate back to normal.

Figure 7-5 shows the relation between the pH inhibition factor and the pH of the feed solution. It should be emphasized that in the anaerobic filter, the methanogens in the interior of the biofilms are not subject to the same pH as the feed solution.

Microbial Kinetics and Toxicity

Industrial wastewaters commonly contain chronic or transient levels of toxicity to methanogens. Consequently, a wastewater treatment system receiving industrial wastes would rarely operate at true steady-state. For instance, under ideal conditions, the microbial regeneration time of methanogens metabolizing acetate would be approximately 2 to 4.2 days (Lawrence and McCarty 1969; Huser et al. 1982). However, when a toxicant is present in the wastewater, the specific utilization of acetate may be reduced (noncompetitive inhibition) or the half-velocity constant of the substrate K_S may be increased (competitive inhibition). The minimum biological SRT is defined as:

$$\theta^c_{min} = (Yk - b)^{-1} \qquad (7\text{-}1)$$

where Y is the growth yield constant, k is the rate of substrate utilization, and b is the endogenous respiration rate. Therefore, anything that decreases the

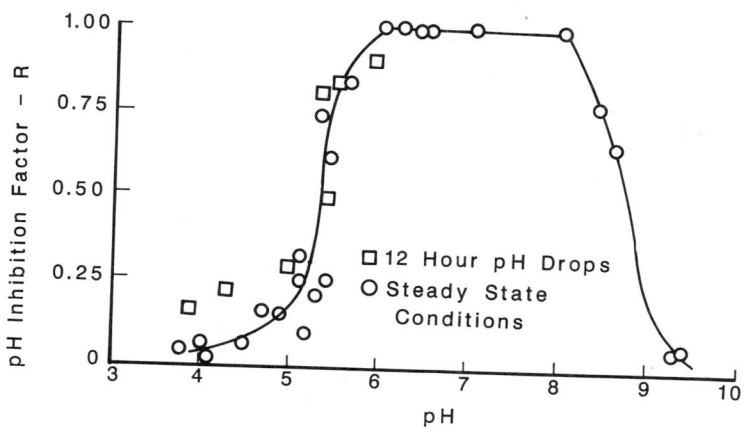

Fig. 7-5. pH inhibition factors observed at different pH levels.

product of the bacterial yield, Y times k, will increase the generation time. Since a toxicant inhibits substrate utilization by definition, it increases the microbial generation time.

The "dynamic" nature of θ^c_{min} with toxicant exposure is demonstrated in Fig. 7-6. After addition of the toxicant, θ^c_{min} increases with concomitant decreases in methane production and process efficiency. Although drawn as a steady state response for each condition, the "elastic" nature of θ^c_{min} under actual field conditions is clear. Actual treatment systems are operated at a reasonably fixed θ_c, and thus temporarily at least the system may be operating under washout conditions after exposure to toxicants. It is also apparent that failure, toxicity, and biodegradability are functions of time. Fig. 7-7 shows the variation in calculated kinetic coefficients and θ^c_{min} with time for acetate utilizing methanogens after exposure to cyanide. A sufficient biological safety factor (defined as θ_c/θ^c_{min}) in the form of a proper design, guards against the above-described failure and allows for acclimation. For example, metabolism of acrylic acid in a suspended-growth acetate-enriched culture was not achieved with a 25-day hydraulic and solids retention time, because the microbial cells washed out before acclimation occurred (Chou et al. 1978a). However, acrylic acid was

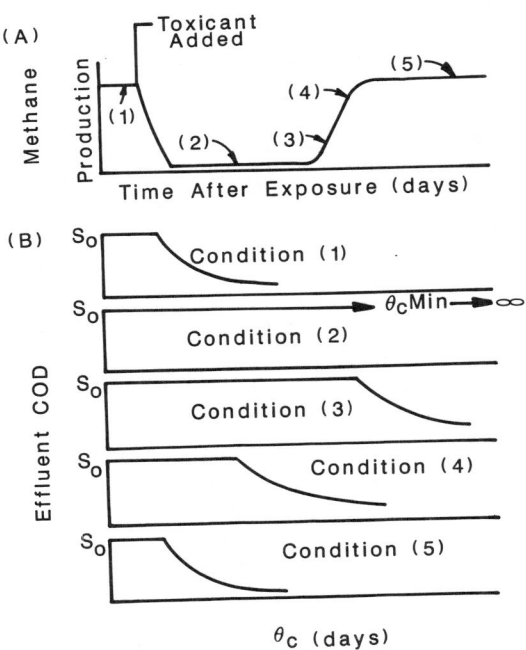

Fig. 7-6. Effect of transient nature of toxicity response on process efficiency.

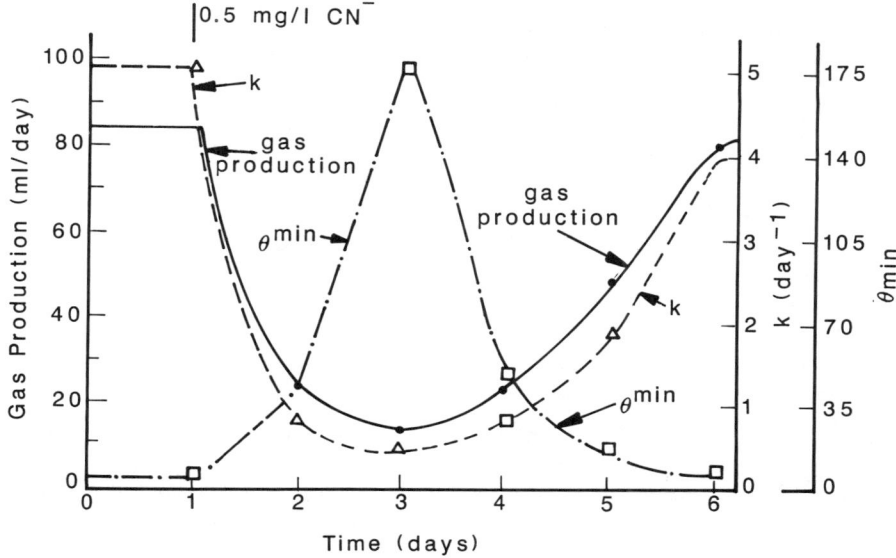

Fig. 7-7. Gas production, k and θ_{min} versus time (0.5 mg/l CN⁻).

metabolized in an anaerobic filter having a solids retention time in excess of 100 days. Thus, the generation time of the culture was apparently in excess of 25 days during the acrylic acid acclimation phase. The need for prolonged solids and microorganism retention times in methane fermentation processes treating industrial wastewaters is not sufficiently recognized.

Toxicity in Anaerobic Digestion Studies

Suspended-growth systems of acetate-enriched methanogens that have received a wide variety of toxicants all show a characteristic recovery pattern when a neutral pH is maintained and the acetate concentration is restored daily to the range of 1,000 to 2,000 mg/l. Fig. 7-8 shows the response of a suspended-growth system to chloroform. The production rate is inhibited in degree and for a time proportional to the toxicant concentration. Even after prolonged periods of no gas production, eventual recovery was very rapid and complete. This indicated that biomass viability was maintained and that the toxicity was bacteriostatic not bacteriocidal (Yang et al. 1980). This response was also shown with cyanide, ammonium, and oxygen. Fig. 7-9 shows the response to repeated exposures to chloroform. After recovery from the first exposure, no inhibition was noted from three subsequent exposures to chloroform.

The reversibility of toxic inhibition was confirmed in two ways (Yang et al.

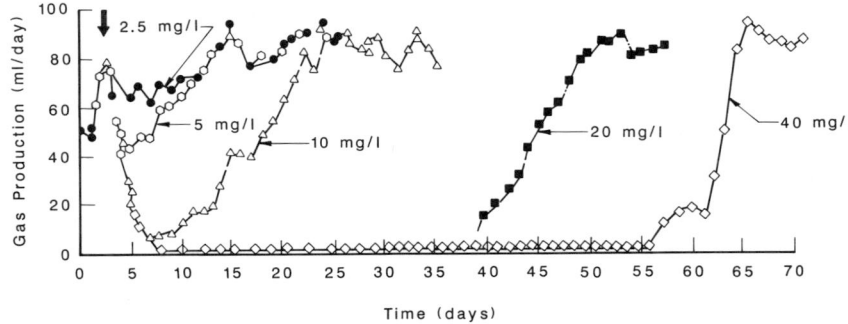

Fig. 7-8. SRT response to toxicant: chloroform 50 days SRT, 35°C.

1980). First, the feed to plug flow anaerobic filters was spiked with various toxicants. The plug flow anaerobic filters resulted in almost complete displacement of the feed and thus of the toxicant spikes in less than two days. Fig. 7-10 indicates that when 2,000 mg/l of sulfide was added to the feed of a filter for a day, gas production almost ceased. However, after 2 days, gas production was back to normal. The second method involved experiments with suspended growth systems. High concentrations of formaldehyde (300, 600, 1,800 mg/l) were injected into duplicate systems. One of the duplicates was centrifuged after exposure for 1 hour. The supernatant was then discarded and replaced with an unadulterated nutrient salt solution. The other duplicate served as a control with no removal of the soluble fraction. Fig. 7-11 shows that gas production in the systems exposed for 1 hour to 300 or 600 mg/l of formaldehyde exhibited only a slight, temporary reduction in gas production. The system exposed to 1,800 mg/l of formaldehyde for 1 hour took about 10 days to recover to 75 percent of the pre-toxicant gas production rate. The gas production ceased in the controls.

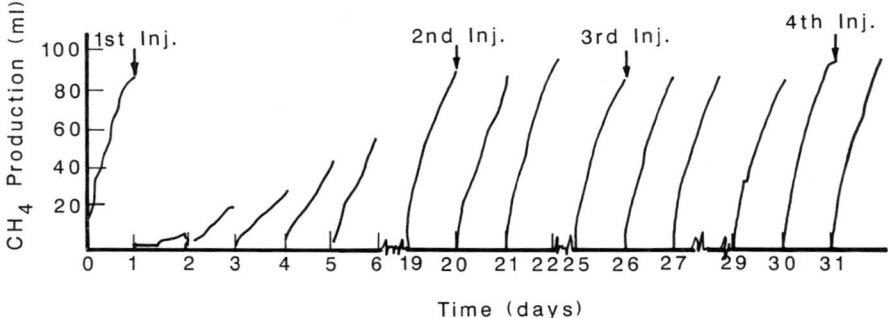

Fig. 7-9. Effect of repeated injections of 2.5 mg/l chloroform on methane production.

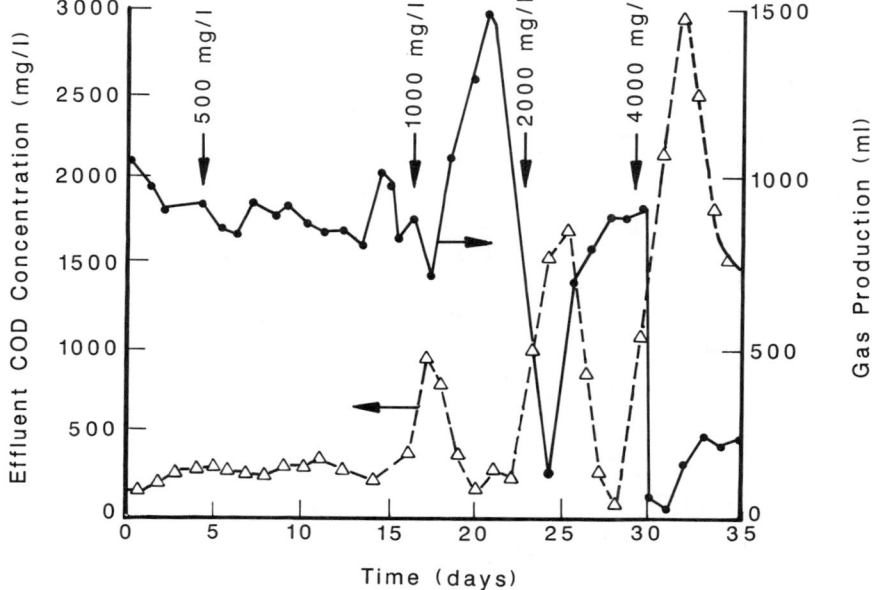

Fig. 7-10. Response of anaerobic filter to toxicant sulfide.

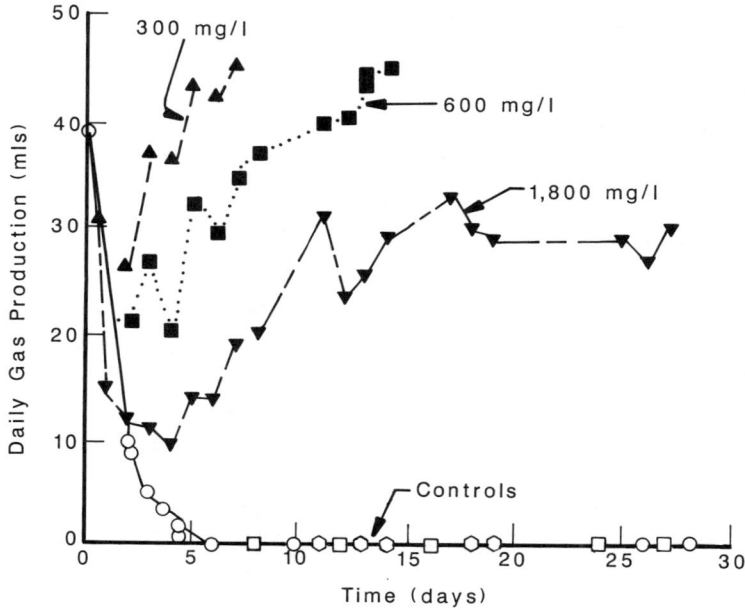

Fig. 7-11. Effect of exposure time on toxicity response: formaldehyde.

High sulfate concentrations will be a common occurrence with some biomass feedstocks and industrial wastewaters. Sulfate reduction in the presence of acetate and hydrogen is energetically favored over methane production. Yet it has been observed that sulfate reduction is not complete in some pilot plant studies of anaerobic digestion. Since excessive hydrogen sulfide concentrations can be toxic, it would be advantageous to be able to inhibit hydrogen sulfide production without inhibiting methane reduction (e.g., molybdenum), would be of academic importance, and, depending on the concentration required and unit cost, it could be of practical value in field scale digesters. Fig. 7-12 shows the relation between feed strength, sulfide precursors, pH, and resulting H_2S in the head gas. Hydrogen sulfide in the head gas in a direct measure of unionized H_2S in the aqueous phase—which is the toxic agent.

Pesticides are used in agriculture to produce greater crop yields, crops of higher quality, and to reduce the input of labor and energy into crop production. If these practices are applied to biomass energy farms, toxicity and inhibition of microorganisms due to pesticide residues may affect digestion process kinetics. A recent review discusses the microbial biodegradability of pesticides along with the toxicity of these compounds to aerobic and anaerobic microorganisms (Hill and Wright 1978).

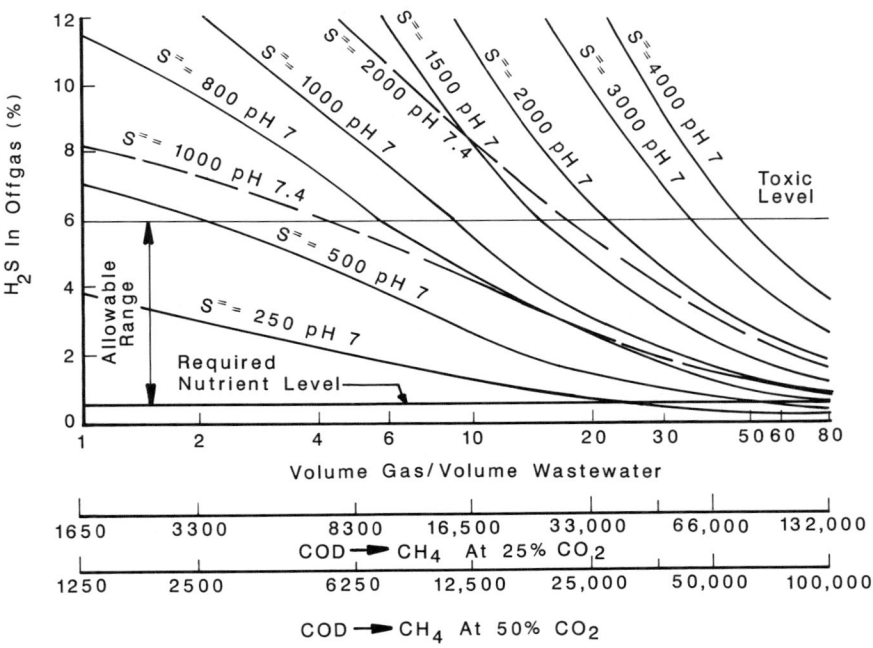

Fig. 7-12. H_2S in offgas versus V_g / V_w (or COD → CH_4), temperature 35°C.

Softwood resin extracts consist of resin acids, fat, and terpenes and also contain a variety of phenolic compounds such as flavonoids, ligands, and stilbenes. Some pine species have been reported to contain up to 14 percent dry weight extractives (Sjostrom 1981). Water soluble tannins are also present in softwoods. Some of these compounds are referred to as ''pathological'' resins as they protect the tree against biological damage. Hardwood resins typically consist of fats, waxes, sterols, but species such as *Eucalyptus* sp. have been reported to be rich in isoprenoid-containing compounds (Sachs et al. 1981). These extractives may be inhibitory to the digestion process and should be studied to determine their effects on biomass degradability.

Some form of pretreatment of biomass feedstocks is desirable to increase yields. Acid or base hydrolysis has shown promise. However, in the process, byproducts can be formed, such as furfural, hydroxymethyl furfural, formic acid, and levulinic acid, which exhibit inhibition (McCarty et al. 1978). Potential acclimation to such inhibition and perhaps eventual metabolism may be possible, but again, process kinetics may be affected.

Summary of Toxic Inhibition

Several general observations on toxicity in anaerobic digestion indicate that methanogens show the same general response pattern to a wide variety of toxicants. However, the toxicity in most cases appears to be reversible. Acclimation to the toxicity is demonstrated with methanogens for some toxicants. These observations demonstrate that methanogens can tolerate toxicity and therefore provide considerable optimism on the potential of anaerobic digestion for treatment of industrial wastewaters.

Next, with most soluble feedstocks for bioconversion to methane, the rate limiting step is not methanogensis. Yet, there is even less information on the effects of various toxicants on the hydrolysis, acid fermentation, or acetogenic steps of anaerobic digestion than for the methanogenesis step. If the digestion process is to be optimized, the rate limiting step must be enhanced.

The mechanism by which a toxic substance inhibits bacterial activity is related to the toxicant. In aerobic bacterial metabolism, general biochemical pathways are defined and the mechanism of toxic inhibition has been studied more extensively. Toxicants have been grouped into classes, depending on the mechanism of toxicity. In anaerobic digestion, the biochemical pathways are not all defined. For instance, the pathway for methane production is not completely known. Therefore, it is more difficult to reclassify toxicants in anaerobic digestion.

There is no conclusive correlation between the toxicity characteristics of an organic compound to an unacclimated acetate-enriched methanogenic culture and the eventual metabolism of that same organic by the acclimated culture. Perhaps this is also true of the entire consortium of bacteria in anaerobic digestion. A clear distinction needs to be drawn between unacclimated and acclimated toxicity. Yet there is essentially no information on the correlation between the exposure time of a toxicant to microorganisms and the degree of resulting acclimation. Neither is there a comparable data base showing the effect of toxicant concentration acclimation potential. In order to utilize the full potential of the consortium of bacteria in anaerobic digestion to accommodate chronic or sporadic toxicity, the relation between type of toxicant, concentration, and duration of exposure must be known.

The acclimation phenomenon is observed, but not understood. Perhaps if it were better understood, it could be more effectively exploited. Of the wide variety of toxicants assayed for methanogenic response, it would seem that the specific site of damage within the cell would be dependent on the toxicant. Yet the gross observation of methanogens based on methane production for the toxicants assayed indicated that the cell is eventually able to repair the damage apparently wherever it is incurred. Full restoration of the original rate of methane production has been observed after toxicant exposure even though the toxicant concentration is unchanged. There appears to be great potential for methanogenic acclimation to toxicity.

Since biomass digestion requires a complete consortium of bacteria, it is relevant to study the effect of a given toxicant and its potential acclimation on the overall process as well as to individual steps within the process. A toxicant can inhibit the rate-limiting step and/or change the step that is rate-limiting. Both manifestations of toxicity can severely affect the overall process.

Propionic acid and, to a lesser extent, some of the higher volatile acids have been reported to accumulate in malfunctioning digestion systems and also commonly during start-up. Acetogenesis is such a key step that specific effort should be concentrated here. Certain acetogens can only be cultured in combination with hydrogen-utilizing methanogens and it is difficult for most laboratories to analytically determine hydrogen at the low concentrations required for acetogenesis to proceed. Therefore, acetogenic toxicity assays will be more complicated to evaluate because of this interdependency, but it must be investigated nevertheless because of its key role.

The reversibility of toxicity strongly affects the choice of digester configuration. Of the limited observations of this phenomena in methanogenic assays, it appears that a plug flow configuration results in less overall reduction in process efficiency than a completely mixed configuration. This needs to be clarified for the hydrolytic, acid fermenting, and acetogenic steps.

There is the need for a more comprehensive viability assay for the microorganisms responsible for all of the steps of anaerobic digestion. Methane production is not sufficiently comprehensive for assaying the viability of methanogens. Cessation of methane production has been shown to not correlate with the viability of methanogens. ATP and F_{420} concentrations also have been shown not to correlate with cell viability. Toxicity response would be greatly enhanced by a comprehensive viability assay for microorganisms in each stage of anaerobic digestion.

Antagonism and synergism have been clearly demonstrated with ion toxicity. Specific treatments have been reported for remedial treatment of toxicity due to anionic detergents, cyanide, and chloroform. There is a need to expand this list of remedial treatments for additional toxicants.

ANAEROBIC TREATMENT ASSAYS

There are two assays used to evaluate biodegradation under anaerobic conditions. One is the biochemical methane potential (BMP) and the other is the anaerobic toxicity assay (ATA). These are very useful assays and are briefly described below.

Biochemical Methane Potential Assay

The biochemical methane potential (BMP) is the anaerobic counterpart of the aerobic BOD (Owen et al. 1979). In the anaerobic treatment of organic pollutants, a reduced product is removed, i.e., methane, whereas aerobically an oxidized product is added, i.e., oxygen.

The assay is conducted in septum sealed bottles incubated at 35°C for a total of 60 days to observe any acclimation trends. The 50 ml of anaerobic seed is taken from an active municipal sludge digester. Gas production is measured daily during the first 10 to 20 days and then at prolonged intervals thereafter. Cumulative gas production is then plotted. Fig. 7-13 shows a BMP curve for an industrial wastewater.

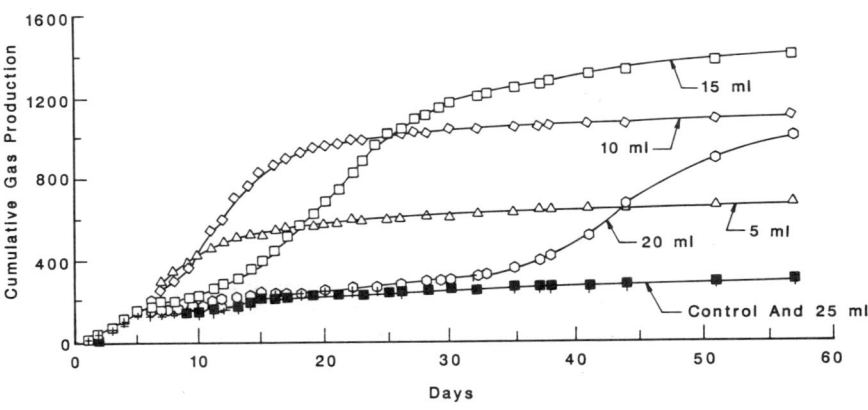

Fig. 7-13. Presentation of results for biochemical methane potential assay.

Anaerobic Toxicity Assay

The anaerobic toxicity assay (ATA) is designed to indicate the presence of toxicity to methanogens in the wastewater being assayed. The anaerobic treatment process utilizes a consortium of microbes to sequentially convert complex materials to simpler, soluble intermediates and finally to methane gas. The final stages in the conversion involve the formation of volatile acids, primarily acetic and propionic, which are the precursors of methane. If the volatile acids are produced at a rate in excess of their conversion to methane, they accumulate in the system and potentially can exceed the alkalinity buffer capacity available, depressing the pH and causing the digestion process to cease. Therefore, the methane forming bacteria which utilize the volatile acids are crucial to the successful digestion of the organic pollutants. In many cases, the organisms utilizing the volatile acids have been shown to be the most sensitive class of microorganisms in the consortium.

Therefore the ATA is conducted to evaluate the relative inhibition of the wastewater sample to the key microorganisms converting volatile acids to methane. Increasing concentrations of the wastewater are incubated with seed replicates to which nonlimiting and noninhibitory concentrations of calcium acetate (10,000 mg/l) have been added. Since substrate is nonlimiting, any toxicity in the wastewater to the methanogens converting acetate to methane will be manifested by a reduced rate of gas production, with higher amounts of wastewater causing more reduction in the gas production rate. Acclimation characteristics of the methanogens to the toxicity will also be reflected in the gas production pattern. Fig. 7-14 shows ATA curves for an industrial wastewater.

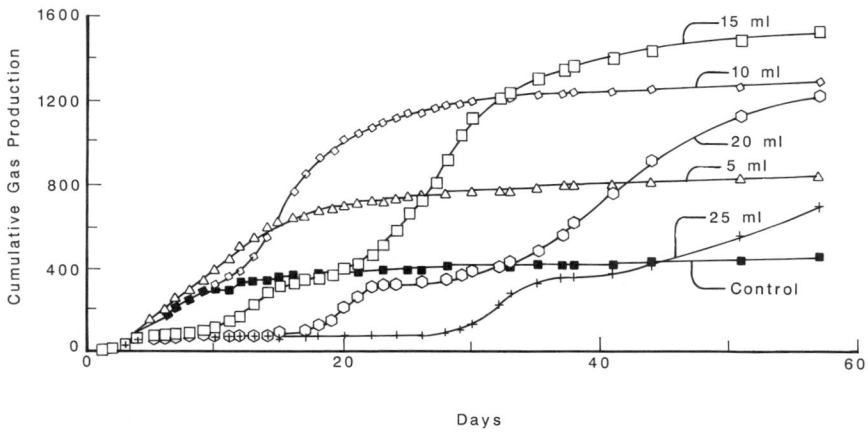

Fig. 7-14. Presentation of results for anaerobic toxicity assay.

REFERENCES

Battersby, N. S. and V. Wilson. 1989. Survey of Anaerobic Biodegradation Potential of Organic Chemicals in Digesting Sludge. *Applied and Environ. Mircobiol.*, **55**: 433–439.

Blum, D. J. W., R. Hergenroeder, G. F. Parkin, and R. E. Speece. 1986. Anaerobic Treatment of Coal Conversion Wastewater Constituents: Biodegradability and Toxicity. *J. WPCF*, **58**: 122–131.

Bower, E. J. 1988. Transformation of Xenobiotics in Biofilms. In: Dahlem Konferenzen Workshop on *Structure and Function of Biofilms*, Berlin.

Boyd, S. A., D. R. Shelton, D. Berry, and J. Tiedje. 1983. *Applied and Environ. Microbiol.*, **48**: 50–54.

Boyd, S. A. and D. R. Shelton. 1984. Anaerobic Biodegradation of Chlorinated Phenols in Fresh and Acclimated Sludge. *Appl. Environ. Microbiol.*, **47**: 272–277.

Chou, W. L., R. E. Speece, and R. H. Siddiqi. 1978a. Acclimation and Degradation of Petrochemical Wastewater Components by Methane Fermentation. *Biotech. and Bioeng. Symp.*, **8**: 391–414.

Chou, W. L., R. E. Speece, R. H. Siddiqi, and K. McKeon. 1978b. The Effect of Petrochemical Structure on Methane Fermentation Toxicity. *Progr. Water Technol.*, **10**: 545–50.

Clark, R. H. and R. E. Speece. 1970. The pH Tolerance of Anaerobic Digestion. *Proc. Fifth Int'l Conf. on Water Pollution Research, San Francisco.*

Guthrie, M. A., E. S. Kirsch, R. F. Wukasch, and C. P. L. Grady. 1984. *Water Research*, **18**: 451–461.

Hakulinen, R. and M. Salkinoja-Salonen. 1982. Treatment of Pulp and Paper Industry Wastewaters in an Anaerobic Fluidized Bed Reactor. *Process Biochem.*, **17**: 18–22.

Hill, I. R. and S. Wright. 1978. *Pesticide Microbiology*. New York: Academic Press.

Huser, B. A., K. Wuhrmann, and A. J. B. Zehnder. 1982. *Methanothrix soehngenii* Non-Hydrogen Oxidizing Methanobacterium. *Arch. Microbiol*, **132**: 1–9.

Krumme, M. L. and S. A. Boyd. 1988. Reductive Dechlorination of Chlorinated Phenols in Anaerobic Upflow Reactors. *Wat. Res.*, **22**(2): 171–177.

Lawrence, A. W. and P. L. McCarty. 1969. Kinetics of Methane Fermentation in Anaerobic Treatment. *J. Water Poll. Cont.*, **41**: R1–R17.

McCarty, P. L., L. Y. Young, J. B. Healy, W. F. Owen, and D. C. Stuckey. 1978. Thermochemical Treatment of Lignocellulosic and Nitrogenous Residuals for Increasing Anaerobic Biodegradability. Presented at Second Symposium Fuels from Biomass, Renssalaer Polytechnical Institute.

Owen, W. F., D. C. Stuckey, J. G. Healy, Jr., L. Y. Young, and P. L. McCarty. 1979. Bioassay for Monitoring Biochemical Methane Potential and Anaerobic Toxicity. *Water Research* **13**: 485–492.

Sachs, R. M., D. W. Gilpin, and T. Mick. 1981. Yields of Short Rotation *Eucalyptus grandis* in High Density Plantings. In: Symposium Papers, *Energy from Biomass and Waters IV.* Chicago: Institute of Gas Technology.

Salkinoja-Salonen, M., R. Paosivino, M. Koistinin, and R. Hakulinen. 1979. *Dechema*, 349–358.

Shelton, D. R., and J. M. Tiedje. 1984. General Method for Determining Anaerobic Biodegradation Potential. *Applied and Environ. Microbiol.*, **47**: 850–857.

Sjostrom, E. 1981. *Wood Chemistry: Fundamentals and Applications*. New York: Academic Press.

Speece, R. E. 1983. Environment Requirements for Anaerobic Digestion of Biomass. *Advances in Solar Energy*, An Annual Review of Research and Development.

Suidan, M. T., P. Fox, and J. T. Pfeffer. 1987. Anaerobic Treatment of Coal Gasification Wastewater. *Water Sci. Technol.*, **19**: 229–236.

Woods, S. L., J. F. Ferguson, and M. M. Benjamin. 1988. Characterization of Chlorophenol and Chloromethoxybenzene Biodegradation During Anaerobic Treatment. *Environ. Sci. Tech.*

Yang, J., R. E. Speece, G. F. Parkin, J. Gosset, and W. Kocher. 1980. The Response of Methane Fermentation to Cyanide and Chloroform. *Progr. Water Techol.*, **12**: 977–89, 1980.

8

Stripping of Volatile Organics

Yerachmiel Argaman

AIR STRIPPING

Air stripping is one of several processes available that removes volatile organic compounds (VOCs) from aqueous streams. More than half of the organics on the U.S. EPA's toxic pollutant list are sufficiently volatile to be effectively removed by stripping. Of the fifty-seven organics controlled by the Organic Chemicals (OCPSF) Industry Categorical Standards, thirty-three are classified as "very easily" stripped, and only six are "very difficult" or "non-strippable." Air stripping processes employ gas-liquid contacting systems that enhance the transfer of VOCs from the liquid phase to the gas phase. Various types of aeration devices have been designed to optimize transfer, and they fall into two general categories: (1) injection of water into air, and (2) injection of air into water. In both systems, mechanical energy creates air-water interfaces across which transfer of the contaminant can occur. Examples of water-into-air systems are spray systems into the open air, spray towers, and tray and packed towers. Air-into-water systems typically use diffused aeration.

Alternative Air Stripping Techniques

Diffused Aeration. In a diffused aeration system, air is injected into the water through a sparging device or through porous diffusers which produce a multitude of fine bubbles. As the bubbles rise, mass transfer occurs across the water-air interface until the bubble either leaves the water column or becomes saturated with the contaminant. The rate of mass transfer and its extent can be enhanced by increasing the depth of the tank, improving bubble dispersion, decreasing bubble size, and increasing the volumetric air to water ratio.

Increasing the depth of the tank may not increase the mass transfer if the air bubble reaches saturation before it exits the liquid surface.

Generally, diffused aeration systems are operated as continuously stirred tank reactors, and are thus inherently less efficient than a counter-flow process. The system is widely used for gas absorption, principally absorption of oxygen in biological treatment systems. Removal of VOCs by stripping during aeration in the activated sludge process will be discussed later in this chapter.

Mechanical Aeration.   Air can also be introduced into water by mechanical means. In this system, the contents of the tank are circulated providing continuous contact between the atmospheric air and the water. Principal design parameters include aeration power, the type of aerator, the depth of the tank, and the residence time. The system has been used with success to supply oxygen for biological treatment.

Sprays and Spray Towers.   High rates of mass transfer can be achieved by pumping water through spray nozzles that break the liquid stream into fine droplets. Sprays can be used to inject water into the open air or into a tower. Principal design parameters for sprays and spray towers include the type of nozzle, nozzle pressure drop, and the process configuration. Because of back mixing in the tower and in the air, true countercurrent transfer does not occur and mass transfer rates are usually modest. As a consequence, very high removal efficiencies may not be economically feasible.

Tray and Packed Towers.   The design principles for air stripping in packed and tray towers have been extensively examined in the chemical engineering literature over the past 30 to 40 years. Most chemical engineering applications generally involve design of systems to treat concentrated solutions. However, the general design procedures developed in the chemical processing industry have recently been extended to cover the full range of concentrations, including the case of dilute solutions typically encountered in water and wastewater treatment applications.

Tray or packed tower air stripping systems similar to those depicted in Fig. 8-1 are generally more effective for the removal of volatile organics from water than the other systems described above. The air is introduced near the base of the tower, while the liquid feed is introduced near the top. As the air rises and comes into contact with the falling water, the volatile components are transferred from the organic rich water to the air. The organic laden air is carried out the top and the stripped water exits the bottom of the tower.

In tray towers, the tower is filled with regularly spaced trays or plates allowing for staged contact between the two phases. The vapor passes through openings in each tray and contacts the liquid flowing across the tray. A quantity of liquid

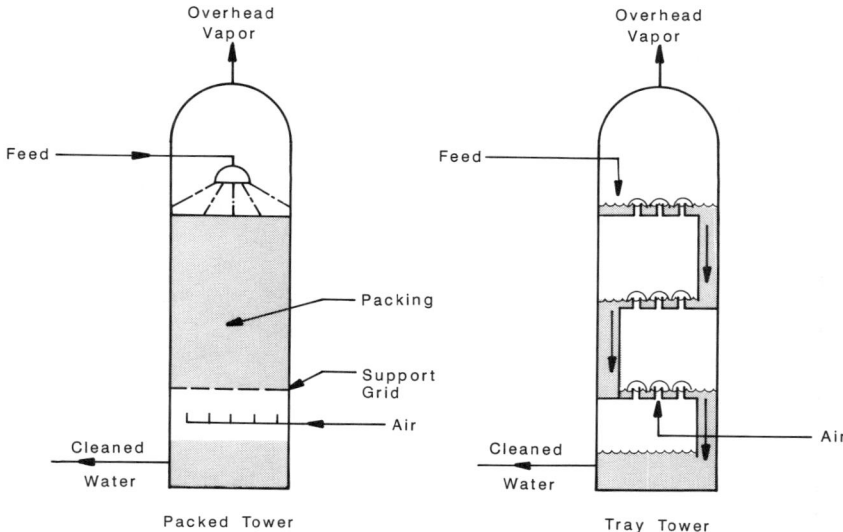

Fig. 8-1. Air stripping towers.

is retained on each tray by a weir. To reach the next stage, the liquid flows over the weir through a downcomer which provides sufficient volume and enough residence time for the liquid to be freed of entrained vapor before entering the next tray. The overall effect is a multiple countercurrent contact of air and water, although each tray is characterized by a crossflow of the two.

Packed towers are simple when compared with tray towers. A typical tower consists of a cylindrical shell containing a support tray for the packing material and a liquid distributor designed to provide effective irrigation of the packing. The water flows down through the packed bed, exposing a large surface area for transfer into the air which enters at the bottom of the tower. Commercially available packing materials come in a variety of shapes and sizes. Most packings are made of either ceramic, metal, or plastic. Depending on the types and size of packing they may be either randomly dumped or carefully stacked in the column.

Theory of Air Stripping

The feasibility of air stripping of a given compound depends on its physical-chemical properties as expressed by Henry's Constant. For an aqueous mixture containing low concentrations of VOCs, the distribution of a particular VOC between air and water under equilibrium is given by Henry's law:

$$P_A = H_A X_A^*$$ (8-1)

where:

P_A = partial pressure of compound A in the gas phase, atm
X_A^* = mole fraction of compound A in liquid phase in equilibrium with the gas phase.
H_A = Henry's constant, atm

When using Henry's law and comparing Henry's constant for various compounds, care must be taken to use a consistent system of units. If the concentration in the liquid is expressed in mol/l (or kmol/m^3) and P_A is in atm, the Henry's constant is in atm · l/mol or atm · m^3/kmol. The Henry's constant is sometimes expressed as a dimensionless number when both the concentration of the compound in the water and the air are in mole fractions or in the same mass/volume units. Table 8-1 summarizes the various forms of the Henry constant and the appropriate conversion factors.

Henry's constant is temperature dependent. The temperature effect on H can be expressed by:

$$\log H = A - \frac{B}{T - T_0} + \frac{C}{T} \qquad (8\text{-}2)$$

where:

T = temperature, °K
A, B, C, T_0 = empirical constants

As a general rule, compounds having a Henry's Constant H_M, greater than 10^{-3} atm · m^3/mol are considered "easy to strip." Those with an H_M between 10^{-4} and 10^{-5} atm · m^3/mol are "difficult to strip." Compounds with an H_M less than 10^{-5} atm · m^3/mol are generally labeled "non-strippable." Some examples of compounds in the three regions are listed in Table 8-2.

Table 8-1. Units and Conversion Factors for Henry's Constant.

Concentration in Gas Phase	Concentration in Liquid Phase	Henry's Constant	
		Symbol	Units
atm	mol fraction	H_A	atm
atm	mol/m^3	H_M	atm · m^3/mol
g/m^3	g/m^3	H_c	dimensionless
mol fraction	mol fraction	H_c'	dimensionless

Conversion factors:
$H_A = H_M \cdot 55.6 \cdot 10^3$
$H_c' = H_M \cdot 55.6 \cdot 10^3$
$H_c = H_M \cdot 44.64 \ (H_c = H_M/RT)$

Table 8-2. Henry's Constants for Selected Compounds.

Compound	Formula	Henry's Constant H_M, atm · m³/mol	Comments
Vinyl chloride	CH_2CHCl	6.38	
Trichloroethylene	$CCHCl_3$	1.0×10^{-2}	
1,1,1-Trichlorethane	CCH_3Cl_3	7.0×10^{-3}	Easy to strip
Toluene	$C_6H_5CH_3$	6.0×10^{-3}	
Benzene	C_6H_6	4.0×10^{-3}	
Chloroform	$CHCl_3$	3.0×10^{-3}	
1,1,2-Trichloroethane	CCH_3Cl_3	7.7×10^{-4}	Difficult to
Bromoform	$CHBr_3$	6.3×10^{-4}	strip
Pentachlorophenol	$C_6(OH)Cl_5$	2.1×10^{-6}	Nonstrippable
Dieldrin	—	1.7×10^{-8}	

Demonstrated Performance

Data on full scale performance of air stripping operations treating industrial wastewater are scarce, because the few full scale plants that exist have not published data. Results from a bench-scale study (Breton et al. 1986) are shown in Fig. 8-2. These were done in a 3.75 inch ID column packed with 1.5 ft of $\frac{1}{4}$ inch berl saddles. In general, the study showed that removal efficiencies increased with increasing Henry's constant and with increasing air/water ratios.

Packed Tower Design

Packed tower air stripping is a well developed chemical engineering process. The theory has been applied to develop design procedures for stripping of organics from water and wastewater (Kavanaugh and Trussell 1980). The design model is summarized herein.

Assuming Henry's law applies, the height of the packing required under steady-state conditions is given by:

$$Z = (HTU)(NTU) \qquad (8\text{-}3)$$

where:

Z = packing height, m
HTU = height of a transfer unit, m
NTU = number of transfer units, dimensionless.

The height of a transfer unit is a somewhat artificial term of the mass transfer theory indicating the packing height that will result in a certain standard removal ratio of a water contaminant. Under ideal stripping conditions, one transfer unit reduces contaminant concentration by a factor of 2.718 (the natural logarithm

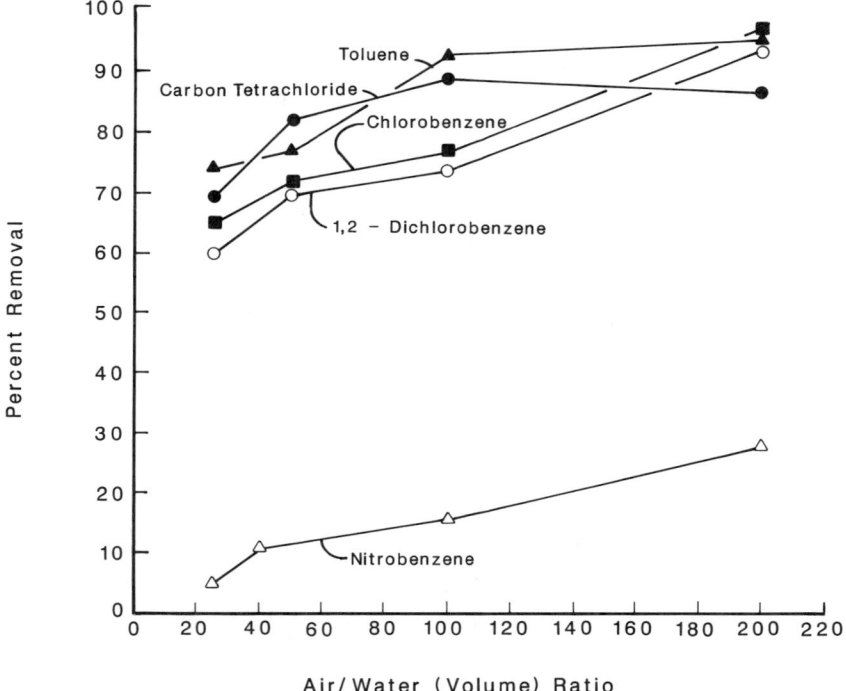

Fig. 8-2. Results of bench-scale air stripping study.

base). Under nonideal conditions, the removal ratio effected by one transfer unit would be lower than 2.718.

The *HTU* is a function of the packing medium, operating conditions and the physical/chemical properties of the solution and the stripped compound. For highly volatile organics, i.e., when mass transfer is controlled by resistance in the liquid phase only, the *HTU* can be expressed by:

$$HTU = \frac{L}{K_L a C_0} \tag{8-4}$$

where:

L = liquid loading rate, kmol/m² · sec
C_0 = molar density of water (55.6 kmol/m³)
$K_L a$ = product of the overall liquid mass transfer coefficient K_L, m/sec, and the specific interfacial area, a, m²/m³, in the packing medium

The $K_L a$ parameter depends on the compound to be stripped, the packing medium, and the air and water loading rates. Values of $K_L a$ or *HTU* may sometimes be obtained from media manufacturers or the literature. In many cases, experimental derivation of this parameter is required.

Several empirical equations for predicting $K_L a$ values for packed media have been proposed (Perry and Chilton 1973). One of those is the Sherwood and Holloway equation:

$$K_L a = \alpha D \left(\frac{L}{\mu}\right)^{1-n} \left(\frac{\mu}{\rho D}\right)^{0.5} \qquad (8\text{-}5)$$

where:

$K_L a$ = overall mass transfer coefficient, hr^{-1}
 D = diffusivity, ft^2/hr
 L = liquid loading rate, $\text{lb}/\text{hr} \cdot \text{ft}^2$
 μ = liquid viscosity, $\text{lb}/\text{hr} \cdot \text{ft}$
 ρ = liquid density, lb/ft^3
α, n = empirical constants

Values of the coefficients α and n are given in Table 8-3. When Eq. (8-5) and Table 8-3 are used to predict $K_L a$, all data must be entered in the specified system of units, i.e., ft, lb, hr.

If the $K_L a$ value of a certain packing is known for a given compound, it can be used in predict values for other compounds under the same operating conditions. This is true if the mass transfer for both compounds is controlled by the liquid film resistance. Under these conditions, $K_L a$ is related to the molecular diffusivity by:

$$K_L a = \alpha \cdot D^n \qquad (8\text{-}6)$$

where:

D = molecular diffusivity, m^2/sec
α = proportionality coefficient
n = empirical constant

Table 8-3. Constants for the Sherwood-Holloway Equation.

Packing	Nominal Size, in.	α	n
Raschig rings	0.5	280	0.35
	1.0	100	0.22
	1.5	90	0.22
	2.0	80	0.22
Berl saddles	0.5	150	0.28
	1.0	170	0.28
	1.5	160	0.28
Tile	3.0	110	0.28

The theoretical value of n is 0.5—see Eq. (8-5)—while experimental data for several chlorinated hydrocarbons yielded a value of 0.66.

Diffusivities of various solutes in water can be predicted by:

$$D = 7.4 \cdot 10^{-8} \frac{T}{\mu} \frac{M^{0.5}}{V^{0.6}} \qquad (8-7)$$

where:

D = diffusivity, cm^2/sec
T = absolute temperature, °K
μ = viscosity, centipoise
M = molecular weight of water, g/mole
V = molar volume of solute, $cm^3/mole$

The molar volume can be calculated by:

$$V = \frac{M}{\rho} \qquad (8-8)$$

where:

M = molecular weight of solute, g/mole
ρ = density of solute, g/cm^3

For water at 20°C:

μ = 1.0 centipoise
= 2.42 lb/ft hr
ρ = 1.0 g/cm^3
= 62.4 lb/ft^3

The number of transfer units required for a specific removal is given by

$$NTU = \frac{R}{R-1} \ln \frac{C_{in}/C_{out}(R-1)+1}{R} \qquad (8-9)$$

where:

R = stripping factor, dimensionless
C_{in}, C_{out} = concentration of contaminant entering and exiting the tower, respectively, mg/l.

The stripping factor R is given by

$$R = \frac{H_A}{P_t} \frac{G}{L} \qquad (8-10)$$

where:

P_t = ambient pressure, atm
G = air loading rate, kmol/m² · sec
L = water loading rate, kmol/m² · sec
H_A = Henry's Constant, atm

The effect of R on removal and *NTU* is illustrated in Fig. 8-3. Three regions of R values can be identified:

- $R < 1$. In this region all NTU curves converge and packing height ceases to influence the percent removed. An air stripper should never be designed for operation at this range.
- $1 < R < 10$. The NTU curves "fan out" over the span of this region, and removal performance grows dramatically for a constant packing height with relatively small increases of R. The stripper height is becoming more important with increasing values of R.
- $R > 10$. At high R values, the required *NTU* ceases to be a function of R

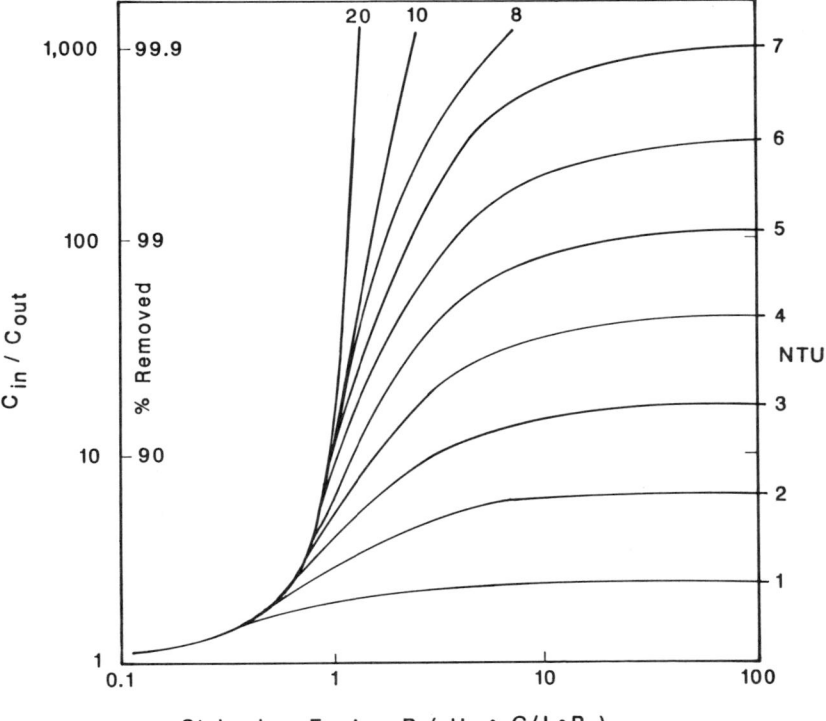

Fig. 8-3. Effect of stripping factor on performance.

and each additional transfer unit increases removal by a factor of 2.718 (the natural logarithm base).

The gas pressure drop through the packing medium is given in Fig. 8-4. It shows the effects of gas and liquid flow rates on pressure drop. Normally, it is not recommended to operate at abscissa values less than 0.02 or above 4.0. Flooding conditions develop at high gas flow rates. This is a condition in which the air velocity is so high that it holds up the water in the column to the point where the water becomes the continuous phase rather than the air.

Packed Tower Design Procedure. Design of a packed tower for a specific application requires selection of the following parameters:

- Packing type
- Packing depth
- Tower diameter
- Air flow

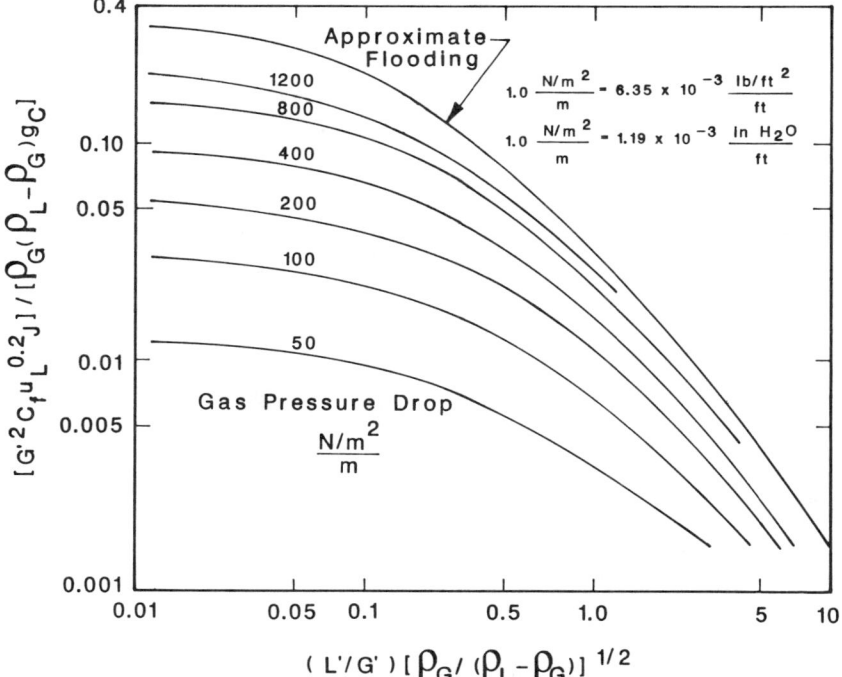

Fig. 8-4. Flooding and pressure drop in packed tower. L' and G' in units of mass/time · area. For SI units, $g_c = 1$ and $J = 1$. For $G' = \text{lb}/\text{ft}^2 \cdot \text{h}$, $\rho = \text{lb}/\text{ft}^3$, $\mu_c = \text{cP}$, $g_c = 4.18 \times 10^8$, $J = 1.502$, C_f for 1 inch plastic saddles $= 33$.

Since there is more than one solution to the problem, an optimization procedure is called for. Several designs can be made and compared for construction and O&M costs. The following design procedure is suggested.

1. Select a packing media of appropriate material, which has a low packing factor and a high relative mass transfer coefficient. The packing factors for several plastic packings and Raschig rings are given in Table 8-4. Values of the relative mass transfer coefficient can be obtained from data in Table 8-5. This table contains $K_G a$ values for several packings derived for CO_2 absorption at a given water loading. K_G is the overall gas phase mass transfer coefficient. For a liquid-phase controlled process, it is related to K_L through

$$K_L = \frac{K_G H_M}{C_0} \qquad (8\text{-}11)$$

Hence for a given contaminant, $K_L a$ is directly proportional to $K_G a$.

Table 8-4. Packing factor C_f for Common Plastic Packings and Raschig Rings.

Packing Type	C_f for Nominal Packing Size Shown				
	1 in.	1.5 in.	2 in.	3 in.	3.5 in.
Super Intalox	33		21	16	
Pall rings	52	32	25		16
Tellerettes	40		20		
Maspak			32	20	
Heil-Pak	45		18	15	
Raschig rings	155	95	65	37	
Berl saddles	110	65	45		

Table 8-5. Relating Mass Transfer Coefficients: Relative $K_G a$ Values for Several Common Packings.[a]

Packing Type	$K_G a$ for Nominal Packing Size (kmol/s · m³ · atm) × 10^2				
	1 in.	1.5 in.	2 in.	3 in.	3.5 in
Super Intalox	1.43		0.94	0.62	
Pall rings	1.10	0.98	0.85		0.62
Tellerettes			0.85		
Maspak			0.76	0.89	
Heil-Pak			1.25	0.53	
Raschig rings (ceramic)	1.03	0.85	0.74		0.62
Berl saddles (ceramic)	1.11	0.80	0.65		

[a]mol/hr · cu ft (atm) = 224 · (kmol/s · m³ · atm). Values for CO_2 absorption by NaOH at $L' = 6.7$ kg/m² · s.

Alternatively, the $K_L a$ value can be estimated from Eq. (8-5), after L has been selected.

2. Select a stripping factor, R. The normal range of practical values is 2 to 20. (Alternatively, it is possible to select a G/L ratio and calculate R from Eq. (8-10). A typical value for G/L is 40.) Obtain NTU for the selected R and desired removal rate C_{out}/C_{in} from Fig. 8-3 or Eq. (8-9). Calculate G/L from Eq. (8-10).

3. Select a desired pressure drop rate, normally between 50 to 400 N/m² · m (0.062 to 0.49 in. H_2O/ft), determine G' from Fig. 8-4, and calculate L' from the known G/L ratio. Note:

$$\frac{G}{L} = \frac{G'/MW_{air}}{L'/MW_{water}}$$

$$= \frac{G'/28.8}{L'/18}$$

$$= 0.625G/L'$$

4. Calculate height of a transfer unit HTU, from Eq. (8-4).
5. Calculate packing height, Z, from Eq. (8-3) and total air pressure drop, $\Delta\rho$, by

$$\Delta\rho = Z \frac{\Delta\rho}{\Delta Z} \tag{8-12}$$

6. Calculate total head loss by

$$\Delta h = \frac{\Delta\rho}{\gamma_{water}}$$

$$= \frac{\Delta\rho}{9810} \frac{(N/m^2)}{(N/m^3)} \tag{8-13}$$

PACKED TOWER DESIGN EXAMPLE

Calculate the height and diameter of a packed tower for removal of 1,1,2-trichloroethane in wastewater from 5 mg/l to 0.01 mg/l. The operating temperature is 20°C, and the flow rate is 1.0 mgd (0.0438 m³/sec or 43.8 kg/sec).

Solution:

1. The selected packing is 2″ ceramic Raschig rings.

2. A stripping factor $R = 3$ will be used. The required removal rate is:

$$\frac{C_{\text{out}}}{C_{\text{in}}} = \frac{0.01}{5.0} = 0.002$$

The required number of transfer units from Eq. (8.9) is:

$$NTU = \frac{3}{3 - 1} \ln \frac{(1/0.002)(3 - 1) + 1}{3}$$

$$= 8.7$$

The air-to-water ratio is, from Eq. (8-10):

$$\frac{G}{L} = \frac{3 \cdot 1.0}{7.7 \cdot 10^{-4} \cdot 55.6 \cdot 10^3}$$

$$= 0.07 \text{ mol/mol}$$

$$\frac{G'}{L'} = \frac{0.07}{1.6}$$

$$= 0.0438, \text{ g/g}$$

3. Select an air pressure drop of 200 N/m²/m. The abscissa value from Fig. 8-4 is:

$$\frac{L'}{G'} \left(\frac{\rho_G}{\rho_L - \rho_G} \right)^{1/2} = \frac{1}{0.0438} \cdot \left(\frac{1.295}{1,000 - 1.295} \right)^{1/2}$$

$$= 0.822$$

The ordinate is 0.0145, hence

$$\frac{G'^2 C_f \mu_L^{0.2} J}{\rho_G (\rho_L - \rho_G) g_c} = 0.0145$$

where

$c_f = 65$ (from Table 8-4)
$\mu_L = 0.001$ kg/m · sec
$\rho_G = 1.295$ kg/m³
$\rho_L = 1,000$ kg/m³
$g_c = 1.0$ (for SI units)
$J = 1.0$ (for SI units)

Hence

$$G'^2 = \frac{0.0145 \cdot 1.295(1,000 - 1.295)1.0}{65 \cdot (0.001)^{0.2} \cdot 1.0}$$

$$= 1.149$$

$$G' = (1.149)^{1/2}$$
$$= 1.072 \text{ kg/m}^2 \cdot \text{sec}$$

$$L' = \frac{G'}{G'/L'} = \frac{1.072}{0.0438}$$
$$= 24.47 \text{ kg/m}^2 \cdot \text{sec}$$

Cross-sectional area of tower:

$$A = \frac{Q_L}{L'}$$
$$= \frac{43.8}{24.47}$$
$$= 1.79 \text{ m}^2$$

Tower diameter:

$$D = \left(\frac{1.79 \cdot 4}{\pi}\right)^{1/2}$$
$$= 1.51 \text{ m } (4.95 \text{ ft})$$

Air flow:

$$G = 1.072 \; (\text{kg/m}^2 \cdot \text{sec}) \cdot 1.79 (\text{m}^2)$$
$$\cdot \frac{35.3 \,(\text{ft}^3/\text{m}^3)\, 60 \,(\text{sec/min})}{1.29 \,(\text{kg/m}^3)}$$
$$= 3150 \text{ cfm}$$

4. Calculate *HTU*. Available data on $K_L a$ for chloroform on 2″-Berl saddle is 238.8 hr^{-1}. To correct for Raschig rings (see Table 8-5):

$$(K_L a)_{\text{Raschig}} = (K_L a)_{\text{Berl}} \cdot \frac{0.74}{0.65}$$

To correct for 1,1,2-trichloroethane, the diffusivities of chloroform and 1,1,2-trichloroethane are as follows:

Chloroform: $0.94 \cdot 10^{-9} \text{ m}^2/\text{sec}$
1,1,2-trichloroethane: $0.83 \cdot 10^{-9} \text{ m}^2/\text{sec}$ (assumed equal to 1,1,1-trichloroethane)

Hence by Eq. (8-6) the $K_L a$ correction will be

$$(K_L a)_{TCE} = \left(\frac{0.83}{0.94}\right)^{1/2} (K_L a)_{\text{chloroform}}$$

Using both corrections, the $K_L a$ for 1,1,2-trichloroethane using 2 in. Raschig rings will be:

$$K_L a = 238.8 \cdot \frac{0.74}{0.65} \cdot \left(\frac{0.83}{0.94}\right)^{1/2}$$

$$= 255.5 \text{ hr}^{-1}$$

$$= 0.071 \text{ sec}^{-1}$$

The *HTU* is obtained by Eq. (8-4):

$$HTU = \frac{L}{K_L a C_0}$$

$$= \frac{24.47/18}{0.071 \cdot 55.6}$$

$$= 0.344 \text{ m}$$

Tower height:

$$Z = (HTU)(NTU)$$

$$= 0.344 \cdot 8.7$$

$$= 3.0 \text{ m } (9.83 \text{ ft})$$

Note: The vent gas concentration Y_{out} is related to influent liquid concentration X_{in} by materials balance:

$$Y_{\text{out}} = X_{\text{in}} \cdot \frac{L}{G}$$

$$= \frac{1}{0.07} X_{\text{in}}$$

$$= 14.3 X_{\text{in}}$$

At equilibrium the vent gas concentration would be:

$$Y_{\text{out}} = H_A \cdot X_{\text{in}}$$

$$= 7.7 \cdot 10^{-4} \cdot 55.6 \cdot 10^3 X_{\text{in}}$$

$$Y_{\text{out}} = 42.8 X_{\text{in}}$$

Hence the air flow is $(42.8/14.3 =)$ 3 times the minimum required.

Surface Aeration Systems Design

For a completely mixed surface aeration system, a mass balance equation yields:

$$C_e = \frac{C_i + K_L a C_s t}{1 + K_L a t} \qquad (8\text{-}14)$$

where

C_i, C_e = influent and effluent concentration of volatile compound, mg/l
C_s = saturation concentration of contaminant in equilibrium with ambient air, mg/l
t = hydraulic residence time, hr
$K_L a$ = overall mass transfer coefficient for the volatile compound, hr^{-1}

In most cases C_s is so low that the second term in the numerator of Eq. (8-14) is neglected.

The overall mass transfer coefficient for a highly volatile compound is related to the oxygen mass transfer by:

$$K_L a = \psi (K_L a)_0 \qquad (8\text{-}15)$$

where:

$(K_L a)_0$ = mass transfer coefficient for O_2, hr^{-1}
ψ = proportionality factor, which according to Eq. (8-5) is proportional to diffusivity raised to the power 0.5 to 0.66

Early studies (Roberts et al. 1984) indicated that ψ of Eq. (8-15) is constant for highly volatile compounds characterized by a Henry's law constant $H_c >$ 0.19. More recently (Munz and Roberts 1989), this has been revised when new data indicated the greater role of the gas layer resistance to mass transfer. Hence, the constant ψ assumption is applicable to compounds with $H_c > 0.55$.

The proportionality factor ψ can be defined in a way that it is applicable for all compounds regardless of their volatility:

$$\psi = \frac{(K_L a)_i}{(K_L a)_0}$$

$$= \left(\frac{D_i}{D_0}\right)^n \left(1 + \frac{1}{k_G/k_L\, H_c}\right)^{-1} \qquad (8\text{-}16)$$

where:

D_i, D_0 = diffusivity of solute i, and oxygen, respectively
k_G/k_L = rate of gas to liquid mass transfer coefficient

For high energy input range (>350 hp/mil gal) the k_G/k_L ratio was estimated as 30 to 40.

Values of $(K_L a)_0$ can be estimated from standard oxygen transfer efficiency. Thus:

$$N_O = (K_L a)_0 C_s \frac{V}{P} \qquad (8\text{-}17)$$

where:

N_O = standard oxygen transfer rate, kg O/hp · hr

C_s = oxygen solubility at standard conditions $\left(9.2 \cdot \dfrac{10^{-3} \text{ kg}}{\text{m}^3}\right)$

V = aeration volume, m³

P = aeration power, hp

Rearranging Eq. (8-17) yields:

$$(K_L a)_0 = \frac{N_O P}{C_s V} \qquad (8\text{-}18)$$

SURFACE AERATION DESIGN EXAMPLE

Design a surface aeration system for the removal of 1,1,2-trichloroethane under the same condition as in the packed tower design example.

Solution:

1. Assume an aeration power level of 200 hp/mil gal, and a standard oxygen transfer capacity of 3.5 lb O_2/hp hr (slow speed surface aerators). Hence, from Eq. (8-18) the oxygen transfer coefficient is:

$$(K_L a)_0 = \frac{3.5 \text{ (lb)}}{\text{(hp hr)}} \cdot \frac{200 \text{ (hp)}}{\text{(mil gal)}} \frac{\text{(mg/l)}}{9.2 \text{ (mg/l)} \ 8.34 \text{ (lb/mil gal)}}$$

$$= 9.12 \text{ hr}^{-1}$$

From literature ψ (1,1,2-trichloroethane) $= 0.61$

$$(K_L a)_{\text{trichloroethane}} = 0.61 \cdot 9.12 \text{ hr}^{-1}$$

$$= 5.57 \text{ hr}^{-1}$$

The required residence time can be derived from Eq. (8-14) (assuming C_s ≈ 0):

$$t = \frac{C_i/C_e - 1}{K_L a}$$

$$= \frac{5/0.01 - 1}{5.57}$$

$$= 89.6 \text{ hr}$$

Required basin volume:

$$V = 1.0 \text{ (mgd)} \cdot \frac{89.6 \text{ (hr)}}{24 \text{ (hr/day)}}$$

$$= 3.73 \text{ mil gal}$$

Power required

$$P = V \cdot P/V$$

$$= 3.73 \text{ (mil gal)} \cdot 200 \text{ (hp/mil gal)}$$

$$= 747 \text{ hp}$$

Diffused Aeration Systems Design

The fractional saturation of an air bubble rising through a water column (when the liquid is assumed to be completely mixed) is given by:

$$\frac{C_{Ge}}{C_G^*} = 1 + \left(\frac{C_{Gi}}{C_G^*} - 1 \right) \exp(-\phi') \tag{8-19}$$

where:

C_{Gi}, C_{Ge} = concentration of volatile compound in air bubble entering and leaving the basin, respectively, g/m^3.

C_G^* = concentration of volatile compound in air which is in equilibrium with the liquid, g/m^3 ($C_G^* = C_L \cdot H_c$),

C_L = equilibrium concentration of volatile compound in the liquid, g/m^3

ϕ' = saturation parameter defined by:

$$= \frac{K_L a V}{H_c Q_G} \tag{8-20}$$

Q_G = gas (air) flow rate, m^3/hr

For a volatile compound that is not present in the ambient air ($C_{Gi} = 0$) Eq. (8-19) reduces to:

$$\frac{C_{Ge}}{C_G^*} = 1 - \exp(-\phi') \tag{8-21}$$

The liquid concentration of the volatile compound at equilibrium is obtained from mass balance (for the case that the influent concentration in the gas is 0):

$$C_{Le} = \frac{C_{Li}}{1 + \dfrac{Q_G}{Q_L} H_c(1 - e^{-\phi'})} \tag{8-22}$$

where:

C_{Li}, C_{Le} = Concentration of volatile compound in the liquid entering and leaving the basin, respectively, g/m^3

Q_L = liquid flow rate, m^3/hr

Note: $C_L = C_{Le}$ for a completely mixed system.

The $(K_L a)_0$ value can be estimated if the oxygen liquid concentration, oxygen uptake, and transfer efficiency are known. The following procedure can be used:

The concentration of oxygen in the air entering the basin is:

$$C_{Gi} = \frac{Y_0 P M_0}{RT} \tag{8-23}$$

where:

Y_0 = mole fraction of oxygen in air (0.21)
P = pressure at mid-depth of basin, atm
M_0 = molecular weight of oxygen (32)
R = gas constant = $8.206 \cdot 10^{-5}$ atm m^3/mol °K
T = temperature, °K

The oxygen concentration in the exiting air is:

$$C_{Ge} = (1 - \mu)C_{Gi} \tag{8-24}$$

where μ is the oxygen transfer efficiency

If R_0 is the oxygen uptake (in g O_2/m^3 treated wastewater), then:

$$R_0 \cdot Q_L = Q_G(C_{Gi} - C_{Ge}) \tag{8-25}$$

from which Q_G can be calculated.

Knowing all terms except ϕ', this parameter can be calculated from Eq. (8-19). The Henry's constant for oxygen, H_c, is 32.

Then $(K_L a)_0$ can be calculated by rearranging Eq. (8-20):

$$(K_L a)_0 = (\phi')_0 Q_G / V(H_c)_0 \tag{8-26}$$

From this $(K_L a)_0$ value, the $K_L a$ of the volatile compound can be calculated using term ψ and Eq. (8-15). The effluent concentration is then obtained using Eq. (8-22).

DIFFUSED AERATION DESIGN EXAMPLE

A diffused air system is used in a 1 mgd plant in which the COD is reduced from 3,000 mg/l to 150 mg/l in a 1 day retention time. The aeration basin DO is 2.0 mg/l and the oxygen utilization efficiency is 5 percent. The basin depth is 14 ft. The oxygen consumption is 0.8 lb O_2/lb COD. Calculate $(K_L a)_0$ for this system and design a diffused air system for removing 1,1,2-trichloro-ethane as in the surface aeration design example.

Solution:

1. The concentration of oxygen in the bubbles released into the aeration basin is obtained from Eq. (8-23):

$$C_{Gi} = \frac{0.21 \cdot 1.21 \ (\text{atm}) \ 32 \ (\text{g/mol})}{8.206 \cdot 10^{-5} \ (\text{atm} \cdot \text{m}^3/\text{mol} \cdot {}^\circ\text{K}) \ 293 \ ({}^\circ\text{K})}$$

$$= 338 \ \text{g/m}^3$$

Based on the 5 percent oxygen utilization, the oxygen concentration in the *exiting* air is:

$$C_{Ge} = (1 - 0.05) \cdot 338 (\text{g/m}^3)$$

$$= 321 \ \text{g/m}^3$$

The air flow rate is obtained from Eq. (8-25):

$$Q_G = \frac{0.8 (\text{gO}_2/\text{gCOD}) \cdot (3000 - 150) \ \text{gCOD/m}^3 \cdot 1.0 \ \text{mgd} \cdot 3785 (\text{m}^3/\text{mil gal})}{(338 - 321) \text{g/m}^3}$$

$$= 507{,}635 \ \text{m}^3/\text{day} \ (12{,}448 \ \text{cfm})$$

From Eq. (8-19):

$$\phi' = \ln\left[\left(\frac{C_{Gi}}{C_G^*} - 1 \right) \middle/ \left(\frac{C_{Ge}}{C_G^*} - 1 \right) \right]$$

$$C_G^* = C_L H_c$$

$$= 2.0 \ \text{g/m}^3 \cdot 32$$

$$= 64 \ \text{g/m}^3$$

$$\phi' = \ln \frac{(338/64) - 1}{(321/64) - 1}$$

$$= 0.064$$

And from Eq. (8-20):

$$(K_L a)_0 = \frac{0.064 \cdot 32 \cdot 507635 \ (\text{m}^3/\text{day})}{1 \ (\text{mil gal}) \cdot 3785 \ (\text{m}^3/\text{mil gal}) \cdot 24 \ (\text{hr}/\text{day})}$$

$$= 11.4 \ \text{hr}^{-1}$$

For 1,1,2-trichloroethane $\psi = 0.61$ and by Eq. (8-15)

$$K_L a = 0.61 \cdot 11.4 \ (\text{hr}^{-1})$$

$$= 6.95 \ \text{hr}^{-1}$$

From Eq. (8-22):

$$\frac{Q_G}{Q_L} H_c (1 - e^{-\phi'}) = \frac{C_{Li}}{C_{Le}} - 1$$

Hence, for various values of Q_G, various values of V (volume of reactor) will be obtained.

(a) Assume

$$Q_G = 10^5 \ Q_L$$

$$= 10^5 \cdot 3785 \ (\text{m}^3/\text{day})$$

$$= 3.785 \cdot 10^8 \ \text{m}^3/\text{day}$$

$$\phi' = -\ln \left[1 - \frac{(C_{Li}/C_{Le}) - 1}{H_c (Q_G/Q_L)} \right]$$

$$= -\ln \left[1 - \frac{5/0.01 - 1}{0.034 \dfrac{3.785 \cdot 10^8}{3785}} \right]$$

$$= 0.1587$$

where $H_c = H_m \cdot 44.64$ (from Table 8-1)

$$= (7.7 \times 10^{-4}) \cdot 44.64 \ (\text{from Table 8-2})$$

$$= 0.034$$

$$V = \frac{\phi' H_c Q_G}{K_L a}$$

$$= \frac{0.1587 \cdot 0.034 \cdot 3.785 \cdot 10^8 (m^3/day)}{6.95 \ (hr^{-1}) \ 24 \ (hr/day)}$$

$$= 12,244 \ m^3$$

$$= 3.23 \ mil \ gal$$

(b) Assume

$$Q_G = 0.5 \cdot 10^5 \ Q_L$$

$$= 0.5 \cdot 10^5 \cdot 3785 \ (m^3/day)$$

$$= 1.892 \cdot 10^8 \ (m^3/day)$$

$$\phi' = -\ln \left[1 - \frac{5/0.01 - 1}{0.034 \cdot \dfrac{1.829 \cdot 10^8}{3785}} \right]$$

$$= 0.347$$

$$V = \frac{0.347 \cdot 0.034 \cdot 1.892 \cdot 10^8 \ (m^3/day)}{6.95 \ (hr^{-1}) \ 24 \ (hr/day)}$$

$$= 13,382 \ m^3$$

$$= 3.53 \ mil \ gal$$

A comparison of the three different air stripping systems in the previous examples is given in Table 8-6. As seen in the table, packed towers are very effective for this stripping application. Although for other stripping problems the difference in unit size may not be as large as in these examples, the general trend seen here still exists.

Table 8-6. Comparison of Results of Air Stripping Examples.

			Diffused Aeration	
	Packed Tower	Surface Aeration	(a)	(b)
Size	Diameter = 4.95 ft	Vol = 3.73 mil gal	Vol = 3.23 mil gal	3.53 mil gal
	Height = 9.83 ft			
Air/Power	Air = 3,150 cfm	Power = 747 hp	Air = 7,450 cfm	3,723 cfm

STRIPPING IN ACTIVATED SLUDGE PLANTS

Volatile organic compounds will air strip in activated sludge plants due to the high air volumes which come in contact with the aerated wastes. Depending on the VOC in question, both air-stripping and biodegradation may occur. The stripping of VOC in biological treatment processes is currently receiving considerable attention in the United States since current legislation severely limits permissible emissions of VOCs.

The fraction of VOCs which will be stripped will depend upon several factors, namely the Henry's Law constant, the biodegradability, the power level in the aeration basin, and in some cases the initial concentration of VOCs. The concentration of nonvolatile organics will affect the composition of the sludge and hence the fraction biodegraded and stripped.

Data for a range of VOC's are shown in Table 8-7. The effect of biodegradability on stripping is shown in Fig. 8-5 (Weber and Jones 1983). As the number of chlorine atoms on the benzene molecule is increased, biodegradability decreases and hence the fraction stripped is increased.

Table 8-7. The Fate of VOC in the Activated Sludge Process .

Compound	SRT days	Stripped	Influent Concentration	Reference
Toluene	3	12–16	100	a
	3	17	0.1	b
Ethylbenzene	3	15	40	a
	12	5	40	a
	6	22	0.1	b
Nitrobenzene	6	< 1	0.1	b
Benzene	6	15	153	a
	6	16	0.1	b
Chlorobenzene	6	20	0.1	b
1,2-Dichlorobenzene	6	59	0.1	b
1,2-Dichlorobenzene	6	24	83	a
1,2,4-Trichlorobenzene	6	90	0.1	b
o-Xylene	6	25	0.1	b
1,2-Dichloroethane	3	92–96	150	a
1,2-Dichloropropane	6	5	180	a
Methyl ethyl ketone	7.1	3	55	c
	7.2	10.2	430	c
1,1,1-Trichloroethane	6	76	141	c

[a]Kincannon and Fazel (1986).
[b]Weber and Jones (1983).
[c]Koczwara et al. (1987).

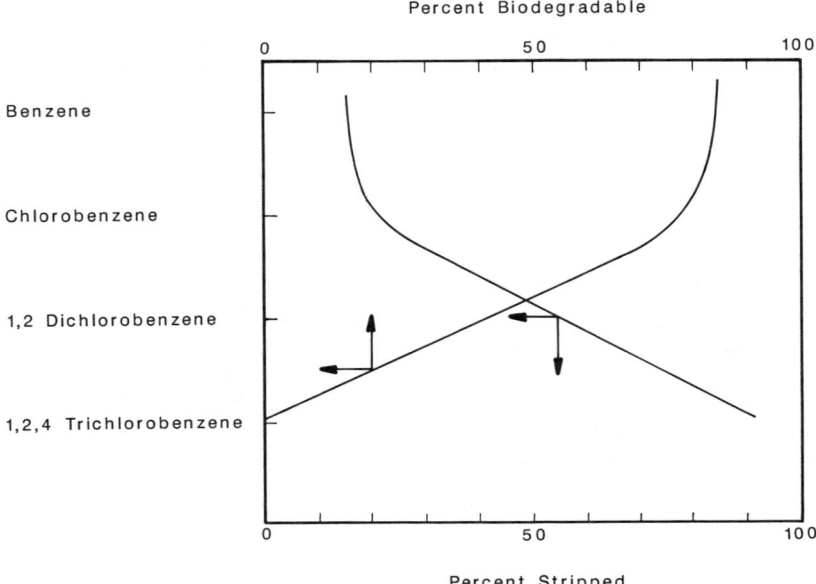

Fig. 8-5. Stripping and biodegradation of organics in the activated sludge process.

Stripping Computation

The fraction emitted to the air of any volatile component entering an activated sludge aeration basin can be derived from the following mass balance equation:

$$QS_{0i} = QS_{ei} + r_i + v_i \qquad (8\text{-}27)$$

where:

Q = flow rate, $1/\text{sec}$
S_{0i} = influent concentration of component i, $g/1$
S_{ei} = effluent concentration of component i, $g/1$
r_i = rate of biodegradation of component i, g/sec
v_i = rate of volatilization of component i, g/sec

For a completely mixed system, the fraction stripped to the air is given by:

$$f_{si} = \frac{v_i}{QS_{ei} + r_i + v_i} \qquad (8\text{-}28)$$

In order to derive the rate of biodegradation r_i, the biomass is considered as a mixed culture with each substrate being the growth limiting factor for a specific microbial population. Under these circumstances, the fraction of biomass dedicated to a specific substrate would be proportional to the biomass yield associated with this substrate. Hence the fraction dedicated to component i will be:

$$\alpha_i = a_i \cdot f_{bi} \cdot S_{0i} \Big/ \sum_{i=1}^{n} (a_i \cdot f_{bi} \cdot S_{0i}) \qquad (8\text{-}29)$$

where:

α_i = fraction of biomass dedicated to component i
f_{bi} = fraction of component i which is degraded
S_{0i} = influent concentration of component i, g/l
a_i = yield coefficient of component i, g VSS/g COD

Under most circumstances encountered in practice, the volatile fraction of the overall influent COD is rather small. Hence Eq. (8-29) can be reduced to:

$$\alpha_i = a_i f_{bi} S_{0i} (a_T S_{0T})^{-1} \qquad (8\text{-}30)$$

where:

a_T = yield coefficient for total waste, g VSS/g COD
S_{0T} = influent total organic concentration, g COD/l

The Monod kinetic model is used to describe biodegradation of the individual compounds:

$$r_i = \frac{\mu_{\max i}}{a_i} \frac{S_{ei}}{K_{si} + S_{ei}} \cdot \alpha_i X_v \cdot V \qquad (8\text{-}31)$$

where:

$\mu_{\max i}$ = maximum growth rate of microorganisms dedicated to component i, day^{-1}
K_{si} = half saturation concentration for component i, g/l
X_v = overall biomass concentration (MLVSS), g/l
V = volume of aeration basin, l

The expression for the rate of stripping depends on the type of aeration used. For surface aeration this expression is:

$$v_i = (K_L a)_i (S_{ei} - S_{si}) V \qquad (8\text{-}32)$$

where:

$(K_L a)_i$ = mass transfer coefficient for component i, sec^{-1}
S_{si} = saturation concentration of component i, g/l

For a diffused air system:

$$v_i = Q_A H_c [1 - \exp(-\phi')] S_{ei} \qquad (8\text{-}33)$$

where:

Q_A = air flow rate, $1/\text{sec}$
H_c = Henry's constant
ϕ' = air saturation parameter

By substituting the approximate expression for v_i and r_i into Eq. (8-28), the fraction of a volatile component that is stripped to the air can be obtained. Hence, for a surface aeration system, assuming $S_{si} = 0$, the fraction stripped is:

$$f_{si} = \frac{(K_L a)_i}{\left[\dfrac{\mu_{\text{maxi}} \alpha_i X_v}{a_i (K_{si} + S_{ei})} + (K_L a)_i + \dfrac{1}{t} \right]} \qquad (8\text{-}34)$$

where:

t = hydraulic residence time, sec

For a diffused air system, if complete saturation is assumed, the stripped fraction is:

$$f_{si} = \frac{Q_A H_c}{\left[\dfrac{\mu_{\text{maxi}} \alpha_i X_v V}{a_i (K_{si} + S_{ei})} + Q_A H_c + Q \right]} \qquad (8\text{-}35)$$

Solving for the stripped fraction f_{si} by either Eq. (8-34) or (8-35) requires a trial and error calculation since α_i is dependent on f_{si}.

STRIPPING IN ACTIVATED SLUDGE EXAMPLE

Calculate the fraction of benzene stripped to the air under the following conditions:

Flow rate, $Q = 178 \ 1/\text{sec}$ (4.0 mgd)
Aeration basin volume, $V = 3.87 \cdot 10^6 \ 1$ (1 mil gal)
Aeration power, $P = 60$ hp
Oxygen transfer capacity = 1.5 lb/hp hr
Dissolved oxygen in aeration basin = 2.0 mg/l
Influent total COD, $S_{0T} = 0.25$ g/l
Influent benzene $S_{0i} = 0.01$ g/l
Effluent benzene $S_{ei} = 0.02 \cdot 10^{-3}$ g/l
Aeration MLVSS, $X_v = 3.0$ g/l

Half saturation concentration for benzene, $K_{si} = 1.0 \cdot 10^{-3}$ g/l
$K_L a$ ratio (benzene/oxygen), $\psi = 0.6$
Maximum growth rate for "benzene organisms," $\mu_{max\,i} = 3.47 \cdot 10^{-6}$ sec^{-1}
Yield coefficient for all organisms, a_i, $a_T = 0.6$
COD for benzene 3.08 g/g
Effluent COD $= 0.02$ g/l

Solution:

1. From the oxygen transfer data:

$$(K_L a)_0 = \frac{1.5 \text{ lb}}{\text{hp} \cdot \text{hr}} \cdot \frac{60 \text{ hp}}{(9.2 - 2.0) \text{ mg}} \frac{1}{3.87 \cdot 10^6 \text{ l}} \frac{1}{3,600} \frac{\text{hr}}{\text{sec}}$$

$$\cdot \frac{454,000 \text{ mg}}{\text{lb}}$$

$$= 0.407 \cdot 10^{-3} \text{ sec}^{-1}$$

$$(K_L a)_{\text{benzene}} = 0.6 \cdot 0.407 \cdot 10^{-3}$$

$$= 0.244 \cdot 10^{-3} \text{ sec}^{-1}$$

Assume 10 percent stripped, hence from Eq. (8-30):

$$\alpha_i = \frac{0.6 \cdot (1 - 0.1) \cdot 0.01 \cdot 3.08}{0.6 \cdot (0.25 - 0.02)}$$

$$= 0.12$$

The fraction emitted to air is calculated from Eq. (8-34):

$$f_{si} = \frac{0.244 \cdot 10^{-3}}{\dfrac{3.47 \cdot 10^{-6} \cdot 0.12 \cdot 3.0}{0.6(1.0 \cdot 10^{-3} + 0.02 \cdot 10^{-3})} + 0.244 \cdot 10^{-3} + \dfrac{178}{3.87 \cdot 10^6}}$$

$$= 0.104 \text{ (10.4 percent, close to the assumed value of 10 percent)}$$

Results of similar calculations, for both diffused and surface aeration are shown in Fig. 8-6. Considerably higher emissions are expected in surface aeration systems when compared with diffused air.

STEAM STRIPPING

In wastewater treatment, steam stripping is normally used to separate and recover volatile organics from concentrated waste streams. It is usually considered a

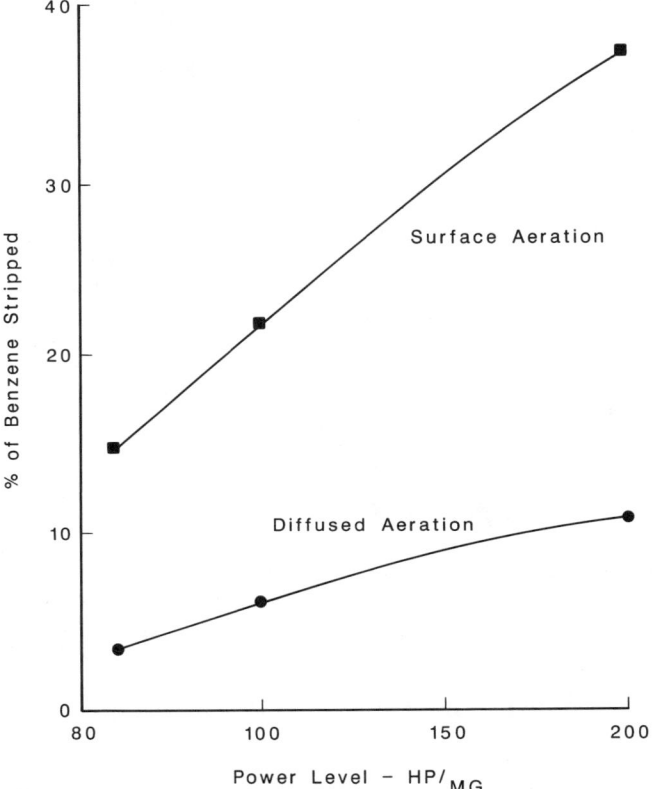

Fig. 8-6. Stripping of benzene in the activated sludge process as related to power level.

pretreatment process to be followed by some other conventional process for effluent polishing.

A typical flow diagram of the steam stripping process is presented in Fig. 8-7. Preheated raw wastewater is introduced at the top of the stripping tower and flows downward, countercurrent to the rising vapors. The overhead vapor is condensed and passed into a decanter where the recovered organics are separated from the aqueous phase. The stripped waste exits at the bottom of the tower. Reflux of the decanted condensate may or may not be practiced depending on the desired composition of the overhead stream.

Many volatile organics form low-boiling azeotropes with water. The azeotrope composition is the limiting attainable composition of the overhead in a steam stripping operation. Miscibility of the organic with water will determine

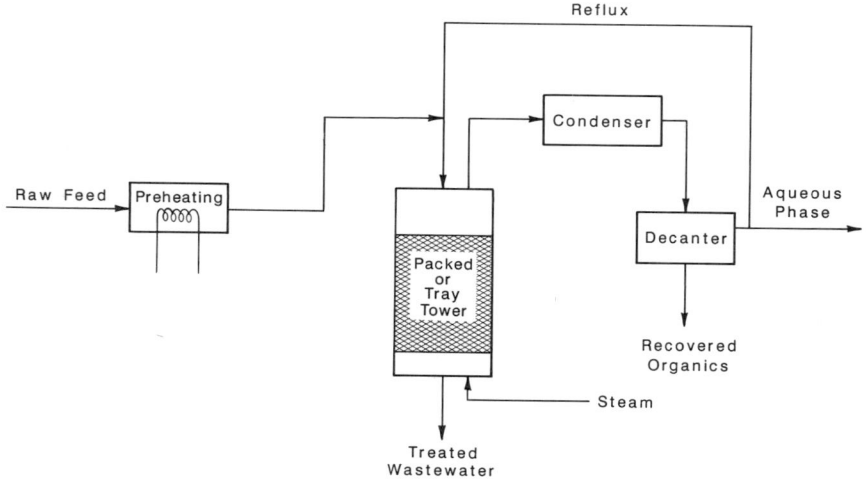

Fig. 8-7. Typical steam stripping process.

its distribution between the organic and aqueous phases. Formation of azeotropes allows separation at lower temperatures, thus avoiding the problems which may be associated with polymerization or destruction of organics.

Design and Operation Parameters

Steam stripping is typically operated isothermally and the feed is preheated to the boiling point, thus minimizing steam requirements and, consequently, the treated waste volume.

Separation can be enhanced by refluxing a portion of the condensed vapor and returning it to the tower. Introducing the feed at a lower level in the tower can also increase the concentration of organics in the overhead vapor.

Packed towers are less expensive, have low pressure drop and are preferred for treatment of corrosive, foaming, or viscous liquids, when compared to tray towers. Tray towers are more predictable in performance and more flexible in operation.

Common tower dimensions range from 1 to 10 feet in diameter, and 10 to 200 feet in height. Throughputs as high as 0.5 mgd have been reported.

Steam requirements range from 10 to 30 percent of feed on a molar basis.

Demonstrated Performance

Results from a U.S. EPA survey of full-scale plants in the pesticides and the organic chemicals industries are presented in Table 8-8. As seen in this table,

Table 8-8.   Full-Scale Industrial Steam Stripper Performance Summary.

Plants Using Steam Stripping	Stripped Compound	Henry's Constant atm	Vapor Pressure mm Hg @ 25°C	Concentration, ppm		Percent Removal
				Influent	Effluent	
Pesticide Industry						
Plant 1	Methylene chloride	177	425	<159	<0.01	99.9
Plant 2	Chloroform	188	180	70.0	<5.0	>92.6
Plant 3	Toluene	370	29	721	43.4	94.0
Organic Chemicals Industry						
Plant 4	Benzene	306	74	<15.4	<0.230	98.5
Plant 5	Methylene chloride	177	425	<3.02	<0.0141	99.5
	Tolune	370	29	178	<52.8	>70.3
Plant 6	Methylene chloride	177	425	1,430	<0.0153	>99.99
	Carbon tetrachloride	1,280	113	<665	<0.0549	>99.99
	Chloroform	188	180	<8.81	1.15	<86.9
Plant 6b	Methylene chloride	177	425	4.73	<0.0021	>99.95
	Chloroform	188	180	<18.6	<1.9	89.8
	1,2-Dichloroethane	62	82	<36.2	<4.36	88.0
	Carbon tetrachloride	1,280	113	<9.7	<0.030	99.7
	Benzene	306	74	24.1	<0.042	>99.8
	Toluene	370	29	22.3	<0.091	>99.6
Plant 7	Methylene chloride	177	425	34	<0.01	>99.97
	Chloroform	188	180	4,509	<0.01	>99.99
	1,2-Dichloroethane	62	82	9,030	<0.01	>99.99

Adapted from: Breton et al. (1986).

not all compounds are removed at the same efficiency, and the same compound is not equally removable in different plants. Toluene, chloroform, and 1,2-dichloroethane appear to be less effectively removed compared with other compounds on the list. This cannot be explained by lowering Henry's constant alone. Vapor pressure may be a better parameter of stripability under certain conditions. This is particularly true in stripping towers that yield high overhead concentrations, where the Henry's law assumption of dilute solution does not apply.

REFERENCES

Breton, M. et al. 1986. Technical Resource Document: Treatment Technologies for Solvent Containing Wastes. EPA/600/2-86/-95.

Kavanaugh, M. C. and R. R. Trussell. 1980. Design of Aeration Towers to Strip Volatile Contaminants from Drinking Water. *J. AWWA*, **72**(12): 684–692.

Kincannon, D. F., and A. Fazel. 1986. Volatilization of Organics in Activated Sludge Reactors. *Proc. 41st Ind. Waste Conf., Purdue University.*

Koczwara, M. K., et al. 1987. Activated Sludge Treatment of Selected Aqueous Organic Hazardous Waste Compounds. *Proc. 42nd Ind. Waste Conf., Purdue University.*

Munz, C. and P. V. Roberts. 1989. Gas- and Liquid-Phase Mass Transfer Resistances of Organic Compounds During Mechanical Surface Aeration. *Water Research*, **23**: 589–601.

Perry, R. H., and C. H. Chilton. 1973. *Chemical Engineer's Handbook*, 5th edition. New York: McGraw-Hill Book Company.

Roberts, R. V. et al. 1984. Modeling Volatile Organic Solute Removal by Surface and Bubble Aeration. *J. WPCF*, **56**(2): 157–163.

Weber, W. J., and B. Jones. 1983. Toxic Substance Removal in Activated Sludge and PAC Treatment Systems. U.S. EPA NTIS No. PB-86-18242J/AS.

9

Granular Carbon Adsorption of Toxics

W. Wesley Eckenfelder, Jr.

PRINCIPLES OF ADSORPTION

In many cases, organics responsible for toxicity are either nonbiodegradable or refractory byproducts. These materials can frequently be removed by adsorption on an active-solid surface. The most commonly used adsorbent is activated carbon.

There are three principle types of adsorption: exchange, chemical, and physical. Exchange adsorption results from electrostatic attraction between absorbate ions and charged absorbent surface sites. Chemical adsorption results in the formation of a monomolecular layer of the adsorbate on the surface through forces of residual valence of the surface molecules. Physical adsorption results from molecular condensation in the capillaries of the solid. All three of these adsorption mechanisms are generally experienced simultaneously in adsorption applications.

There are essentially three consecutive steps in adsorption: film diffusion, pore diffusion, and adsorptive reaction. The adsorbate passes through the surface film to the adsorbent surface (film diffusion). The adsorbate then passes through the adsorbent pores (pore diffusion). Lastly, the adsorbate binds to the adsorbent surface (adsorptive reaction). In continuous flow carbon columns, film diffusion or pore diffusion may be the rate limiting step of adsorption. The rate of adsorption:

- Varies reciprocally with the square of the particle diameter if pore diffusion controls
- Increases with increasing concentration of solute
- Increases with increasing temperature

- Decreases with increasing molecular weight of the solute if pore diffusion controls
- Increases with decreasing pH because of changes in surface charges of the carbon.

The adsorptive capacity of a carbon for a solute will likewise be dependent on both the carbon and the solute.

Most wastewaters are highly complex and vary widely in the adsorbability of all the compounds present. Molecular structure, solubility, etc. all affect the adsorbability as follows:

- An increasing solubility of the solute in the liquid carrier decreases its adsorbability
- Branched chains are usually more adsorbable than straight chains; an increasing length of the chain decreases solubility and thereby increases adsorbability
- As adsorbate polarity increases, water solubility increases and therefore adsorbability decreases
- Constituent chemical groups affect adsorbability (see Table 9-1)
- Generally, strong ionized solutions are not as adsorbable as weakly ionized ones; i.e., undissociated molecules are in general preferentially adsorbed
- The amount of hydrolytic adsorption depends on the ability of the hydrolysis to form an adsorbable acid or base
- Unless the screening action of the carbon pores intervene, large molecules are more sorbable than small molecules of similar chemical nature; this is attributed to more solute carbon chemical bonds being formed, making desorption more difficult
- Molecules with low polarity are more adsorbable than highly polar ones
- The adsorptive capacity of a compound is decreased in the presence of competing compounds

Table 9-1. Effect of Constituent Groups on Adsorbability.

Constituent Group	Nature of Influence
Hydroxyl	Generally reduces adsorbability. Extent of decrease depends on structure of host molecule.
Amino	Effect similar to that of hydroxyl but somewhat greater. Many amino acids are not adsorbed to any appreciable extent.
Carbonyl	Effect varies according to host molecule. Glyoxylic acid more adsorbable than acetic, but similar increase does not occur when introduced into higher fatty acids.
Double bonds	Variable effects as with carbonyl.
Halogens	Variable effects.
Sulfonic	Usually decreases adsorbability.
Nitro	Often increases adsorbability.

Source: Eckenfelder (1989).

The effects of specific chemical groups on carbon adsorbability are shown in Table 9-1. The effectiveness of a carbon in removing selected contaminants can be predicted using equilibrium adsorption isotherms developed from batch tests. The Freundlich isotherm is usually used to describe adsorption phenomena involving industrial wastewater applications:

$$X/M = KC^{1/n}$$

$$\log X/M = \log K + 1/n \log C \qquad (9\text{-}1)$$

where:

   X = amount of impurity adsorbed
   M = weight of carbon
   C = equilibrium concentration of impurity in solution
   K, n = constants

This relationship may be graphically linearized by plotting the logarithm of the quantity of contaminant adsorbed (X/M) as a function of the logarithm of the equilibrium contaminant concentration (c), such as BOD, COD, TOC, a pure chemical, or color. The presentation of data obtained from the Freundlich isotherm is shown in Fig. 9-1. The use of this relationship may be limited for concentrated and complex wastewaters where a significant portion of the organic impurities may not be amenable to adsorption. This results in a constant residual regardless of the carbon dosage.

In Eq. (9-1), the constants K and n are indicative of the adsorbability of the wastewater constituents. Generally:

- n and K decrease with increasing wastewater complexity
- High K and high n values indicate high adsorption throughout the concentration range studied
- A low K and high n indicates a low adsorption throughout the concentration range studied
- A low n value, or steep slope, indicates high adsorption at strong solute concentrations and low adsorption at dilute solute concentrations

Freundlich parameters for several priority pollutants are summarized in Table 9-2. The adsorbability of various organic compounds is shown in Table 9-3. In general, carbon capacity (which increases with the concentration) is higher when determined from a continuous flow carbon column study than when determined from batch isotherms, as shown in Fig. 9-2. This can be attributed to a continuously high concentration gradient at the interface of the adsorption zone in the column versus the decreasing concentration gradient over time in the batch isotherm test. In addition, biodegradable organics may degrade after adsorption in the column, thus increasing the apparent capacity of the carbon.

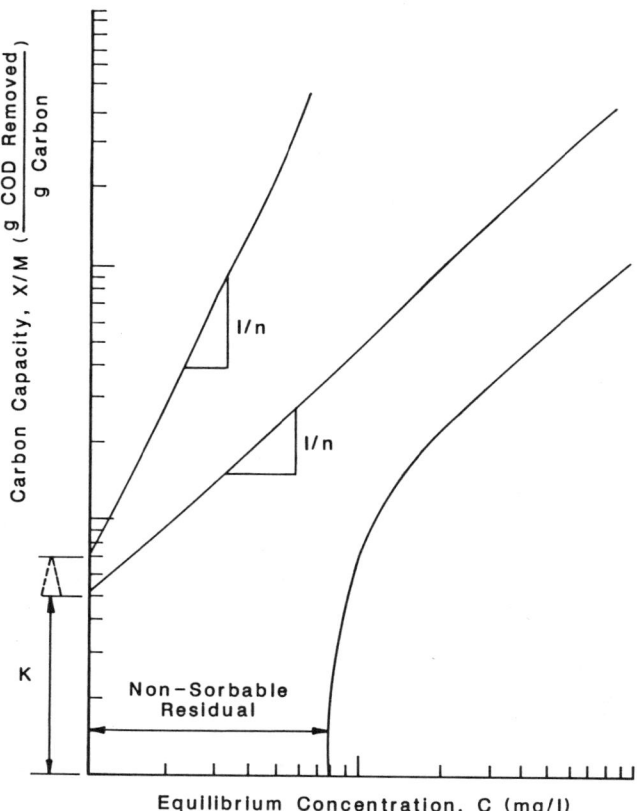

Fig. 9-1. Freundlich adsorption isotherm.

Table 9-2. Summary of Freundlich Parameters at Neutral pH.

Compound	K, mg/g	$1/n$
Hexachlorobutadiene	360	0.63
Anethole	300	0.42
Phenyl mercuric acetate	270	0.44
p-Nonylphenol	250	0.37
Acridine yellow	230	0.12
Benzidine dihydrochloride	220	0.37
n-Butylphthalate	220	0.45
N-Nitrosodiphenylamine	220	0.37
Dimethylphenylcarbinol	210	0.33
Bromoform	200	0.83

Table 9-2.   (*Continued*)

Compound	K, mg/g	$1/n$
β-Naphthol	100	0.26
Acridine orange	180	0.29
α-Naphthol	180	0.31
α-Naphthylamine	160	0.34
Pentachlorophenol	150	0.42
p-Nitroaniline	140	0.27
1-Chloro-2-nitrobenzene	130	0.46
Benzothiazole	120	0.27
Diphenylamine	120	0.31
Guanine	120	0.40
Styrene	120	0.56
Dimethyl phthalate	97	0.41
Chlorobenzene	93	0.98
Hydroquinone	90	0.25
p-Xylene	85	0.16
Acetophenone	74	0.44
1,2,3,4-Tetrahydronaphthalene	74	0.81
Adenine	71	0.38
Nitrobenzene	68	0.43
Dibromochloromethane	63	0.93

Source: U.S. EPA (1973).

Table 9-3.   Amenability of Selected Organic Compounds to Activated Carbon Adsorption.

Compound	Molecular Weight	Aqueous Solubility, %	Concentration, mg/l Initial, C_0	Concentration, mg/l Final, C_f	Adsorbability[a] g Compound/ g carbon	Adsorbability[a] Percent Reduction
Alcohols						
Methanol	32.0	∞	1,000	964	0.007	3.6
Ethanol	46.1	∞	1,000	901	0.020	10.0
Propanol	60.1	∞	1,000	811	0.038	18.9
Butanol	74.1	7.1	1,000	466	0.107	53.4
n-Amyl alcohol	88.2	1.7	1,000	282	0.155	71.8
n-Hexanol	102.2	0.58	1,000	45	0.191	95.5
Isopropanol	60.1	∞	1,000	874	0.025	12.6
Allyl alcohol	58.1	∞	1,010	789	0.024	21.9
Isobutanol	74.1	8.5	1,000	581	0.084	41.9
t-Butanol	74.1	∞	1,000	705	0.059	29.5
2-Ethyl butanol	102.2	0.43	1,000	145	0.170	85.5
2-Ethyl hexanol	130.2	0.07	700	10	0.138	98.5
Aldehydes						
Formaldehyde	30.0	∞	1,000	908	0.018	9.2
Acetaldehyde	44.1	∞	1,000	881	0.022	11.9

Table 9-3.   (*Continued*)

Compound	Molecular Weight	Aqueous Solubility, %	Concentration, mg/l Initial, C_0	Concentration, mg/l Final, C_f	Adsorbability[a] g Compound/ g carbon	Adsorbability[a] Percent Reduction
Aldehydes						
Propionaldehyde	58.1	22	1,000	723	0.057	27.7
Butyraldehyde	72.1	7.1	1,000	472	0.106	52.8
Acrolein	56.1	20.6	1,000	694	0.061	30.6
Crotonaldehyde	70.1	15.5	1,000	544	0.092	45.6
Benzaldehyde	106.1	0.33	1,000	60	0.188	94.0
Paraldehyde	132.2	10.5	1,000	261	0.148	73.9
Amines						
Di-N-Propylamine	101.2	∞	1,000	198	0.174	80.2
Butylamine	73.1	∞	1,000	480	0.103	52.0
Di-N-Butylamine	129.3	∞	1,000	130	0.174	87.0
Allylamine	57.1	∞	1,000	686	0.063	31.4
Ethylenediamine	60.1	∞	1,000	893	0.021	10.7
Diethylenetriamine	103.2	∞	1,000	706	0.062	29.4
Monethanolamine	61.1	∞	1,012	939	0.015	7.2
Diethanolamine	105.1	95.4	996	722	0.057	27.5
Triethanolamine	149.1	∞	1,000	670	0.067	33.0
Monoisopropanolamine	75.1	∞	1,000	800	0.040	20.0
Disopropanolamine	133.2	87	1,000	543	0.091	45.7
Pyridines and Morpholines						
Pyridine	79.1	∞	1,000	527	0.095	47.3
2-Methyl-5-ethyl pyridine	121.2	slight	1,000	107	0.179	89.3
N-Methyl morpholine	101.2	∞	1,000	575	0.085	42.5
N-Ethyl morpholine	115.2	∞	1,000	467	0.107	53.3
Aromatics						
Benzene	78.1	0.07	416	21	0.080	95.0
Toluene	92.1	0.047	317	66	0.050	79.2
Ethyl benzene	106.2	0.02	115	18	0.019	84.3
Phenol	94	6.7	1,000	194	0.161	80.6
Hydroquinone	110.1	6.0	1,000	167	0.167	83.3
Aniline	93.1	3.4	1,000	251	0.150	74.9
Styrene	104.2	0.03	180	18	0.028	88.8
Nitrobenzene	123.1	0.19	1,023	44	0.196	95.6
Esters						
Methyl acetate	74.1	31.9	1,030	760	0.054	26.2
Ethyl acetate	88.1	8.7	1,000	495	0.100	50.5
Propyl acetate	102.1	2	1,000	248	0.149	75.2
Butyl acetate	116.2	0.68	1,000	154	0.169	84.6
Primary amyl acetate	130.2	0.2	985	119	0.175	88.0
Isopropyl acetate	102.2	2.9	1,000	319	0.137	68.1
Isobutyl acetate	116.2	0.63	1,000	180	0.164	82.0
Vinyl acetate	86.1	2.8	1,000	357	0.129	64.3

Table 9-3. (*Continued*)

Compound	Molecular Weight	Aqueous Solubility, %	Concentration, mg/l Initial, C_0	Concentration, mg/l Final, C_f	Adsorbability[a] g Compound/ g carbon	Adsorbability[a] Percent Reduction
Esters						
Ethylene glycol monoethyl ether acetate	132.2	22.9	1,000	342	0.132	65.8
Ethyl acrylate	100.1	2.0	1,015	226	0.157	77.7
Butyl acrylate	128.2	0.2	1,000	43	0.193	95.9
Isopropyl ether	102.1	1.2	1,023	203	0.162	80.0
Butyl ether	130.2	0.03	197	nil	0.039	100.0
Dichloroisopropyl ether	171.1	0.17	1,008	nil	0.200	100.0
Glycols and Glycol Ethers						
Ethylene glycol	62.1	∞	1,000	932	0.0136	6.8
Diethylene glycol	106.1	∞	1,000	738	0.053	26.2
Triethylene glycol	150.2	∞	1,000	477	0.105	52.3
Tetraethylene glycol	194.2	∞	1,000	419	0.116	58.1
Propylene glycol	76.1	∞	1,000	884	0.024	11.6
Dipropylene glycol	134.2	∞	1,000	835	0.033	16.5
Hexylene glycol	118.2	∞	1,000	386	0.122	61.4
Ethylene glycol monomethyl ether	76.1	∞	1,024	886	0.028	13.5
Ethylene glycol monoethyl ether	90.1	∞	1,022	705	0.063	31.0
Ethylene glycol monobutyl ether	118.2	∞	1,000	441	0.112	55.9
Ethylene glycol monohexyl ether	146.2	0.99	975	126	0.170	87.1
Diethylene glycol monoethyl ether	134.2	∞	1,010	570	0.087	43.6
Diethylene glycol monobutyl ether	162.2	∞	1,000	173	0.166	82.7
Ethoxytriglycol	178.2	∞	1,000	303	0.139	69.7
Halogenated						
Ethylene dichloride	99.0	0.81	1,000	189	0.163	81.1
Propylene dichloride	113.0	0.30	1,000	71	0.183	92.9
Ketones						
Acetone	58.1	∞	1,000	782	0.043	21.8
Methylethyl ketone	72.1	26.8	1,000	532	0.094	46.8
Methylpropyl ketone	86.1	4.3	1,000	305	0.139	69.5
Methylbutyl ketone	100.2	very slight	988	191	0.159	80.7
Methylisobutyl ketone	100.2	1.9	1,000	152	0.169	84.8
Methylisoamyl ketone	114.2	0.54	986	146	0.169	85.2
Diisobutyl ketone	142.2	0.05	300	nil	0.060	100.0
Cyclohexanone	98.2	2.5	1,000	332	0.134	66.8
Acetophenone	120.1	0.55	1,000	28	0.194	97.2
Isophorone	138.2	1.2	1,000	34	0.193	96.6

Table 9-3. (*Continued*)

| Compound | Molecular Weight | Aqueous Solubility, % | Concentration, mg/l | | Adsorbability[a] | |
			Initial, C_0	Final, C_f	g Compound/ g carbon	Percent Reduction
Organic Acids						
Formic acid	46.0	∞	1,000	765	0.047	23.5
Acetic acid	60.1	∞	1,000	760	0.048	24.0
Propionic acid	74.1	∞	1,000	674	0.065	32.6
Butyric acid	88.1	∞	1,000	405	0.119	59.5
Valeric acid	102.1	2.4	1,000	203	0.159	79.7
Caproic acid	116.2	1.1	1,000	30	0.194	97.0
Acrylic acid	72.1	∞	1,000	355	0.129	64.5
Benzoic acid	122.1	0.29	1,000	89	0.183	91.1
Oxides						
Propylene oxide	58.1	40.5	1,000	739	0.052	26.1
Styrene oxide	120.2	0.3	1,000	47	0.190	95.3

Source: Adams et al. (1981).
[a]Dosage: 5 m/l carbon.

PROPERTIES OF ACTIVATED CARBON

Activated carbons are made from a variety of materials including wood, lignin, bituminous coal, lignite, and petroleum residues. Granular carbons produced from medium volatile bituminous coal or lignite have been most widely applied to the treatment of wastewater. Activated carbons have specific properties depending on the material source and the mode of activation. Property standards are helpful in specifying carbons for a specific application. In general, granular carbons from bituminous coal have a small pore size, a large surface area, and the highest bulk density as compared to lignite carbon, as shown in Table 9-4.

- Adsorptive capacity is the effectiveness of the carbon in removing desired constituents such as COD, color, phenol, etc., from the wastewater; several tests have been employed to characterize adsorptive capacity
- The phenol number is used as an index of a carbon's ability to remove taste and odor compounds
- The iodine number relates to the ability of activated carbon to adsorb low-molecular-weight substances (micropores having an effective radius of less than 2 μm)
- The molasses number relates to the carbon's ability to adsorb high-molecular substances (pores ranging from 1 to 50 μm)
- In general, high iodine numbers will be most effective on wastewaters with predominantly low-molecular-weight organics, while high molasses numbers will be most effective for wastewaters with a dominance of high-

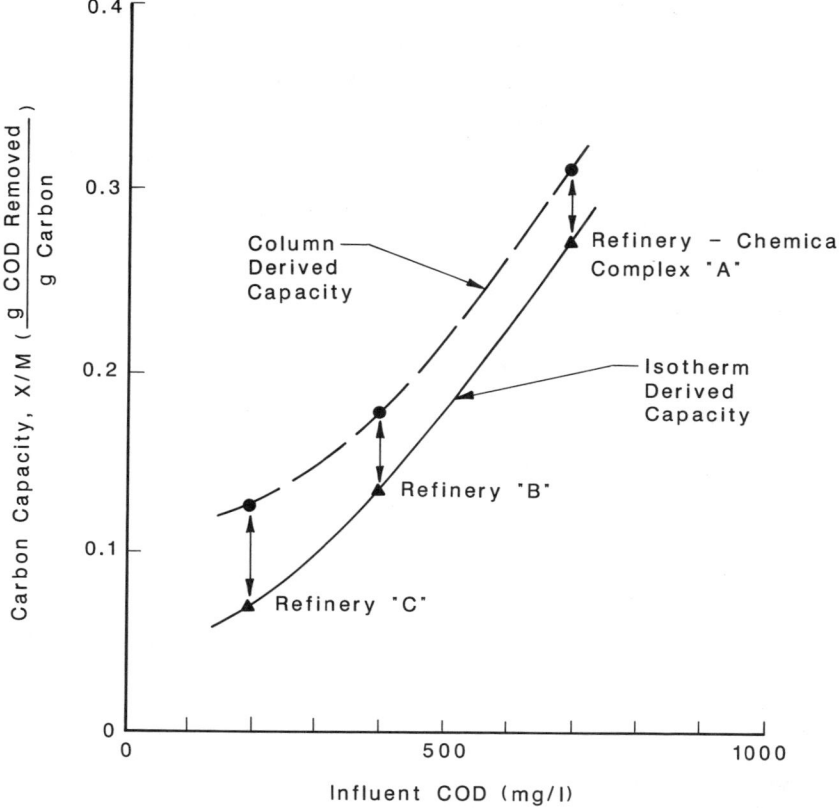

Fig. 9-2. Carbon capacity from batch and column systems. (*Source:* Adams, Ford and Eckenfelder 1981.)

molecular-weight organics; properties of commercial carbons are shown in Table 9-4

Depending on the characteristics of the wastewater, one type of carbon may be superior to another (Fig. 9-3) since the capacity is greater at equilibrium effluent concentrations. For TOC, carbon 2 is better for carbon column operation since the capacity is greater in equilibrium with the influent (C_0) at column exhaustion, while carbon 1 would be better for batch treatment.

Laboratory Evaluation of Alternative Carbons

In order to evaluate the feasibility and economics of adsorption, a laboratory adsorption study should be conducted. If granular carbon is to be evaluated, the

Table 9-4. Properties of Commercially Available Carbons.

	American Norit Hydrodarco 3,000 (lignite)	Calgon Filtrasorb 300 (8 × 30) (bituminous)	Westvaco Nuchar WV-L (8 × 30) (bituminous)	Witco 517 (12 × 30) (bituminous)
Physical properties				
Surface area, m^2/g (BET)	600–650	950–1,050	1,050	1,050
Apparent density, g/cm^3	0.43	0.48	0.48	0.48
Density, backwashed and drained, lb/ft^3	22	26	26	30
Real density, g/cm^3	2.0	2.1	2.1	2.1
Particle density, g/cm^3	1.4–1.5	1.3–1.4	1.4	0.92
Effective size, mm	0.8–0.9	0.8–0.9	0.85–1.05	0.89
Uniformity coefficient	1.7	≥ 1.9	≥ 1.8	≥ 1.44
Pore volume, cm^3/g	0.95	0.85	0.85	0.60
Mean particle diameter, mm	1.6	1.5–1.7	1.5–1.7	1.2
Specifications				
Sieve Size (U.S. standard series)				
Larger than No. 8 (max %)	8	8	8	[a]
Larger than No. 12 (max %)	[a]	[a]	[a]	5
Smaller than No. 30 (max %)	5	5	5	5
Smaller than No. 40 (max %)	[a]	[a]	[a]	[a]
Iodine No.	650	900	950	1,000
Abrasion No., minimum	[b]	70	70	85
Ash, %	[b]	8	7.5	0.5
Moisture as packed (max. %)	[b]	2	2	1

Source: U.S. EPA (1973).
[a]Not applicable to this size carbon.
[b]No available data from the manufacturer.
Note: $lb/ft^3 = 16\ kg/m^3$.

carbon must first be ground to pass a 325-mesh screen. Grinding the carbon will not significantly affect its adsorptive capacity, but will increase the rate of adsorption.

The time of contact required to approach equilibrium should first be evaluated. A carbon dosage of 500 mg/l should be contacted with the wastewater for various periods of time, and the degree of adsorption determined at selected

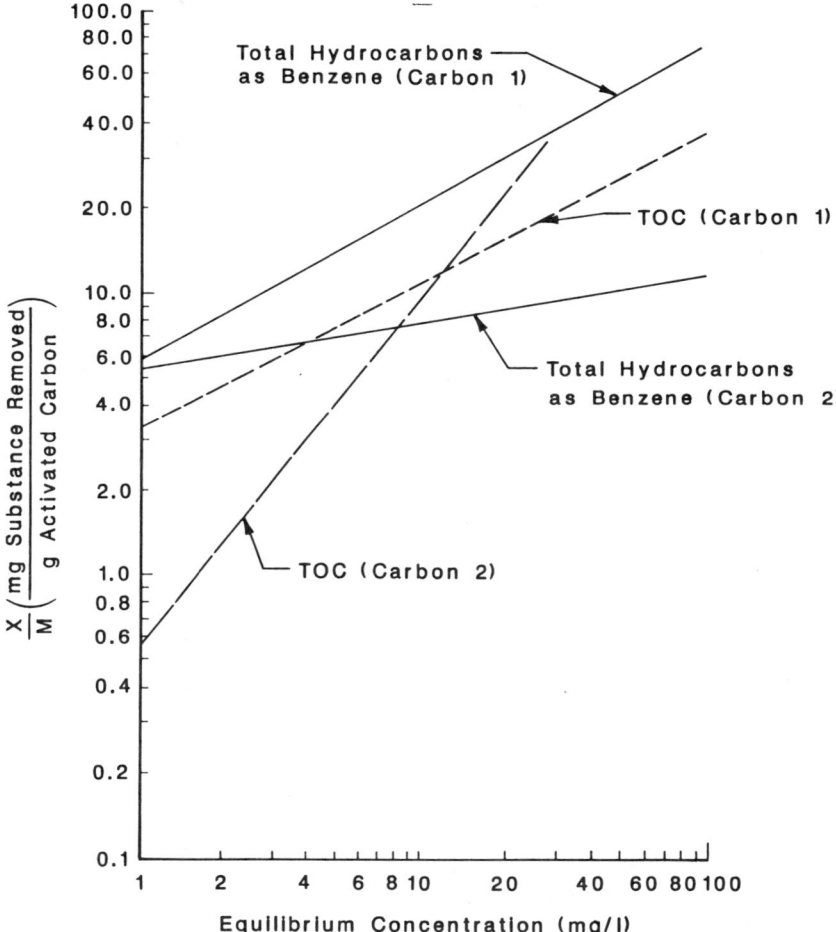

Fig. 9-3. Freundlich isotherms for total hydrocarbons and total organic carbon, contact time 48 hours. (*Source:* Eckenfelder 1989.)

time intervals. A mixing time sufficient to achieve 90 percent or more of equilibrium should be used for subsequent studies. Usually, a 2-hour contact is sufficient to attain greater than 90 percent of equilibrium, though in some cases a longer contact time is required. The initial testing should include a 24-hour contact time. If the equilibrium value after 2 hours is greater than 90 percent of the 24-hour value, the 2-hour test can be used. Depending on the wastewater in question, adsorption over a range of pH values may be of interest.

Laboratory-Scale Isotherm Procedure

Equipment:

- Shaker assembly or jar test apparatus (see Fig. 9-4)
- Selected powdered or crushed granular carbon
- Beakers, pipettes, and other glassware

Methodology:

1. Fill six to eight flasks (for shaker assembly) or beakers (for jar test assembly) with a fixed volume of wastewater. The initial concentration of organics in terms of BOD, COD, TOC, or TOD and specific pollutants of interest should be accurately determined.
2. Place varying weights of prewashed dried test carbon into each container, corresponding to 0.05 to 5.0 g TOC applied/g carbon. If any volatile compounds are present in the wastewater which might be air stripped during agitation, then a ''blank'' should be included to correct for this organic removal, consisting of one container with the wastewater but no activated carbon.
3. Agitate the flasks or beakers and monitor the organic concentration after the contact time selected above. The total sample withdrawal during the test should not exceed more than 5 percent of the total liquid volume. The best approach is to use micro-samples for TOC or TOD determination. All samples withdrawn for organic analysis should first be filtered.

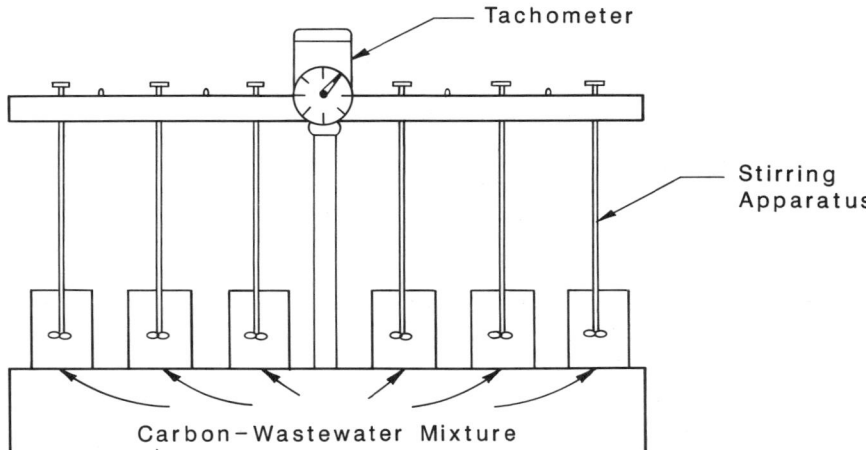

Fig. 9-4. Laboratory scale jar test assembly for carbon isotherm determination.

4. Once the contaminant level in the flask or beaker has stabilized, note the equilibrium concentration.

5. The equilibrium value C as a function of the value is plotted on log-log paper, with X/M as the ordinate and C as the abscissa. Each flask or beaker will represent one point on the plot. Calculations are shown in the following example.

ISOTHERM DETERMINATION EXAMPLE

Given the following data for the adsorption of naphthalene on Filtersorb 400 carbon:

Carbon Dose, mg/l	Effluent, mg/l
0	9.94
11.2	5.3
22.3	3.0
56.1	0.71
168.3	0.17
224.4	0.06

a. Develop the adsorption isotherm and estimate the Freundlich parameters.
b. Determine the adsorption capacity at an initial concentration of 10 mg/l?
c. Determine the carbon dose required to reduce the concentration from 1 mg/l to 0.01 mg/l?

Solution:

a. Freundlich adsorption isotherm: To correlate the adsorption data, the following table was developed in order to plot X/M vs. C on log-log paper:

Carbon M, mg/l	Adsorbed X, mg/l	Capacity X/M, mg/l	Equilibrium C, mg/l
0	0	—	9.94
11.2	4.64	0.414	5.3
22.3	6.94	0.311	3.0
56.1	9.23	0.165	0.71
168.3	9.77	0.058	0.17
224.4	9.88	0.044	0.06

From Figure 9-5, the constants K and n are obtained as follows:

$$\text{Slope} = 0.524 = 1/n, \quad \text{then } n = 1.91$$
$$\text{Intercept} = -0.755, \quad \text{then } K = 0.176$$

b. Adsorption capacity at an initial concentration of 10 mg/l: The isotherm can be written as:

$$(C_0 - C)/M = 0.176C^{0.524}$$

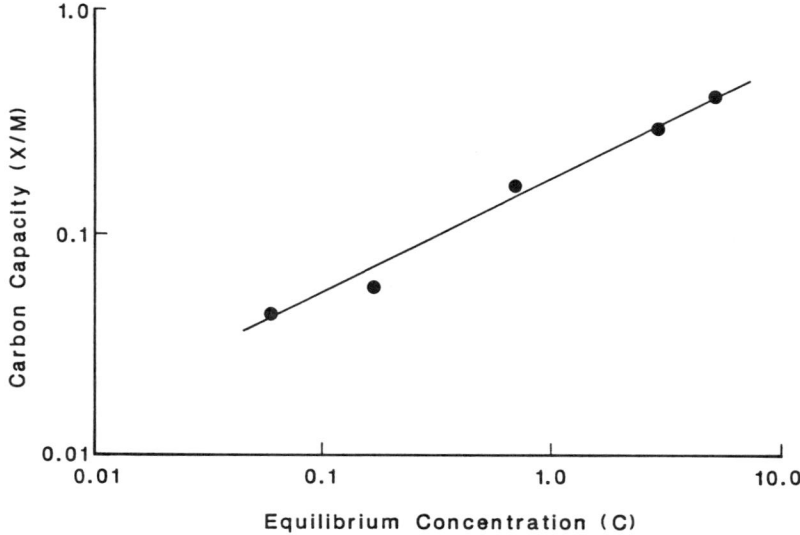

Fig. 9-5. Freundlich adsorption isotherm for example.

If $C_0 = 10$ mg/l and $M = 1$ mg/l, then C is 9.43 mg/l, and

$$X/M = (10 - 9.43)/1 = 0.57 \text{ mg/mg}$$

c. Carbon dose required to reduce concentration from 1 to 0.01 mg/l. From

$$X/M = 0.176C^{0.524}$$

$$= 0.176 \times 0.01^{0.524} = 0.0158 \text{ mg/mg}$$

M can be obtained as follows:

$$M = X/0.0158 = (1 - 0.01)/0.0158 = 64 \text{ mg/l}$$

ADSORPTION COLUMN SYSTEMS

The carbon column can be considered as a non-steady-state process in which, as an increasing quantity of water is passed through the bed, the adsorbent is removed in an increasing amount. As water initially passes through the upper-most layers, rapid adsorption occurs in equilibrium with the effluent concentration. As this water passes through the bed, the equilibrium shifts with the decreasing concentration of remaining solute; resulting in a substantially solute-free effluent. With continuing flow of water the adsorption zone, in equilibrium with the influent concentration, moves downward through the bed. As this zone approaches the bottom of the bed, the concentration of solute in the effluent increases. The breakthrough point is defined as the volume of water which can be passed through the bed before maximum effluent concentration is reached. As the adsorption zone falls to the bottom of the bed, the effluent concentration increases until it equals the influent concentration.

As a general rule, the breakthrough point:

- Decreases with decreased bed height
- Decreases with increasing particle size of adsorbent
- Decreases with increased flow rate
- Decreases with increased initial solute concentration

In granular carbon columns, the carbon capacity at breakthrough as related to exhaustion is a function of the waste complexity, as shown in Fig. 9-6. A single organic will yield a sharp breakthrough curve such that the column is greater than 90 percent exhausted when breakthrough occurs. By contrast, a multicomponent wastewater shows a drawn-out breakthrough curve due to varying rates of sorption and desorption.

Depending on the nature of the wastewater one of several column configurations may be employed:

1. Downflow. These are fixed beds in series. When breakthrough occurs in the last column, the first column is in equilibrium with the influent concentration (C_0) in order to achieve a maximum carbon capacity. After carbon replacement in the first column, it is reconfigured to become the last column in the series, etc. (Fig. 9-7). Downflow columns provide both adsorption and filtration. Consequently, these columns may require backwash capabilities.

2. Multiple units. These are operated in parallel with the effluent blended to achieve the final desired quality. The effluent from a column ready for regeneration or replacement, which is high in organics, is blended with the other effluents from fresh carbon columns to achieve the desired quality (Fig. 9-7). This mode of operation is most adaptable to waters in which the ratio of the capacity at breakthrough to the capacity at exhaustion is near 1.0.

3. Upflow. Expanded beds are used when suspended solids are present in the influent or when biological action occurs in the bed (Fig. 9-7).

4. Continuous counterflow. These are column or pulsed beds in which the spent carbon from the bottom (in equilibrium with influent solute concentration) sent to regeneration. Since this design cannot be backwashed, residual biodegradable organic content in the influent should be very low to avoid plugging. Regenerated and makeup carbon is fed to the top of the reactor (Fig. 9-7).

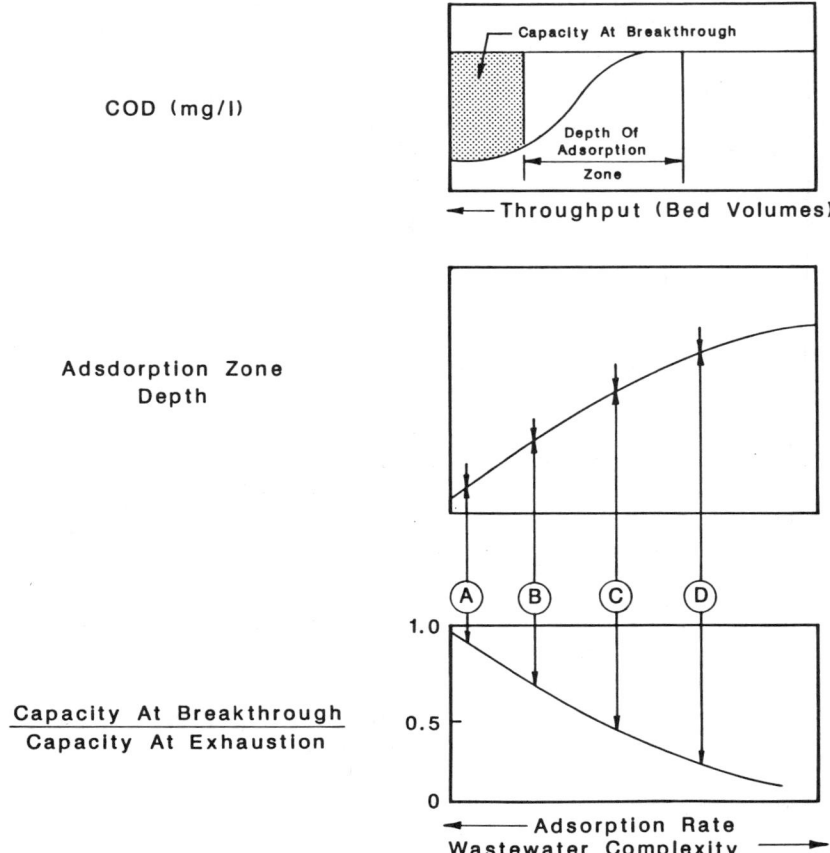

Fig. 9-6. Influence of wastewater complexity on carbon capacity
   A — one pure organic compound.
   B — sewage, activated sludge effluent.
   C — refinery activated sludge effluent.
   D — Refinery primary effluent.

5. Countercurrent two-bed series. The two beds are arranged so that the gravity, open-top structures are operated in a series upflow "roughing" contactor and a downflow "polishing" contactor. Once breakthrough occurs, the pair of columns are taken off line, the spent upflow column regenerated, and the unused capacity of the downflow column is used by reversing the flow and employing it as the upflow reactor, using the former upflow column containing regenerated carbon as the downflow polishing unit.

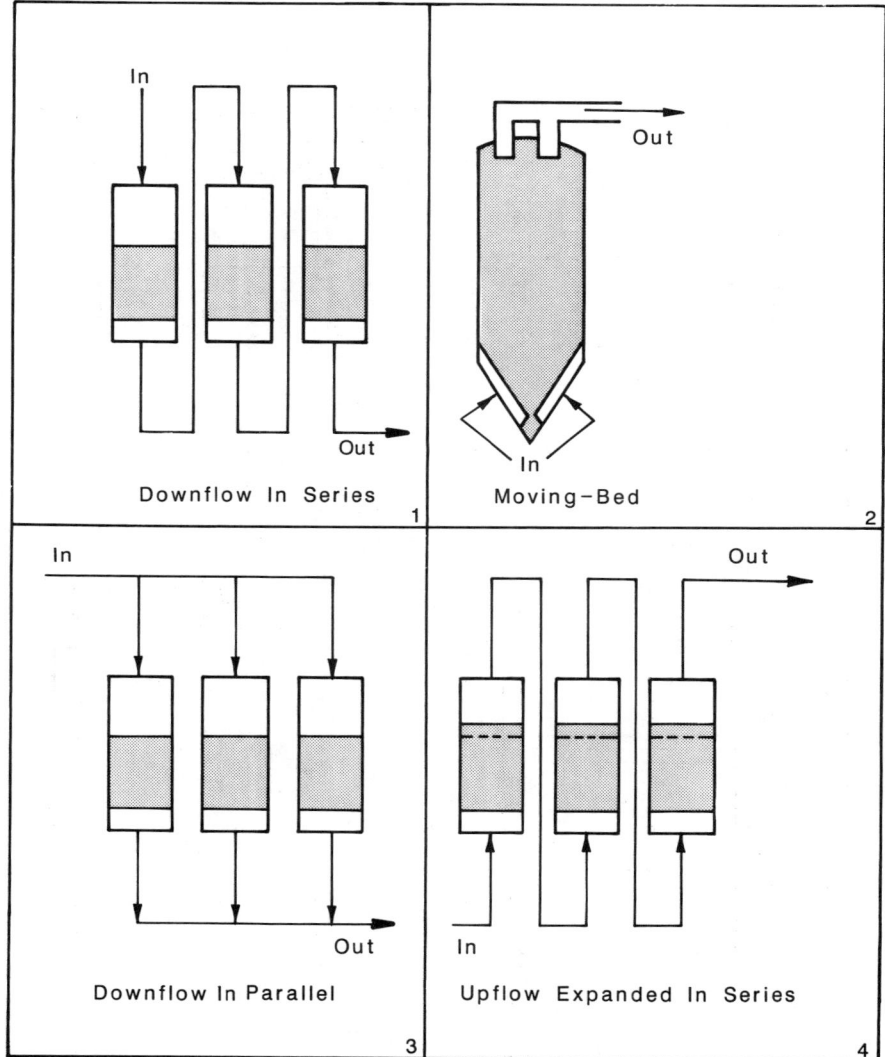

Fig. 9-7. Alternative GAC column configurations. (*Source:* Calgon Carbon Corporation.)

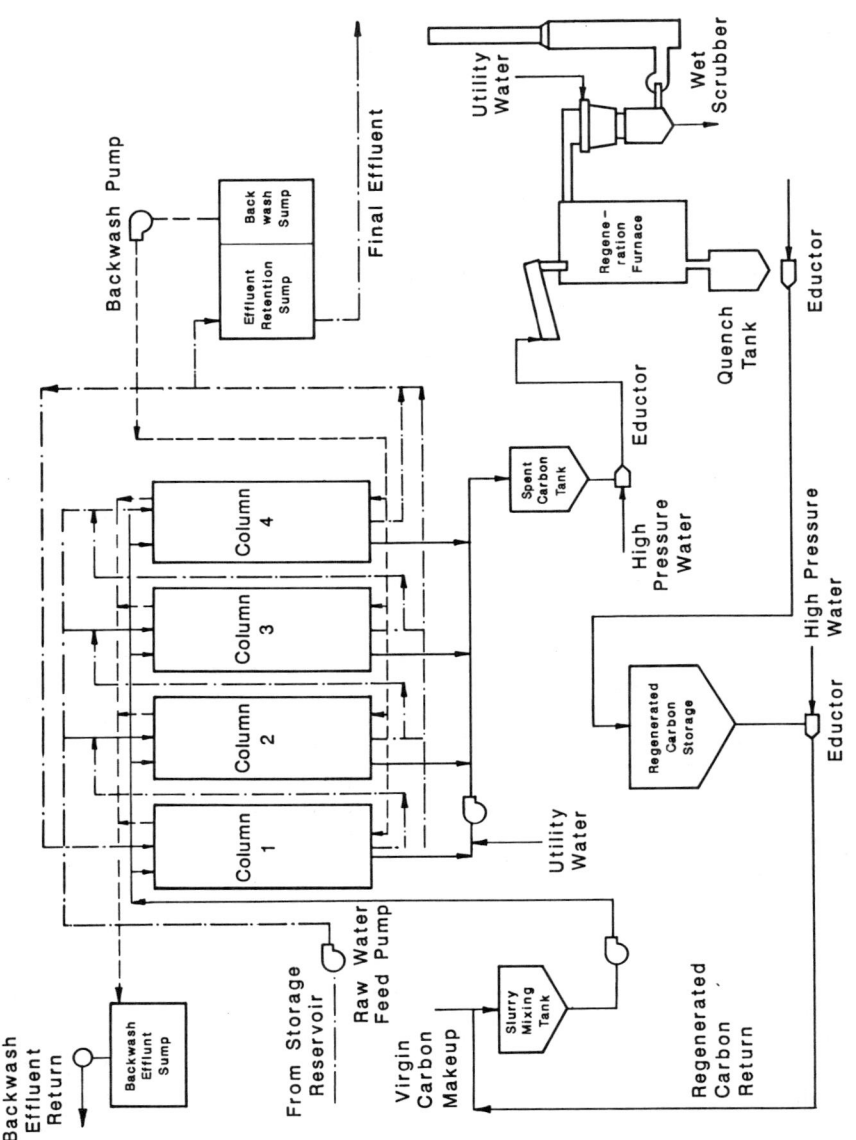

Fig. 9-8. GAC process flowsheet.

Bohart and Adams (Dale et al. 1946) developed a relationship, based on a surface-reaction-rate theory, which can be used to predict the performance of continuous carbon columns. The relationship can be expressed in a simplified form:

$$t = \frac{N_0 X}{C_0 v} - \frac{1}{KC_0} \ln \frac{C_0}{C_B} - 1 \tag{9-2}$$

where:

t = service time, hr
v = linear flow rate, ft/hr
X = depth of bed, ft
K = rate constant, 1/mg hr
N_0 = adsorptive capacity, mg/l
C_0 = influent concentration
C_B = allowable effluent concentration

The bed depth, which theoretically is just sufficient to prevent penetration of a concentration in excess of C_B at time zero, is defined as the critical bed depth and is determined from Eq. (9-2) when $t = 0$:

$$X_0 = \frac{v}{KN_0} \ln \frac{C_0}{C_B} - 1 \tag{9-3}$$

$$b = -\frac{1}{C_0 K} \ln \frac{C_0}{C_B} - 1 \tag{9-4}$$

Hutchins (1973) presented a modification of a Bohart-Adams equation which requires only three column tests to collect the necessary data. This is called the bed depth-service time (BDST) approach. From Eq. (9-2) it can be shown that the adsorptive capacity N_0 can be determined from the slope of a linear plot of t versus X. The rate constant K is then calculated from the intercept of this plot. The Bohart-Adams equation can be expressed as

$$t = aX + b \tag{9-5}$$

where:

a = slope = $N_0/C_0 v$
b = intercept = $-(1/KC_0) \ln (C_0/C_B - 1)$

If the value of a is determined for one flow rate, values for other flow rates can be computed by multiplying the original slope by the ratio of the original and new flow rates assuming C_0 remains unchanged. The b value change is insignificant with respect to changing flow rates. Adjustment for changing initial a and b values can be made as follows:

$$a_2 = a_1 \frac{Q_1}{Q_2}$$

$$b_2 = b_1 \frac{C_1 \ln (C_2/C_F - 1)}{C_2 \ln (C_1/C_B - 1)}$$

C_F and C_B are the effluent concentrations at C_2 and C_1, respectively. In order to develop a BDST correlation, a number of pilot columns of equal depth are operated in series and breakthrough curves plotted for each, as shown in Fig. 9-9. These data are then used to plot a BDST correlation by recording the operating time required to reach a certain removal at each depth. A BDST plot of the data from Fig. 9-9 is shown in Fig. 9-10.

The slope of the BDST line is equal to the reciprocal velocity of the adsorption zone and the X intercept is the critical depth defined as the minimum bed depth required to obtain the desired effluent quality at time zero.

If the adsorption zone is arbitrarily defined as the carbon layer through which the liquid concentration varies from 90 percent of the feed concentration to the desired effluent concentration, then this zone is defined by the horizontal distance between these two lines on the BDST plot.

In order to design an adsorption system with maximum carbon utilization, the carbon removed should be near saturation (e.g., in equilibrium with the

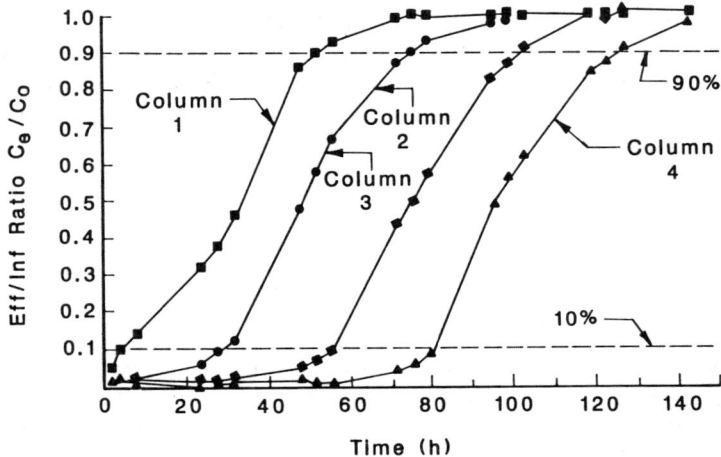

Fig. 9-9. Column breakthrough curves.

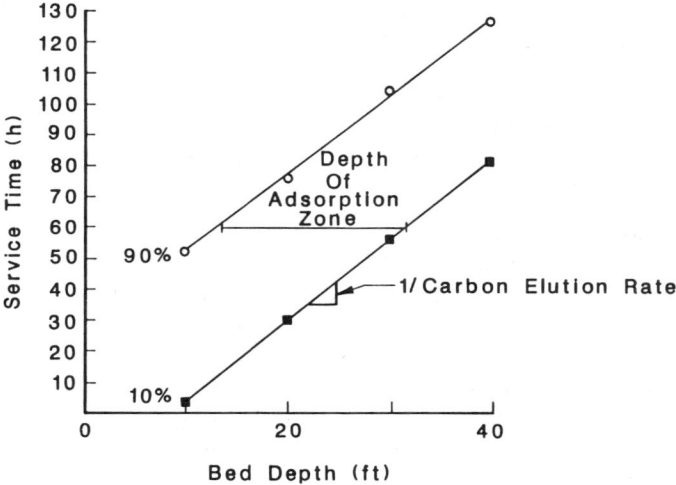

Fig. 9-10. Bed depth-service time design curves.

influent concentration). In multistage columns, the first run of three columns will be made with all fresh carbon. When the third column breaks through, the first column should be exhausted and a fourth column with fresh carbon is placed at the end of the train. This process is repeated each time the last column breaks through.

The BDST curve should be developed from the breakthrough curves after the third column breaks through. Assuming that 90 percent removal represents exhaustion, the horizontal distance between 90 and the desired breakthrough concentration is taken as the depth of the adsorption zone. This is also the minimum depth in a pulsed or moving-bed system. For a multistage system, the number of stages and the bed depth in each stage are related to the depth of the adsorption zone:

$$n = \frac{D}{d} + 1 \qquad (9\text{-}6)$$

where:

n = number of stages in series
D = depth of adsorption zone
d = depth of single stage

Selection of d should be based on practical considerations and result in D/d being a whole number. Selecting a small d will result in small-size equipment with lower carbon inventory, but a high number of stages and consequently more costly equipment.

Pilot-Scale Column Procedure

Equipment Required:

- Three to six acrylic plastic or glass carbon columns depending on the wastewater constituent, three to six in. in diameter by approximately six ft in depth; connector piping, valves, pressure gauges, and pumps designed for series column operation with backwash capability; pressure gauges should be attached to each column (see Fig. 9-11)
- Selected granular carbon
- Filtration column (if required) using graded sand
- Sample collection capability from each column
- Flow measuring equipment

Methodology:

1. Set up the carbon columns in a manner similar to that shown in Fig. 9-11. They should be located close to the source of feed water to minimize pumping losses. A variable speed centrifugal pump is desirable for flexibility of operation. If the suspended solids or oil is high in the feed wastewater (>50 mg/1), then use one column as a prefilter.
2. The units should be piped and valved so that samples can be taken from each column and each column can be backwashed individually. Down-

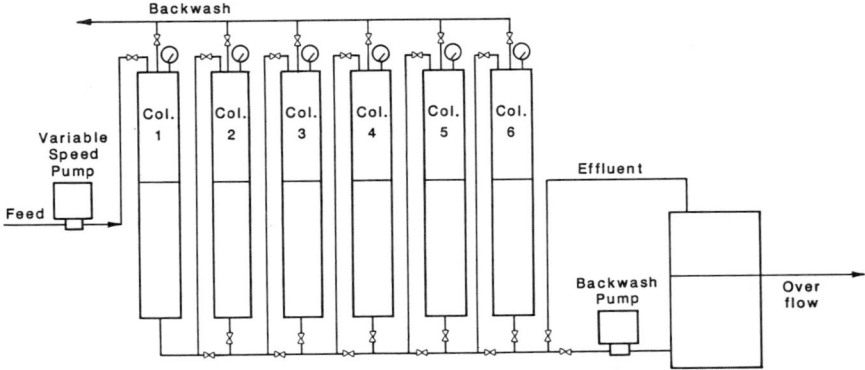

Fig. 9-11. Pilot scale carbon column assembly.

flow operations are normally used, although the upflow mode of operation can be investigated if desired.

3. Load the columns with the required amount of test carbon. Each column should have an equal amount. It may be helpful to paint the columns black or to cover them to prevent sunlight penetration during the test run. This will minimize algal growth in the columns and reduce backwash frequency.
4. Begin the run at the prescribed linear flow velocity and initial conditions. Monitor the effluent quality of each column by withdrawing samples at least once daily. If composite samples are required, electrically operated solenoid valves can be used to deliver the samples to compositing containers. The collected samples can be analyzed for any or all of the following parameters:
   a. BOD$_5$
   b. COD
   c. TOD
   d. TOC
   e. Phenolic materials
   f. Color
   g. Aquatic toxicity (bioassay)
   h. Total Kjeldahl nitrogen (TKN)
   f. Specific organic or inorganic constitutents
5. The effluent from the last column should be collected in a 55-gal drum which can serve as a reservoir for backwash water. Each column can be backwashed on a routine basis or at a specified pressure drop through the column.
6. The test should be continued until the breakthrough curves are fully developed: for example, the final column effluent concentration of the most slowly removed constituent approaches that of the first column influent concentration.

BED DEPTH SERVICE TIME DESIGN EXAMPLE

A petrochemical wastewater with a flow of 85,000 gal/d (322 m^3/d) has to be treated to 90 percent removal. A four-column pilot plant was operated with a carbon that had a density of 30 lb/ft^3 (482 kg/m^3). The columns were 10 ft (3 m) long and loaded at a hydraulic rate of 5 gal/(min · ft^2) [0.20 m^3/(min · m^2)]. The pilot plant was operated in series: the effluent from column 1 passed to the top of column 2 and sequentially to columns 3 and 4. Calculate the depth of the adsorption zone, the required number of columns, the time required to exhaust a column, the column diameter, the daily carbon use, and the carbon adsorption loading.

Solution:
The data should be plotted as sequential breakthrough curves (Fig. 9-9) of the bed effluent concentration as a ratio of the influent concentration of the system (630 mg/l).

The breakthrough curves for the four columns are symmetrical. The adsorption zone is defined as the carbon layer through which the concentration varies from 10 percent to 90 percent of the feed concentration. The breakthrough times are plotted as the service times and are a function of total carbon bed depth (Fig. 9-10).

In this case, the lines are almost parallel and the depth of the adsorption zone is between 18 and 19 ft. The total bed depth will be the adsorption zone plus one additional column:

$$\text{Number of columns} = \frac{19}{10} + 1 = 2.9, \text{ round up to } 3$$

The bed depth service time (Fig. 9-10) data fits well in the Bohart-Adams equation:

$$t = \frac{N_0}{C_0 v}(X) - \frac{1}{C_0 K} \ln \frac{C_0}{C_B} - 1$$

At 10 percent,

$$2.57(X) - 21.5$$

with correlation coefficient $(r) = 0.9999$; at 90 percent,

$$2.50(X) - 27.0$$

with correlation coefficient $(r) = 0.9981$.

Adsorption velocity $\qquad = \dfrac{1}{\text{slope}} = \dfrac{1}{2.57 \text{ h/ft}}$

$\qquad\qquad\qquad\qquad = 0.39$ ft/hr or 9.36 ft/d (2.85 m/d)

Time to exhaust a column $\quad = \dfrac{10 \text{ ft}}{9.36 \text{ ft/d}} \times 24$ hr/d $= 26$ hr

Area required $\qquad\qquad = \dfrac{Q}{A}$

$\qquad\qquad\qquad\qquad = \dfrac{85,000 \text{ gal/d}}{5 \text{ gal/(min} \cdot \text{ft}^2)} \times \dfrac{\text{d}}{1,440 \text{ min}}$

$\qquad\qquad\qquad\qquad = 11.8 \text{ ft}^2 \ (1.1 \text{ m}^2)$

Diameter $\qquad\qquad\quad = 3.88$ ft (1.2 m)

Carbon use $\qquad\qquad = 11.8 \text{ ft}^2 \times 9.34 \text{ ft/d} \times 39 \text{ lb/ft}^3$

$\qquad\qquad\qquad\qquad = 3,300$ lb/d (1,500 kg/d)

$$\text{Carbon adsorption loading} = \frac{(630 - 63) \text{ mg}/1 \times Q}{\text{carbon use}}$$

$$= \frac{567 \text{ mg}/1 \times 85,000 \text{ gal}/d}{3,300 \text{ lb}/d} \times \frac{8.34 \text{ lb}/\text{gal}}{10^6 \text{ mg}/1}$$

$$= 0.122 \text{ lb}/\text{lb carbon}$$

CARBON REGENERATION

It is generally feasible to regenerate spent carbon for economic reasons. In the regeneration process, the object is to remove from the carbon pore structure the previously adsorbed materials. The modes of regeneration are thermal, steam, solvent extraction, acid or base treatment, and chemical oxidation. The latter methods are usually to be preferred when applicable, since they can be accomplished in situ whereas thermal regeneration normally requires that the carbon be transported. The difficulty arises that adsorption from multicomponent wastewaters usually does not lend itself to high-efficiency regeneration by these methods. Exceptions are (i) phenol, which can be regenerated with caustic in which the phenol is converted to the more soluble phenate, and (ii) a single chlorinated hydrocarbon which can be removed with steam. In most wastewater cases, thermal regeneration is required. Thermal regeneration is the process of drying, thermal desorption, and high-temperature heat treatment (1200 to 1800°F, 650 to 980°C) in the presence of limited quantities of water vapor, flue gas, and oxygen. Multiple-hearth furnaces or fluidized-bed furnaces can be used.

Weight losses of carbon result from attrition and carbon oxidation. Depending on the type of carbon and furnace operation, this usually amounts to 5 to 10 percent by weight of the carbon regenerated. There is also a change in carbon capacity through regeneration that may be caused by a change in pore size (usually an increase resulting in a decrease in iodine number) and a loss of pores by deposition of residual materials. In the evaluation of carbons for a wastewater-treatment application, the change in capacity through successive regeneration cycles should be evaluated. In most cases, three regeneration cycles will define the maximum capacity loss.

PERFORMANCE OF GRANULAR ACTIVATED CARBON COLUMNS

Activated carbon columns are employed for the treatment of toxic or nonbiodegradable wastewaters and for tertiary treatment following biological oxidation.

When degradable organics (BOD) are present in the wastewater, biological action provides biological regeneration of the carbon, thus increasing the

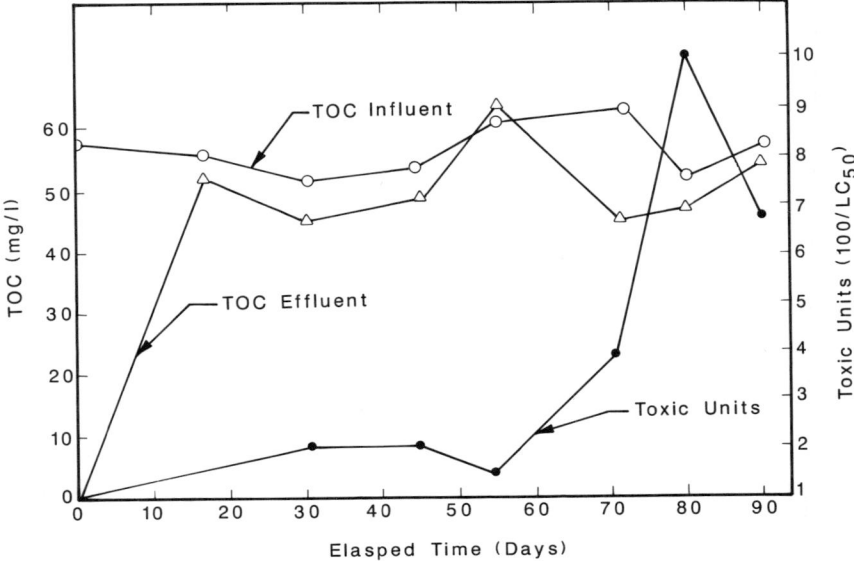

Fig. 9-12. TOC and toxicity reduction using granular carbon columns.

apparent capacity of the carbon. Biological activity may be an asset or a liability. When the applied BOD is in excess of 50 mg/l, anaerobic activity in the columns may cause serious odor problems while aerobic activity may cause plugging due to biomass generation.

Most heavy metals are removed through carbon columns. In order to avoid reduced capacity after regeneration, the carbon should be acid-washed prior to reuse.

Depending on the nature of the organics contributing to the toxicity, toxicity reduction efficiency may be considerably superior to TOC removal efficiency in a given system. Fig. 9-12 depicts a carbon column operation in which TOC breakthrough occurred after 12 days of operation, while breakthrough of toxicity did not occur for 60 days. One possible explanation for these results is the replacement of the less strongly adsorbed nontoxic molecules by more strongly adsorbed toxic molecules.

REFERENCES

Adams, C. E., D. L. Ford, and W. W. Eckenfelder, 1981. *Development of Design and Operational Criteria for Wastewater Treatment*. Boston, MA: Butterworth Publishing.

Dale, J., J. Malcolm, and I. M. Klotz. *Ind. Eng. Chem.*, **38**(1): 289.

Eckenfelder, W. W. 1989. *Industrial Water Pollution Control*. New York: McGraw-Hill.

Hutchins, R. A. 1973. *Chemical Engineering*, **80**(19): 133.

U.S. EPA. 1973. *Process Design Manual for Carbon Adsorption*. Technology Transfer.

10

Powdered Activated Carbon Treatment (PACT®)

Perry W. Lankford

PROCESS DESCRIPTION

The fundamental principles of the adsorption phenomenon are applicable to all solid adsorbents including man-made and natural materials. The basic principles and evaluation techniques for carbon adsorption are not only applicable to granular carbon, but also to powdered carbon. These principles and techniques are described in detail in the literature and in Chapter 9 and will not be repeated here. The principle differences between the granular and powered carbon treatment techniques are the type of contactor and the method of regeneration.

PACT® is simply the application of powered activated carbon (PAC) to the activated sludge process as shown in Fig. 10-1 (Eckenfelder 1989). Activated carbon is mixed with the influent wastewater or fed directly to the aeration basin. The carbon-biosludge mixture is settled and the sludge recycled in the same manner as in the conventional activated sludge process. The waste activated sludge similarly contains the carbon and biosludge mixture.

PAC offers the advantage of being able to be integrated into existing biological treatment facilities at lower capital cost than other carbon adsorption processes might involve. Since the addition of PAC enhances sludge settleability, conventional secondary clarifiers will usually be adequate, even with high carbon dosages.

The total MLSS in a PAC aeration basin will normally be roughly half biological and half carbon, the exact ratio depending on the nature of the substrate (or adsorbate), the biochemical synthesis coefficient, and the biological sludge age. However, this ratio is affected by several variables and makes it difficult to independently predict powdered carbon and biomass responses in a PACT® system. For example, the biological sludge age:

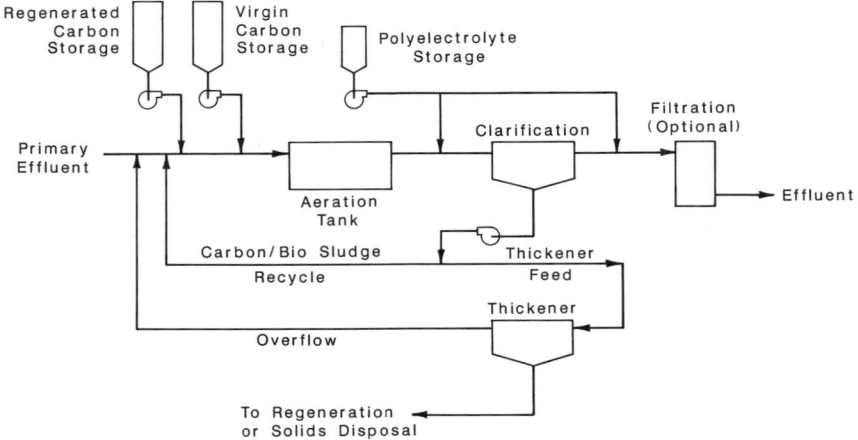

Fig. 10-1. Typical PACT® process diagram. (*Source*: Eckenfelder 1989.)

- Affects the powdered carbon efficiency; higher sludge ages enhance the organic removal per unit of carbon
- Affects the molecular configuration of the adsorbate based on varying biological uptake patterns and end-products
- Establishes the equilibrium biological MLSS level in the aeration basin

Moreover, carbon adsorptive capacities are influenced by:

- The characteristics of the carbon used
- The nature of the adsorbate
- The degree of biological regeneration
- The influent concentration of the organic substrate

Because of the many variables involved in the carbon and biomass components of the total MLSS, it is realistic to compute the sludge age in terms of the total mass only.

In the PACT® process, both low and high molecular weight compounds are adsorbed on the carbon. There is some evidence that the attached biomass degrades some of the adsorbed low molecular weight compounds as demonstrated by TOC removal rates which exceed isotherm predictions. Fig. 10-2 was developed by evaluating the difference in TOC removal for biological units operated in parallel, with and without PAC (Eckenfelder 1989). As can be seen, the performance of the carbon in the bioreactor was significantly greater than predicted by the isotherm only.

The mechanisms felt to be responsible for this phenomenon include:

1. Additional biodegradation of organics due to reduced biological inhibition through adsorption

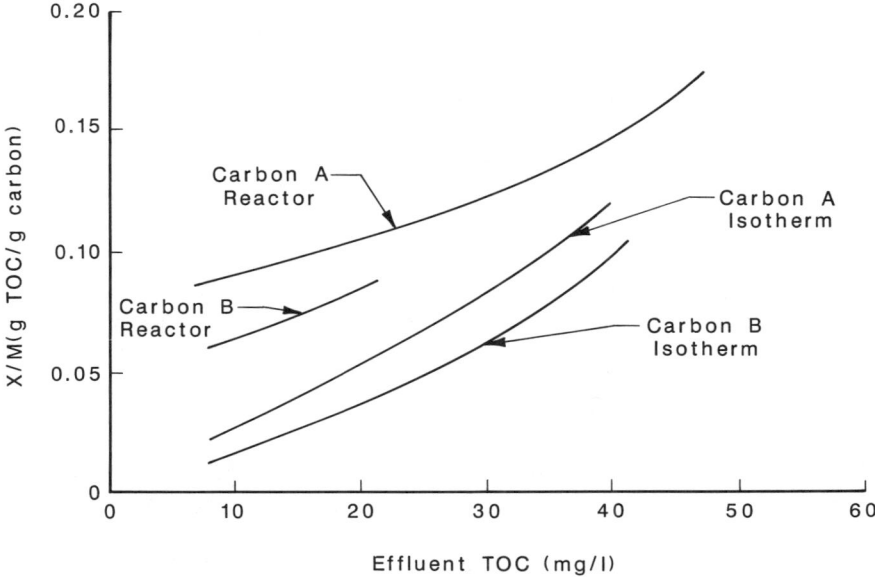

Fig. 10-2. Performance relationship of PACT® reactors with isotherm data. (*Source*: Eckenfelder 1989.)

   2. Degradation of normally nondegradable substances due to increased exposure time to the biomass through adsorption on the carbon; the carbon with adsorbed materials remains in the system for one sludge age, typically 10 to 30 days, while without carbon the substances would remain in the system for only one hydraulic retention time, typically 6 to 36 hours

   3. Substitution/adsorption phenomena, replacement of low molecular weight compounds with high molecular weight compounds, resulting in improved adsorption efficiency

It should be noted that toxicity reduction may or may not correlate with TOC removal by carbon adsorption depending on the specific organics responsible for the toxicity. Relationships between toxicity and the TOC can be developed for specific conditions, but must be periodically verified due to changes in influent character and treatment effectiveness.

COMPARISON WITH ACTIVATED SLUDGE

Through this relatively simple, although sometimes expensive, conversion from secondary to advanced treatment, dramatic improvements can be observed in removal of organics, color, and metals. These improvements can sometimes be realized without substantial physical changes to the treatment system. Some operating variables may require changes as will be described later.

Direct side-by-side comparisons of activated sludge and PACT® for three industrial cases are provided in Table 10-1.

COMPARISON WITH GRANULAR CARBON COLUMNS

As discussed previously, the principles of adsorption are the same for the powdered and granular carbon column processes. If used in the same or a similar contactor, there would be no intrinsic advantage of the granular or powdered form. The granular form lends itself more toward a column contactor where the initial carbon is at equilibrium with the influent feed concentration of contaminants. As more columns are used in series, the efficiency of granular carbon increases. However, this obviously will increase capital cost to reduce operating cost.

Powdered carbon, on the other hand, lends itself to a completely mixed contactor where the carbon is at equilibrium with the effluent concentration. When the number of granular columns in series is minimized to save capital cost, the efficiency of powered carbon can become very competitive with granular. In numerous side-by-side comparisons of PACT® with a single GAC column, the carbon requirement to achieve the same effluent organic concentration has been determined to be approximately the same for both options.

Hutton (1981) of DuPont, compared PACT® with granular carbon columns and indicated:

Adding powdered carbon at 100 ppm dosage to the activated sludge process (DuPont PACT® process) increased performance so that 85 percent removal was achieved for 30 priority [organic] compounds.

The combined activated sludge plus granular carbon [columns] also achieved 85 percent removal for 30 priority [organic] compounds.

However, Hutton's work described the chromatographic effect in granular carbon column removal of specific priority pollutants. He indicated:

Twelve of the compounds gave carbon column effluent concentrations for a period of time at breakthrough higher than ever fed to the columns.

Twenty of the compounds exhibited some bleed-through of the carbon columns from day one.

This chromatographic effect has been predicted for multi-component mixtures like this and constitutes a significant problem in the use of carbon columns.

Table 10-1. Comparison of PACT® with Activated Sludge.

	Case 1[a]			Case 2[b]			Case 3[c]		
	A.S. (without PAC)	Influent	PACT	A.S. (without PAC)	Influent	PACT	A.S. (without PAC)	Influent	PACT
HRT, days	1.6		1.6	0.36		0.32	6		6
SRT, days	30		30	30		30	45		25
PAC Dosage, mg/l	0		250	0		100			
Influent									
BOD, mg/l		320			160			4,035	
TOC, mg/l		245			153			2,965	
TSS, mg/l		70			36			—	
Color, APHA		5,365			1,640			—	
Copper, mg/l		0.41			—			—	
Chrome, mg/l		0.09			—			—	
Nickel, mg/l		0.52			—			—	
Effluent									
BOD, mg/l	3		2	5.4		3.0	17		11
TOC, mg/l	81		29	47		21	65		25
TSS, mg/l	50		34	15		47	—		—
Color, A.P. H.A.	3,830		323	1,280		315	820		94
Copper, mg/l	0.36		0.07	—		—	—		—
Chrome, mg/l	0.06		0.02	—		—	—		—
Nickel, mg/l	0.35		0.24	—		—	—		—

[a]Source: Lankford et al. (1988).
[b]Source: Hutton (1981).
[c]Source: Eckenfelder (1989).

Hutton concluded that PACT® was the more desirable process for that application.

> The carbon was more effectively used by the PACT® process than the carbon columns. At the point of breakthrough, the carbon columns had expended a dosage of 140 ppm. The PACT® process was achieving goals at the carbon dosage of 100 ppm used.

DESIGN CRITERIA

PACT® is a patented process originally developed by DuPont in the 1970s at its Chambers Works Plant in Deepwater, New Jersey. Currently, Zimpro/Passavant of Rothschild, Wisconsin holds the license.

Although there are other design criteria, the more significant parameters are solid retention time (SRT), hydraulic retention time (HRT), mixed liquor suspended solids (MLSS), and, of course, powdered activated carbon (PAC) dosage. The design criteria for industrial applications in Table 10-2 were excerpted from Zimpro's most recent design manual (Zimpro 1984). Zimpro recommends that for industrial wastewater applications, the SRT should be from 5 to 20 days, the HRT should be in the range of 0.17 days (4 hours), the MLSS should be 15 to 20 g/l, and the powdered carbon makeup should be greater than 100 g/l. Some of these criteria will, of course, be dictated more by the biological design of the activated sludge system than by the PACT® design.

A number of case studies are reported on the application of PACT® to industrial wastewater. Five such case studies are provided in Table 10-3, indicating a range of design criteria.

From the case studies, it can be observed that the application of PACT® in such cases coincides fairly well with the criteria provided previously. The hydraulic retention time varies from 0.32 to 4.2 days, the solids retention time varies from 5.8 to 54 days, and the MLSS varies from 10 to 35 g/l. The PAC dosage varies from 114 to 2,270 mg/l.

Table 10-2. PACT® System Design Guidelines for Industrial Waste.

Aeration system	
Hydraulic retention time	0.17 days[a]
Solids retention time	5–20 days
Mixed liquor suspended solids	15–20 g/l
Clarification	
Overflow rate, average	800 gpd/sq ft
Powdered carbon makeup	over 100 mg/l

Source: Zimpro (1984).
[a]Expected to vary with waste strength and treatability.

Table 10-3. PACT® Process Case Studies.

Parameter	Case Study No.				
	1[a]	2[b]	3[c]	4[c]	5[c]
Aeration System					
HRT, Days	0.32	0.75	2.3	3.8	4.2
SRT, Days	54	40	5.8	20	19.3
MLSS, g/l	34.8	10–12	—	—	—
PAC Dosage, mg/l	114	170	2,270	850	1,140
TOC					
Influent, mg/l	174	—	2,470	2,330	2,490
Removal, %	81	—	85.3	98.9	98.4

[a]*Source*: Zimpro (1984).
[b]*Source*: Zimpro (1987).
[c]*Source*: Dietrich et al. (1988).

The MLSS level and mixing requirements in a PACT® system are dictated by the PAC dosage, the biological MLSS concentration and the SRT (θ_c). By definition, the SRT is:

$$\theta_c = (X_a V)/(Q_w X_u + (Q - Q_w)X_e) \qquad (10\text{-}1)$$

where:

θ_c = biological sludge age
X_a = average MLSS concentration
V = aeration volume
Q_w = sludge wastage flow
X_u = sludge underflow concentration
Q = total flow
X_e = effluent TSS concentration

The equilibrium PACT® MLSS concentration can then be defined, based on the following relationship:

$$X_P = \frac{X_i \theta_c}{\text{HRT}} \qquad (10\text{-}2)$$

where:

X_P = equilibrium PAC concentration
X_i = PAC dosage
θ_c = sludge age
HRT = hydraulic retention time

In order to determine the total mass concentration, the biological solids level must be determined for the selected HRT and SRT. The total MLSS is the sum of the carbon and biological components. At the higher MLSS levels, mixing can become critical in the aeration basin; low mixing levels and some diffused air systems may not be acceptable.

DEVELOPMENT OF DESIGN CRITERIA

Prior to beginning testwork, it is important to develop an experimental design which deals with all parameters of concern. The obvious parameters are HRT, SRT, PAC type, and PAC dosage. Additional tests may be warranted for operation at low temperature and use of regenerated PAC.

The first step involves screening of alternate carbons to select the most cost-effective carbon to attain the required performance. This can be done using batch isotherm tests on biologically-treated wastewater. Following screening tests, bench-scale continuous-flow reactors, as shown in Fig. 10-3 (Eckenfelder 1989), can be used to develop basic design criteria. Parallel reactors should be operated including a control unit (with no PAC addition) and other units which control operating variables such as temperature, pH, MLSS, and organic loading. PAC relationships can then be developed to select the desired carbon dosage and PAC mixed liquor concentration required to achieve the stipulated effluent quality.

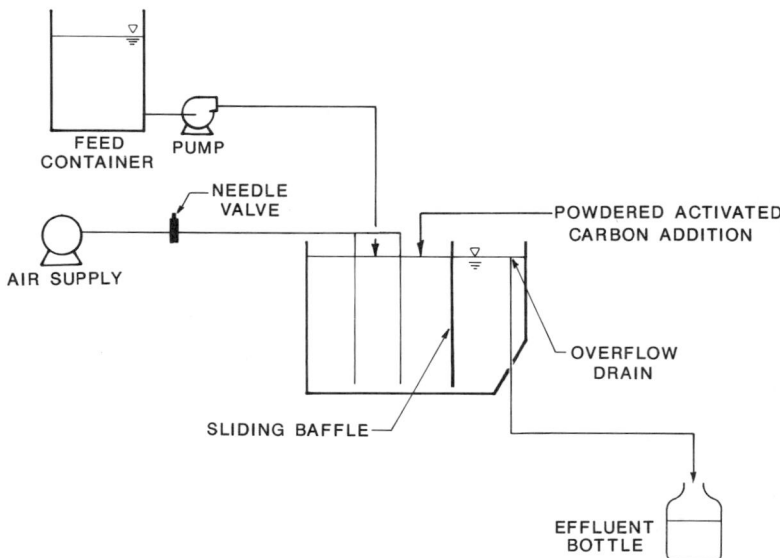

Fig. 10-3. Biological reactor with PAC addition. (*Source*: Eckenfelder 1989.)

The analytical requirements are similar to operation of a standard activated sludge experiment. However, analysis of carbon-containing MLSS is unique, since you must distinguish between biological and carbon solids. This is typically done by nitric acid digestion of biological solids rather than using the volatile solids procedure (DuPont 1981).

PERFORMANCE OF POWDERED ACTIVATED CARBON TREATMENT

TOC Removal

For the purpose of understanding the mechanism for removal of non-specific organics (TOC) it is helpful to describe the PACT® reactor as a two component process, as presented in Fig. 10-4 (Lankford and Miller 1987). The first component is the biological degradation of the organics present in the influent and the formation of adsorbable byproducts. The second component is the adsorption of microbial byproducts by the carbon that is being added to the aeration basin. Equations such as the following can be used to estimate the effluent TOC that can be achieved by the addition of a predefined dose of carbon, C:

$$S_x = S_b \exp\left(-kC\right) \tag{10-3}$$

where:

S_x = Continuous PACT® reactor effluent TOC (mg/l)

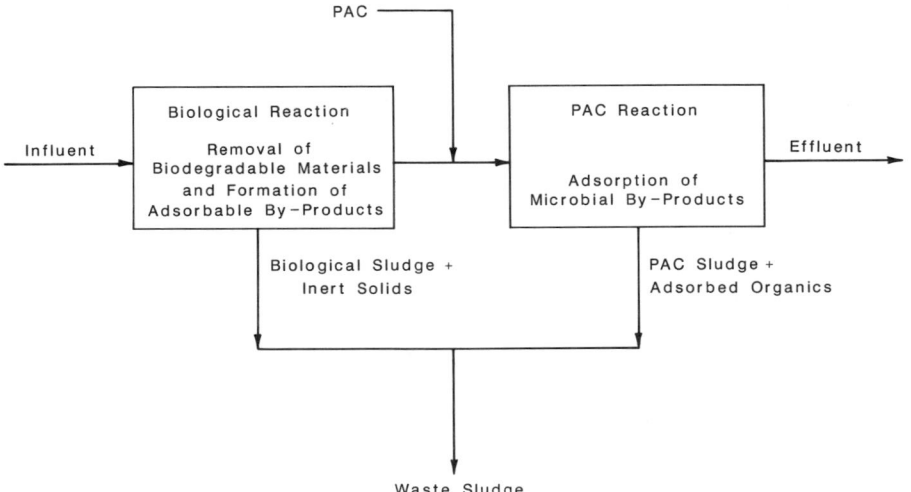

Fig. 10-4. Conceptual model for PACT® reactor. (*Source*: Lankford and Miller 1987.)

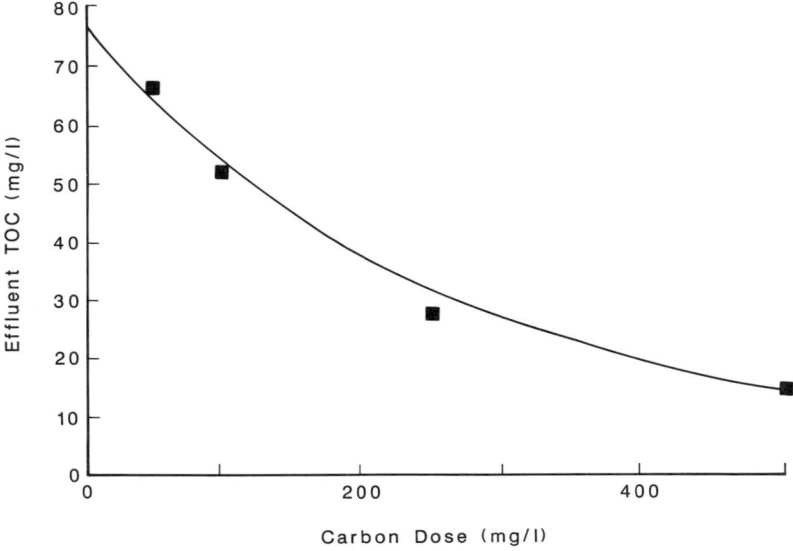

Fig. 10-5. Effluent TOC versus carbon dose. (*Source*: Lankford and Miller 1987.)

S_b = Continuous biological control reactor (no PAC) effluent TOC (mg/l)
C = Carbon dosage (mg/l)
k = Constant (1/mg)

The constant k can be developed from continuous reactor experimental data arranged as shown in Figs. 10-5 and 10-6. The values of k for different carbons, under the continuous operating conditions used, usually do not correspond to the differences predicted by isotherms. As indicated previously in Fig. 10-2, isotherm values do not reflect actual continuous PACT® removals and efficiencies.

Ammonia Removal

Nitrification is very sensitive to the presence of inhibitory chemicals such as heavy metals or amino acids. Addition of powered activated carbon to the activated sludge system may reduce the inhibitory effects, at least partially, of some components of the wastewater, and consequently promote nitrification (Dunn et al. 1981).

This was demonstrated when the effect of powered activated carbon on nitrification was studied using five identical reactors with different amounts of PAC and a full-scale PACT® system (Lankford and Miller 1988). The bench-scale results of this study are presented in Fig. 10-7. Although the biological control

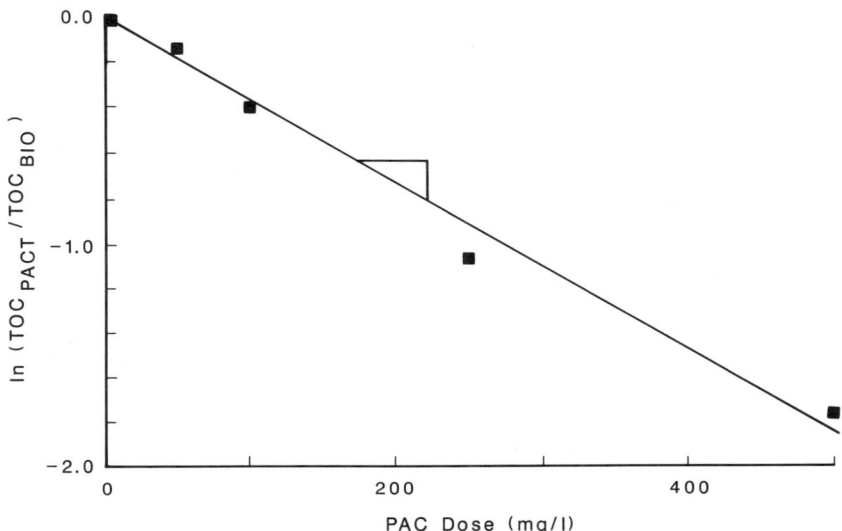

Fig. 10-6. PACT® residual TOC ratio versus PAC dosage. (*Source*: Lankford and Miller 1987.)

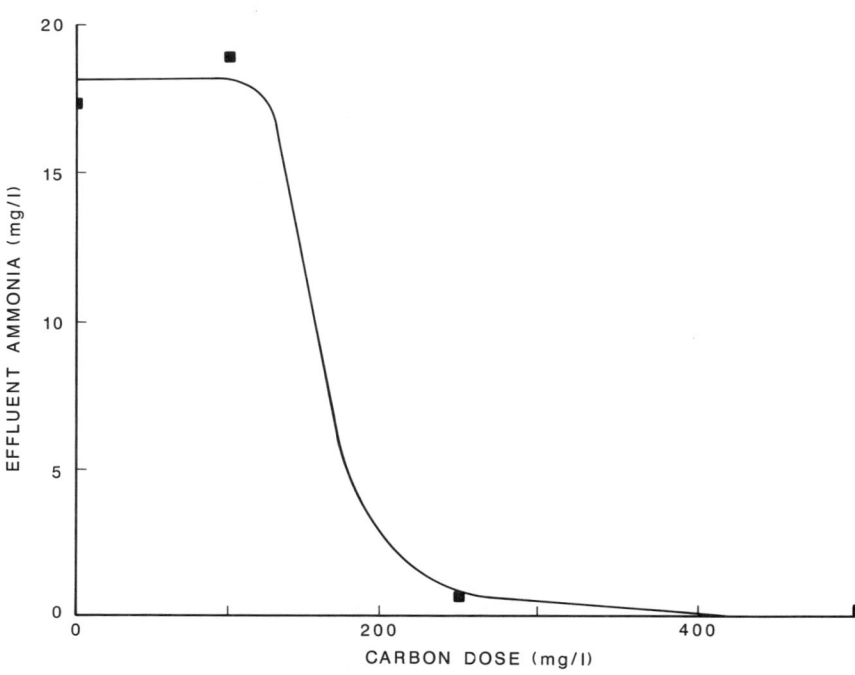

Fig. 10-7. Effect of carbon dose on occurrence of nitrification. (*Source*: Lankford and Miller 1988.)

unit (no PAC) was operated at conditions conducive to nitrification, (an SRT of 30 days and a temperature of 20°C), the ammonia level in the reactor effluent was relatively high, ranging from 8 to 36 mg/l. Under the conditions in the reactor, the ammonia level should have been below 1 mg/l. Five months of acclimation under these conditions did not yield a nitrifying population. This inability of the control reactor to nitrify was explained by the presence of chemical inhibition, which was eliminated by the addition of 250 mg/l of Calgon RB carbon.

Similar observations can be made from the inspection of the full-scale plant performance data presented in Fig. 10-8 (Lankford and Miller 1988). Before the addition of carbon to the aeration basin, the treatment plant was operating strictly as a biological system for removal of carbonaceous materials. The reactor did not nitrify and ammonia levels in the effluent were averaging 43 mg/l. In week five, four weeks after the addition of approximately 50 mg/l of carbon to the system, the effluent ammonia concentration was reduced to below 1 mg/l. During this entire period, the temperature was above 22°C.

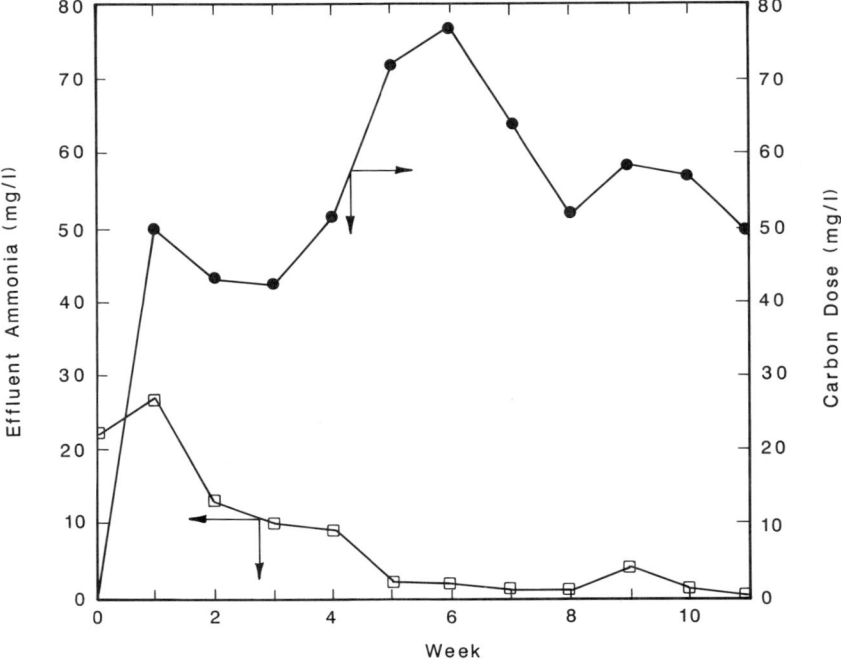

Fig. 10-8. Effect of carbon dose on effluent ammonia concentration in a full-scale PACT® system. (*Source*: Lankford and Miller 1988.)

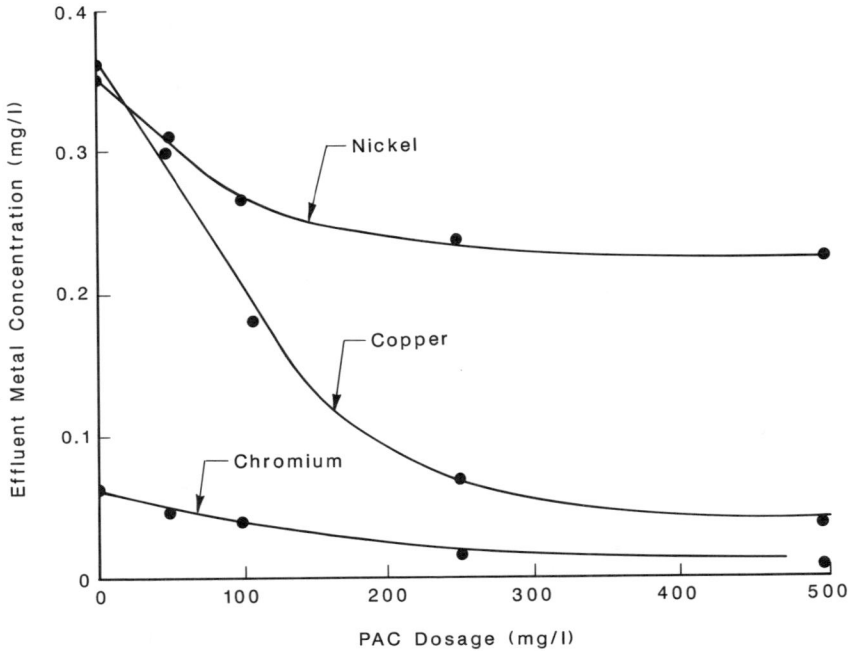

Fig. 10-9. Typical metal removal through the PACT® system. (*Source*: Lankford, Eckenfelder, and Torrens 1987.)

Metals Removal

Although not typically utilized as a standard process in removal of metal contaminants from wastewater, carbon adsorption does at times provide excellent incidental treatment as described in Chapter 5. Although there is no clear explanation of the phenomenon of metals adsorption by activated carbon, there is clear empirical evidence of its occurrence. Theories range from (i) adsorption of an organic which has complexed the metal, to (ii) surface precipitation by sulfide occurring on the carbon due to its high-sulfur content. In the PACT® process, there is the additional possibility of coprecipitation with the biological floc. Typical results using the PACT® process on metals-containing wastewaters are presented in Fig. 10-9 (Lankford et al. 1988).

Color Removal

Powdered activated carbon has been heavily relied upon for color removal in water treatment for many years. PACT® is certainly an excellent configuration in which to accomplish color removal from wastewaters. Fig. 10-10 (Lankford

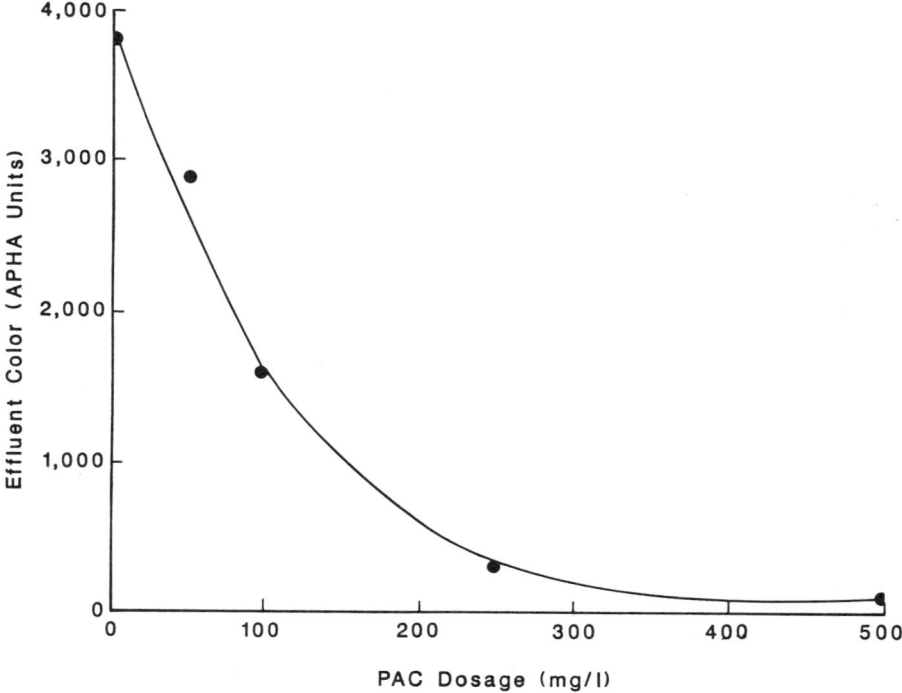

Fig. 10-10. Typical color removal through a PACT® system. (*Source*: Lankford, Ecken-felder, and Torrens 1987.)

et al. 1988) provides typical results of color removal utilizing the PACT® process. The case histories in Table 10-1 also provide color removal data.

Volatiles Control

The addition of PAC to activated sludge not only results in improved removals of such contaminants as TOC and NH_3, through improved biodegradation, and metals and color through adsorption, but it also reduces the losses of some volatile organics to the air phase. Stripping of organics is greatly diminished through the PACT® process. In 1984, Weber (Weber and Jones 1984) concluded:

> In general, the addition of carbon affected significant reductions in effluent and off-gas concentrations of the non-biodegradable compounds lindane and 1,2,4-tri-chlorobenzene, and the poorly degraded compound 1,2-dichlorobenzene.

He further concluded:

> Addition of carbon resulted in approximately equal percent reductions in both effluent and off-gas concentrations of volatile compounds.

Toxic Compounds Removal

In the process of gaining approval of PACT® for demonstration by the 1988 U.S. EPA Superfund Innovative Technology Evaluation program, Zimpro (1988) provided the removal data for priority pollutants shown in Table 10-4. There are also many examples of PACT® systems attaining below detection limit results on all priority pollutants.

Toxicity Reduction

Although there is no universal relationship between reduction of toxicity and degree of treatment for parameters such as TOC, NH_3, or metals, there is a relationship between improved treatment and toxicity reduction. In other words, a lower TOC or NH_3, level does not necessarily provide the toxicity reduction required, but improvements in treatment overall will generally reduce toxicity. PACT® has become a leading process in toxicity reduction over the last few years due to its broad effect and simple means of implementation. Typical results are provided in Fig. 10-11.

REGENERATION OF POWDERED CARBON

PAC is most effectively regenerated by wet air regeneration (WAR). In this process, the biosludge-carbon mixture is placed in a reactor at approximately 260°C and 750 psi in the presence of a slight excess of oxygen. Under these conditions, the biosludge is oxidized and solubilized, and the carbon is regenerated. The COD of the return liquor from the reactor will vary from 3,000 mg/l to 8,000 mg/l, and will constitute an additional organic load on the aeration process. The return liquor also can contain 1,000 mg/l or more of ammonia nitrogen. There will be some buildup of ash through WAR which must be blown down at periodic intervals.

Virgin carbon is required to make up for regeneration system losses of 5 to 20 percent, depending on the severity of regeneration conditions. The regeneration efficiency and losses will vary with both the carbon type and the specific organics involved. In general, as one strives for a higher quality regenerated carbon the losses increase; therefore, a trade-off exists between quality and losses.

Table 10-4.   PACT® Performance Results.

Compounds (all values in $\mu g/l$	PACT® Performance[a]		USEPA OCPSF BAT Limits[b]
	In	Out	
Acenaphtene	520	nil	22
Acrylonitrile	11,700	nil	96
Benzene	290	1	37
Carbon tetrachloride	860	1	18
Chlorobenzene	31	5	15
1,2,4-Trichlorobenzene	210	0.21	68
1,2-Dichloroethane	210	1	68
1,1,1-Trichloroethane	4,970	1	21
1,1-Dichloroethane	640	1	22
1,1,2-Trichloroethane	30	5	21
Chloroethane	667	0.65	104
Chloroform	1,470	1	21
2-Chlorophenol	31.9	0.05	31
1,2-Dichlorobenzene	30	5	77
2,4-Dichlorophenol	19	1.33	39
2,6-Dinitrotoluene	1,100	55	255
Ethylbenzene	185	1	32
Methylene chloride	84	20	40
Methyl chloride	138	0.41	86
Naphthalene	191	1	22
Nitrobenzene	330	0.33	27
2-Nitrophenol	216	13	41
4-Nitrophenol	1,100	33	72
2,4-Dinitrophenol	140	1.4	71
Phenol	2,400	2	15
bis(2-Ethylhexyl)phthalate	561	2	103
Diethyl phthalate	88	1	81
Dimethyl phthalate	332	1	19
Tetrachloroethylene	304	1	22
Toluene	2,730	1	26
Trichloroethylene	326	1	21
Total cyanide	60.1	0.65	420

Source: Zimpro (1988).
[a]Actual results of full-scale or pilot testing of the PACT® system on wastewater. Full reports available.
[b]Maximum monthly average.

As previously discussed, activated carbons differ in their effectiveness for the removal of TOC and toxicity. In addition, regeneration affects adsorption efficiency of various carbons differently. Results from an organic chemicals wastewater are shown in Table 10-5 (Lankford et al. 1988). Type RB carbon was very effective in its virgin state, but lost much of its effectiveness after wet air regeneration. By contrast, HDC carbon showed relatively poor results in its virgin state, but was not dramatically affected by regeneration.

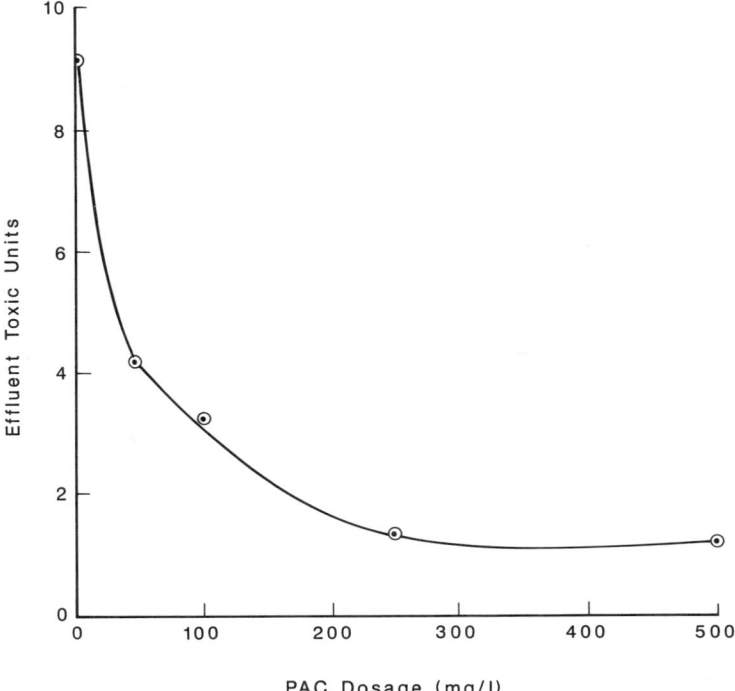

Fig. 10-11. Typical toxicity reduction through the PACT® system. (*Source*: Lankford, Eckenfelder, and Torrens 1987.)

Table 10-5.   Carbon Dosage Comparisons for Toxicity Reduction.

Carbon	Contactor	Dosage,[a] mg/l
Calgon RB virgin	Isotherm	118
Hydrodarco HDC virgin	Isotherm	222
Calgon WPX virgin	Isotherm	111
Calgon RB virgin	PACT® reactor	31
Calgon RB regenerated	PACT® reactor	200
Hydrodarco HDC virgin	PACT® reactor	182
Hydrodarco HDC regenerated	PACT® reactor	230

Source: Lankford et al. (1987).
[a]Influent dosage required to achieve LC_{50} of 50 percent.

REFERENCES

Dietrich, M. J., W. M. Copa, A. K. Chowdhury, and T. L. Randall. 1988. Removal of Pollutants from Dilute Wastewater by the PACT® Treatment Process. *Environmental Progress*, **7**(2): 143–149.

Dunn, G. F., D. G. Hutton, and W. W. Eckenfelder. 1981. Improved Nitrification with the PACT® Process. In *Application of Adsorption to Wastewater Treatment* (W. Wesley Eckenfelder, Jr., ed.). Nashville: Enviro Press, Inc.

DuPont de Nemours, E. I. 1981. *PACT® Technology Design Manual*. Wilmington, DE.

Eckenfelder, W. W. 1989. *Industrial Water Pollution Control*, 2nd edition. New York: McGraw-Hill.

Hutton, D. G. 1981. Priority Pollutant Removal: Comparison of the DuPont PACT® Process with Activated Sludge Followed by Granular Activated Carbon Columns. In *Application of Adsorption to Wastewater Treatment* (W. W. Eckenfelder, Jr., ed.), pp. 435–72. Nashville: Enviro Press, Inc.

Lankford, P. W., W. W. Eckenfelder, and K. D. Torrens. 1988. Reducing Wastewater Toxicity. *Chemical Engineering*, **95**(16).

Lankford, P. W. and E. Miller. 1987. Predictive Design Approach for PACT® Systems. Presented at Annual Meeting Water Pollution Control Federation, Philadelphia, PA.

Lankford, P. W. and E. Miller. 1988. Ammonia Removal in the PACT® Process and its Impact on Effluent Toxicity. *Water Pollution Control Federation*. Presented at Annual Meeting Water Pollution Control Federation, Dallas, TX.

Weber, W. J. and B. E. Jones. 1984. Toxic Substance Removal in Activated Sludge and PAC Treatment Systems. Presented at Annual Meeting Water Pollution Control Federation, New Orleans, LA.

Zimpro, Inc. 1984. Wastewater Treatment Systems Manual 200. Rothschild, Wisconsin.

Zimpro, Inc. 1987. Contaminated Groundwater.

Zimpro, Inc. 1988. Organic Chemicals Wastewater.

11

Chemical Oxidation of Toxic, Inhibitory, and Refractory Organics

Alan R. Bowers

INTRODUCTION

Chemical oxidation for treatment of organic compounds has traditionally been considered uneconomical compared to biological treatment processes. This is principally due to the continuing operating cost of the oxidants, while microorganisms grow without the addition of chemicals, as long as sufficient substrates and nutrients are present. However, many organic compounds are toxic, inhibitory, or refractory to microorganisms; and additionally, particular organic compounds may result in the production of soluble microbial products that produce effluent aquatic toxicity in violation of discharge permits (Grady 1986; Brandes et al. 1987; Eckenfelder 1989). These difficulties have resulted in new regulations for pretreatment of toxic organics prior to discharge into publicly owned treatment works (POTWs). Therefore, chemical oxidation is being considered as a pretreatment alternative for toxic organics, prior to discharge into the biological treatment system or POTW, and in certain instances, as a complete treatment process to meet effluent guidelines for BOD/COD reduction and toxicity.

OXIDATION ALTERNATIVES

When toxic, inhibitory and/or refractory organic compounds are reacted with a chemical oxidant, the extent of degradation of those organic compounds may be characterized as follows (Lyman et al. 1982):

247

- Primary degradation—a structural change in the parent compound(s); bio-degradability may be improved
- Acceptable degradation—degradation to the extent that toxicity is reduced
- Ultimate degradation—complete destruction to CO_2, H_2O, and other inorganics

A primary or acceptable degradation will occur at a much lower consumption of oxidants, and therefore, when coupled with a biological treatment process, the partial destruction will be much more economical than ultimate or complete destruction of the compounds.

A variety of chemical oxidants has been used in water and wastewater treatment processes for disinfection and/or destruction of undesirable chemical species (organic or inorganic). The ideal oxidant is one which is inexpensive, readily available, reactive with the compounds to be treated, and produces no undesirable residual byproducts. The availability and economy are readily determined by commercial suppliers of oxidants, but the reactivity and the characteristics of the by-products may or may not be known. In particular, data on byproducts may not be available except for obvious cases, such as the reduction of Cr(VI) to Cr(III) or the generation of undesirable chlorinated organics as a result of partially oxidized organic species contacted by chlorine. Several of the most common oxidants and their primary applications are presented in Table 11-1.

For most situations, a treatment scheme using chemical oxidation must be developed on a case-by-case basis. Selection of the right oxidant depends not only on the reactivity of the oxidant, but the rate of reaction (catalysts may be required), the reaction conditions (particularly pH) and the characteristics of the residual by-products. No predictive models are available at this time, that address all of these factors, and a considerable amount of laboratory work may be required.

CHEMISTRY OF CHEMICAL OXIDATION

Stoichiometry

It is important to define the stoichiometric relationship between an oxidant and the compounds to be treated so that required oxidant dosages may be estimated and experiments may be designed within reasonable limits. A general approach may be taken so that one can easily convert the stoichiometry for one particular compound from one oxidant to another. For convenience, the half reactions for each oxidant can be expressed in terms of free reactive oxygen, [O], derived from each oxidant, or, using simple oxygen as an example:

$$O_2 \rightarrow 2[O] \tag{11-1}$$

Table 11-1. Common Oxidants, Current Applications and Potential for Wastewater Treatment.

Oxidant	Phase[a]	Major Current Applications	Wastewater Treatment Potential
Cl_2 or $HOCl$	g, l[b]	Disinfection, oxidation of cyanide, bleaching pulp and paper	Suspect due to formation of chlorinated organics as residual species
ClO_2	g	Disinfection, bleaching pulp and paper	Does not produce chlorinated organics
O_3	g	Alternate water disinfectant, destruction of trihalomethane precursors in drinking water, bleaching pulp and paper	Highly reactive, no production of undesirable byproducts, must be generated on site, high capital costs
H_2O_2	l	Addition of O_2 to wastewater systems, odor control in sewers bleaching pulp and paper	Can be stored on site, highly reactive with a variety of organics, no well known undesirable byproducts, economical, generally requires acid and catalyst
$KMnO_4$	s	Removal of taste and odor producing compounds from drinking water	Can be stored on site, highly reactive with a variety of organics, no well known undesirable byproducts, may produce solids (MnO_2) which present handling and disposal problems
FeO_4^{-2}, $S_2O_8^{-2}$,	s	Laboratory studies only	Insufficient laboratory data or case studies to be considered at this time

[a]Refers to the typical form(s) supplied for water and wastewater treatment; g = gas, l = liquid, s = solid
[b]Cl_2 typically supplied as a gas, $HOCl$ supplied as aqueous $NaClO_4$ solutions

Then, any reaction with an organic may be expressed in terms of the equivalent free oxygen and related back to any oxidant by the stoichiometry of the oxidant half reaction. The half reactions for a variety of oxidants are presented in Table 11-2.

Based on the ultimate conversion of an organic compound to CO_2 and H_2O, a general stoichiometric equation may be derived using free reactive oxygen, or:

$$C_aH_bO_c + d[O] \rightarrow aCO_2 + (b/2)H_2O \qquad (11-2)$$

Table 11-2.   Oxidant Half-Reactions.

	Equivalent Reactive Oxygen	
Half Reaction	Moles [O] per Mole Oxidant	Moles [O] per kg Oxidant
$Cl_2 + H_2O \rightarrow [O] + 2Cl^- + 2H^+$	0.5	14.1
$HOCl \rightarrow [O] + Cl^- + H^+$	1.0	19.0
$2ClO_2 + H_2O \rightarrow 5[O] + 2Cl^- + 2H^+$	2.5	37.0
$O_3 \rightarrow [O] + O_2$	1.0	20.8
$H_2O_2 \rightarrow [O] + H_2O$	1.0	29.4
$2MnO_4^- + H_2O \rightarrow 3[O] + 2MnO_2 + 2OH^-$	1.5	9.5
$2FeO_4^{-2} + 2H_2O \rightarrow 3[O] + Fe_2O_3 + 4OH^-$	1.5	7.6

where the number of free reactive oxygens required is represented by the coefficient d which equals $(2a + b/2 - c)$.

Then, the equation may be balanced for any oxidant by adding the half reaction for the oxidant times the number of free reactive oxygens required (d) divided by the stoichiometric number of free reactive oxygens produced (see half-reactions in Table 11-2). Using phenol as an example, from Eq. (11-2):

$$C_6H_5OH + 14[O] \rightarrow 6CO_2 + 3H_2O$$

and, considering H_2O_2 and MnO_4^- as oxidants:

H_2O_2:

$$\frac{\begin{aligned}C_6H_5OH + 14[O] &\rightarrow 6CO_2 + 3H_2O \\ + (14/1)(H_2O_2 &\rightarrow [O] + H_2O)\end{aligned}}{= C_6H_5OH + 14H_2O_2 \rightarrow 6CO_2 + 17H_2O}$$

and, MnO_4^-:

$$\frac{\begin{aligned}C_6H_5OH + 14[O] &\rightarrow 6CO_2 + 3H_2O \\ + (14/3)(2MnO_4^- + H_2O &\rightarrow 3[O] + 2MnO_2 + 2OH^-)\end{aligned}}{= C_6H_5OH + (28/3)MnO_4^- + (5/3)H_2O \rightarrow 6CO_2 + (28/3)MnO_2 + (28/3)OH^-}$$

In many cases, the wastewater will consist of a wide variety of compounds rather then one or a few specific known compounds. Therefore, the use of the theoretical stoichiometry cannot be applied. However, the general approach developed using the free reactive oxygens can be adapted to the surrogate chemical oxygen demand, COD. This readily measurable parameter can be converted to the stoichiometric requirements of an arbitrary wastewater as follows:

$$\text{Oxidant Demand (mg oxidant/L)} = (2/n)(MW/32)\,COD \quad (11\text{-}3)$$

where:

n = moles [O] per mole oxidant (see Table 11-2)
MW = molecular weight of oxidant (g/mole)
COD = chemical oxygen demand of mixed systems (mg O_2/L)

using H_2O_2 and $KMnO_4$ as examples:

H_2O_2 stoichiometric demand = $(2/1)(34/32)$ COD = $2.13 \cdot$ COD

$KMnO_4$ stoichiometric demand = $(2/1.5)(158/32)$ COD = $6.58 \cdot$ COD

Based on the stoichiometric requirements and the yield of free reactive oxygens, i.e., moles [O] per kg (from Table 11-2), the cost for the stoichiometric dosage of each oxidant may be determined (based on $/kg oxidant). Then, a ranking of candidate oxidants based on cost can be established and their actual effectiveness and cost determined from laboratory data.

Thermodynamics

Thermodynamically, the common oxidants used for water and wastewater treatment are theoretically capable of oxidizing any soluble organics found in aqueous solution. However, the reaction rates may be extremely slow in some cases. In addition, the extent of reaction, and therefore the oxidizing potential, will be pH and concentration dependent. For purposes of comparison, the standard electric potentials may be used. Standard potential is expressed as a reduction and is based on 25°C, 1 atm pressure and 1 M activity of all constituents. The higher the standard potential, the greater the oxidant strength. Considering permanganate as an example:

Acidic conditions:

$$MnO_4^- + 4H^+ + 3e^- \rightarrow MnO_2(s) + 2H_2O; \quad E_H = 1.70 \text{ volts}$$

Basic conditions:

$$MnO_4^- + 2H_2O + 3e^- \rightarrow MnO_2(s) + 4OH^- \quad E_H = 0.59 \text{ volts}$$

This illustrates the differences in standard oxidizing potential as a function of pH, i.e., E_H under acidic conditions is approximately triple the E_H under basic conditions for permanganate. A summary of the standard oxidation potentials for the most common oxidants is presented in Table 11-3.

Although the standard electrical potentials are useful for determining the strength of an oxidant, these data do not indicate the performance of an oxidant in wastewater treatment. In fact, the kinetics or rates of reaction are far more important but have little or no relationship with the thermodynamic properties.

Table 11-3.   Standard Potential for Common Oxidants.

Oxidant	Conditions	E_H, volts[a]
Cl_2	acidic	1.36
HOCl	acidic	1.49
	basic	0.89
ClO_2	acidic	1.95
	basic	1.16
O_3	acidic	2.07
	basic	1.25
H_2O_2	acidic	1.72
$KMnO_4$	acidic	1.70
	basic	0.59
K_2FeO_4	acidic	0.74
	basic	2.20

[a]Adapted from Weast (1987/88), except for K_2FeO_4, which is based on values from Wood (1952).

In addition, the oxidant byproducts are a function of the reaction mechanisms and environmental conditions rather than the thermodynamic properties.

Kinetics/Catalysis

The rates of oxidation reactions are perhaps the most important considerations in wastewater applications. The reactions of many species which are thermo-dynamically favorable are in fact unacceptable under practical environmental conditions.

The rates of oxidation reactions are dependent upon many variables which must be considered simultaneously. These include:

- Temperature. All reactions proceed more quickly at elevated temperature (Froment and Bischoff 1979). In addition, oxidation reactions are highly exothermic, which, in concentrated solutions, may lead to an autothermal effect and thus dangerously fast reaction rates (Hoffman and Moser 1985).
- Concentration. Oxidation reactions proceed more quickly as the concentration of the oxidant is increased.
- Catalysts. Catalysts increase the rate of oxidation and typically include metals in a variety of forms as well as certain spectra of light, i.e., photocatalysis.
- pH. Most oxidation reactions are highly complex, involving the transfer of several electrons and the consumption of protons (H^+) or hydroxide ions (OH^-). Typically, the reaction rates will have a sharp optimal pH or a preferred pH region.

- Reaction byproducts. Due to the complex reaction mechanisms involved in oxidation reactions, a wide variety of intermediate organic compounds and final byproducts are formed. These intermediates are interactive and may have an autocatalytic or autoretardant effect on the reaction rate.
- System impurities. A wide variety of common chemical species can interact with the complex oxidation reaction mechanisms. These may have a catalytic or retardant effect, and/or may shift the optimum pH. Examples of typical impurities include soluble metals, complex-forming agents or particulate species.

Because of the wide variety of chemical matrices that may exist in the wastewater environment, it is difficult to translate results from one particular wastewater to another. Therefore, a certain amount of laboratory work will be required before any kinetic relationships may be developed for a specific wastewater.

Typical oxidizing conditions depend upon the oxidant used and, to some extent, the organics to be oxidized. The following subsections discuss these conditions. These are presented for the most common oxidants, H_2O_2, O_3, $KMnO_4$ and ClO_2. Chlorine gas and NaOCl are omitted since their applicability in wastewater treatment is limited due to the formation of chlorinated organics.

Hydrogen Peroxide—H_2O_2. Hydrogen peroxide is available commercially as 30 percent or 50 percent (by weight) aqueous solutions. These solutions may be stored in bulk quantities for substantial time periods without difficulty. Most commercial grades are preserved with a chemical inhibitor to prolong storage time.

The rates of reaction are slow unless catalyzed by high pH (alkaline peroxide) or by a variety of metal species, (Fe, Cu, Mn) at reduced pH. (Fenton 1894; Baral et al. 1985; Klibanov et al. 1980; Francis et al. 1985; Lim et al. 1984). In addition, the enzyme horseradish peroxidase has been successfully used to catalyze reactions with phenols and anilines (Klibanov et al. 1980). These reactions are complex. However, it is believed that the production of free radicals, $OH\cdot$ or $HO_2\cdot$, are responsible for initiating the reactions with organic species.

By far, the most common catalyst is ferrous sulfate, $FeSO_4$, sometimes referred to as Fenton's reagent. Therefore, this discussion will focus on the Fe(II) catalysis system. Two reactions are hypothesized between Fe and H_2O_2 (Merz and Waters 1949):

$$Fe^{+2} + H_2O_2 \rightarrow Fe^{+3} + OH^- + \cdot OH \qquad (11\text{-}4a)$$

and,

$$Fe^{+3} + H_2O_2 \rightarrow Fe^{+2} + HO_2\cdot + H^+ \qquad (11\text{-}4b)$$

A chain reaction is then initiated between the hydroxyl radical and an organic species, which may be summarized as:

$$RH + \cdot OH \rightarrow R\cdot + H_2O \qquad (11\text{-}5a)$$

$$R\cdot + O_2 \rightarrow ROO\cdot \qquad (11\text{-}5b)$$

and

$$ROO\cdot + RH \rightarrow ROOH + R\cdot \qquad (11\text{-}5c)$$

This implies a very complex overall reaction mechanism, and that oxygen may play a role [see Equation (11-5b)] if present in sufficient quantity. The pertinent reaction parameters, H_2O_2 concentration, Fe(II) concentration, and reactive organic concentrations are extremely interactive. Typically, good reaction conditions exist when a stoichiometric amount of H_2O_2 is used and about a 10 to 1 ratio of H_2O_2 to Fe(II) (Merz and Waters 1949).

The optimal conditions for treatment will be the minimum H_2O_2 and Fe(II) dosages required to achieve the desired extent of reaction in an acceptable time. A statistically representative set of laboratory data should be provided around the reactive conditions outlined above, i.e., recommended conditions within an order of magnitude for H_2O_2/Fe(II) and from 10 percent to 300 percent of the stoichiometric H_2O_2. This should insure the selection of optimal reaction conditions. However, variability in wastewaters must also be considered and representative samples, over the range of variability, must be tested.

Hydrogen peroxide reactivity is a strong function of pH. Optimal reaction conditions occur in the acidic pH region and the typical optimum pH is between 2.0 and 4.0 (Merz and Waters 1949). The role of pH is further complicated by the formation of Fe(III) in the catalytic mechanism [Eq. (11-4a)]. If the pH is too high, Fe(III), which is very insoluble as $Fe(OH)_3$, precipitates out, resulting in a two-phase (liquid-solid) reaction, or heterogeneous catalysis. The reaction rate then becomes much slower than homogeneous catalysis. Usually when the pH is greater than 3.0 the reaction becomes heterogeneous.

Rate expressions for reactions of organics and H_2O_2 are fairly empirical and not well documented due to the complexity of the mechanism and the reactivity of intermediate oxidation products. Typically, under optimal conditions, the rates are quite fast and empirical relationships based on laboratory data must be used for reactor design.

Ozone—O_3. Ozone is an unstable gas which must be generated on site. One of the most important considerations in the process design is the mass transfer of ozone from the gaseous to the aqueous phase.

Ozone is fairly soluble in water and, due to the pH-dependent aqueous decomposition of ozone, Henry's Law solubility constants will be a function of pH as well as temperature. The true Henry's Law constants have recently been reported (Roth and Sullivan 1981) as:

$$H_A = 3.84 \times 10^7 [OH^-]^{0.035} \exp(-2428/T) \qquad (11\text{-}6)$$

where:

H_A = Henry's Law solubility constant, atm/mole fraction
$[OH^-]$ = hydroxide ion concentration, M
T = temperature, °K

Considering a relatively pure water system, the ozone solubility may be written as:

$$[O_3]^* = 1.80 \times 10^{-2} P_{O_3}/H_A \qquad (11\text{-}7)$$

where:

$[O_3]^*$ = ozone solubility, M
P_{O_3} = partial pressure of ozone in the saturating air, atm

Mass transfer considerations of ozone are complicated by its self-decomposition in water. The rate of self-decomposition is pH dependent, with ozone becoming increasingly stable at low pH values. Although some disagreement exists, researchers have shown that self-decomposition is first order with respect to ozone concentration.

The following rate expression has been reported (Sullivan and Roth 1980) for a pH range of 0.5 to 10.1 and a temperature of 3.5 to 60°C:

$$r_d = -9.811 \times 10^7 [OH^-]^{0.123} [O_3] \exp(-5606/T) \qquad (11\text{-}8)$$

where:

r_d = rate of O_3 self-decomposition, M $-$ min^{-1}
T = temperature, °K

The exponent on the hydroxide ion, 0.123 in Eq. (11-8), represented the largest discrepancy among the data, ranging from 0.123 to 1.0. This difference has been attributed to the ozone detection methods used (Sullivan and Roth 1980).

The ozone decomposition rates are somewhat dependent on solution composition. Various inorganic ions can catalyze or retard the decomposition rate (Rice and Netzer 1982). Decomposition of ozone results in production of the hydroxyl radical, ·OH. However, both ozone and hydroxyl actively participate in reactions with organic molecules in aqueous solution. Since the hydroxyl radical reacts less selectively than ozone, pH plays a major role in the rate of ozone reactions and in the reaction products that are formed. The complex reaction mechanism that results is summarized schematically in Fig. 11-1.

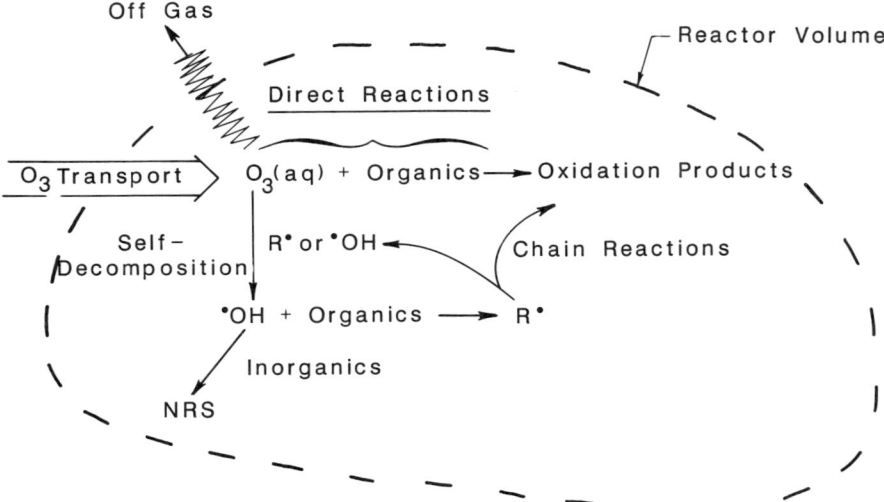

Fig. 11-1. Schematic of ozone reactions in aqueous solution. (*Source:* Rice and Netzer 1982.)

The reactions shown in Fig. 11-1 indicate the complexity of the aqueous ozone/organic system. First, the mass transfer rate of ozone into the water is important, and some O_3 if stripped out with the off-gas. Second, the self-decomposition reactions yield hydroxyl free radicals and organic free radicals, R, which can catalyze the self-decomposition of ozone. Third, inorganics may react with the free radicals to form nonreactive species, NRS. In addition, a variety of secondary reactive species may be formed. These include the perhydroxyl radical, $\cdot HO_2$, as well as Mn(VII) and BrO_3^-, which are dependent upon the initial presence of reduced inorganic species in solution, i.e., Mn^{+2} or Br^- (Rice and Netzer 1982).

Although the kinetics of reaction with organics are complex, a variety of relationships have been reported. Since mass transfer is important and difficult to separate from other processes, such as self-decomposition and reaction with organics, most experiments are performed under conditions where mass transfer limitations are not significant. For example, organics may be added to pre-saturated ozone solutions or gas flow rates are maintained at high enough levels so that solution reaction rates are much slower than the mass transfer.

Ozone mass transfer rates have frequently been modeled using conventional theory (Roth and Sullivan 1981; Sullivan and Roth 1980; Rice and Netzer 1982; Gurol and Nekouinaini 1984):

$$r_{mt} = K_L a \left([O_3]^* - [O_3] \right) \qquad (11\text{-}9)$$

where:

r_{mt} = mass transfer rate
$K_L a$ = lumped mass transfer rate coefficient
$[O_3]^*$ = saturation concentration of ozone, from Eq. (11-7)
$[O_3]$ = instantaneous ozone concentration

While typical values of $K_L a$ depend on the physical configuration of the reactor and the gas transfer equipment, it is also important to note that the identity of the organic substrates has an important impact on the mass transfer rates. Hydrophobic compounds have been reported to increase mass transfer rate coefficients by up to 300 percent, while hydrophilic compounds had little effect on the mass transfer rates (Gurol and Nekouinaini 1984).

Rates of reaction may be written for both ozone and the reacting species in solution (Gurol and Nekouinaini 1985; Eisenhauer 1968; Gould and Weber 1976):

$$dO_3/dt = r_{mt} - r_d - r_{si} \qquad (11-10)$$

where:

r_{mt} = rate of ozone mass transfer
r_{si} = rate of reaction with solute i
r_d = rate of ozone self-decomposition, see Eq. (11-8)

and, considering first order kinetics with respect to each solute:

$$r_{si} = -k_i[S_i][O_3] \qquad (11-11)$$

where:

k_i = rate coefficient
$[S_i]$ = concentration of solute i

Then, considering that the residual ozone concentration $[O_3]$ will depend upon the mass transfer rate, a given reactor configuration will yield the following (Eisenhauer 1968; Gould and Weber 1976)

$$[O_3] = k'D \qquad (11-12)$$

where:

D = ozone dose rate or application rate
K' = lumped constant depending on mass transfer and decomposition rate

This approach has been further refined in terms of a global rate constant (Roth, et al. 1982):

$$dS_i/dt = k[S_i] \qquad (11-13)$$

where:

k = global rate constant = $f(T, G, V, \text{pH})$
T = temperature
G = gaseous ozone flow rate
V = reactor volume

Additionally, the reaction with organics can be broken down into the direct reaction between the solutes and ozone, which is reportedly highly selective, and the indirect reactions of the solutes with the hydroxyl free radical, which are reportedly not very selective (Rice and Netzer 1982; Gurol 1985; Mehta et al. 1989). This implies that a highly interactive relationship exists between pH (controls rate of self-decomposition and thus ·OH production) and the mass transfer of ozone with the solute reactions and the overall selectivity in a mixture of organics. This may be summarized by three reactive regions (Mehta et al. 1989).

Region 1   High pH and/or low ozone dose rate implying that the major reaction mechanism with the solutes entails ·OH. Ozone will exhibit little selectivity.

Region 2   Approximately equal rates of decomposition and mass transfer. Both direct O_3 and ·OH reactions are important.

Region 3   Low pH and/or extremely high ozone dose rate implying that direct oxidation of solutes by O_3 is the controlling reaction. Ozone will exhibit a high degree of selectivity.

Rate constants and selectivity of ozone with a variety of solutes must then be determined through laboratory experiments. Values of pseudo-first order rate constants for direct reactions with ozone reportedly range from about 0.1 to 10^5 per mole-second for a variety of solutes (Rice and Netzer 1982).

Many reactive organic substrates are volatile and are removed from aqueous solution by a combination of air-stripping and reaction with ozone. Although care should be used in all cases where volatile organics are present, the problem is particularly exacerbated by the need to transfer ozone into the aqueous system from the gas phase. This mass transfer also provides a direct route for the transfer of volatile organics from the aqueous phase into the gas phase and out of the system through the off-gas. The problems with volatile organics have been fairly well documented in aerobic biological waste treatment processes, where oxygen must be transferred into the system, which results in the simultaneous air stripping of volatile organics. Roberts et al. (1984) reported that air stripping accounted for 20 to 80 percent removal for a variety of volatile compounds (chlorinated aliphatics) in aerobic biological systems having a 5 to 15 percent oxygen transfer efficiency. However, their analysis indicated that the removal due to air stripping decreased as the oxygen transfer efficiency increased (Roberts et al. 1984). Therefore, due to the expected high transfer efficiency for ozone (~90 percent), air-stripping for most compounds should be kept to a minimum. In addition, those systems which use pure oxygen rather than air for ozone generation should experience a more than fivefold decrease in the air-stripping potential due to the subsequent decrease in off-gas volume.

The main difficulty with volatile organics occurs during laboratory evaluation of ozonation. The typical experimental apparatus rarely matches the actual equipment and differences in mass transfer characteristics frequently exist. When preozonation of the aqueous solution is done before the organics are added, then the air-stripping potential will be underestimated. On the other hand, if an inefficient mass transfer apparatus is used to ozonate the wastewater directly, then the air-stripping potential will be overestimated. Therefore, proper experimental procedure should always include blank runs (air flow without ozone) to determine the contribution of air-stripping to the removal of volatile organics. In addition, the gas transfer characteristics of the experimental apparatus should closely match the actual reactor set-up.

Potassium Permanganate—KMnO$_4$. Potassium permanganate is available as a solid (about 99 percent pure), which has an aqueous solubility limit of about 6 percent by weight at 20°C (Carus Chemical Co., Inc. 1979). Reactions then occur in homogeneous aqueous solutions with the molecular reaction rates controlling. In general, permanganate reactions are quite complex and detailed rate equations and mechanisms for a wide variety of compounds are not available.

Reactions with permanganate are pH dependent, due to changes in the reaction species and the kinetic medium. The half-reactions for permanganate and the applicable pH region are as follows (Throop 1977; Reides et al. in press):
pH < 3.5:

$$MnO_4^- + 8H^+ + 5e^- \rightarrow Mn^{+2} + 4H_2O; \qquad E_H = 1.51 \text{ volts} \qquad (11\text{-}14a)$$

3.5 < pH < 7.0:

$$MnO_4^- + 4H^+ + 3e^- \rightarrow MnO_2 + 2H_2O; \qquad E_H = 1.70 \text{ volts} \qquad (11\text{-}14b)$$

7.0 < pH < 12.0:

$$MnO_4^- + 2H_2O + 3e^- \rightarrow MnO_2 + 4OH^-; \qquad E_H = 0.59 \text{ volts} \qquad (11\text{-}14c)$$

12.0 < pH < 13.0:

$$MnO_4^- + e^- \rightarrow MnO_4^{-2} \qquad E_H = 0.56 \text{ volts} \qquad (11\text{-}14d)$$

Most aqueous reactions are carried out in the pH range between 3.5 and 12.0. Then, the oxidations are carried out in three electron steps under acidic or basic conditions as described above. While the two half-reactions covering this pH range indicate that acid conditions favor the forward reaction, in practice, the rates of reaction are faster in the alkaline region. For example, the optimum pH for phenol oxidation by permanganate was found to be 8.5 to 9.5 (Throop 1977).

Generally, reactions of permanganate do not require the addition of a catalyst. However, Cu^{+2} and Ag^+ have been found to catalyze the conversion of cyanide to cyanate. Additionally, under extremely alkaline conditions ($12 <$ pH < 13), Ca^{+2} has been found to catalyze the disproportionation of MnO_4^{-2} to MnO_4^- and MnO_2, which in turn regenerates MnO_4^- for further oxidation.

Without specific guidelines, the rates of reaction with permanganate will be highly waste specific and must be evaluated in a series of laboratory tests. Variables include temperature, organic concentration, permanganate concentration, metal species present, and pH. Although the stoichiometric dosage of permanganate is rarely required, like peroxide, this is an excellent starting point for laboratory testing. The stoichiometric dosage of permanganate can be calculated from COD values of the waste; see Eq. (11-3). However, n is a function of pH, where:

$$pH < 3.5, \quad n = 2.5$$

$$3.5 < pH < 12, \quad n = 1.5$$

$$12 < pH < 13, \quad n = 0.5$$

Chlorine Dioxide—ClO_2. Chlorine dioxide is an unstable gas which must be generated on site. The generation is performed in aqueous solution by combining $NaClO_2$ with HCl (Masschelein 1984):

$$5NaClO_2 + 4HCl \rightarrow 2ClO_2 + NaCl \qquad (11\text{-}15)$$

Yields of chlorine dioxide are on the order of 95 percent. Once generated, the concentrated aqueous solutions are stored and used for dosing. This eliminates gas phase mass transport as a consideration. However, the aqueous solubility of chlorine dioxide is still important, because of its explosiveness above a maximum concentration in the air of 10 percent. Henry's Law constants may be used to determine solubility, as:

$$[ClO_{(aq)}]^* = K_H P_{ClO_2} \qquad (11\text{-}16)$$

where:

$[ClO_{2(aq)}]^* =$ aqueous saturation limit of ClO_2, g/L
$P_{ClO_2} =$ partial pressure of ClO_2 in the atmosphere, atm
$K_H =$ Henry's Law constant, g/L-atm

For temperatures between 5°C and 40°C, K_H was fit from data presented by Masschelein (1984):

$$K_H = 3.65 \times 10^{-4} \exp(3,650/T) \qquad (11\text{-}17)$$

where:

T = Temperature, °K
K_H = Henry's constant, g/L-atm

Rates of reaction of chlorine dioxide in aqueous solution are reportedly "pseudo-first order," in other words, the reactions follow a first-order law with respect to chlorine dioxide concentration and organic concentration only up to a certain point (60 to 90 percent decomposition of the original organic), and then the reaction rate slows down dramatically. The pseudo-first order rate expression may be presented as (Rav-Acha and Blits 1985; Rav-Acha and Chosen 1987):

$$dS_i/dt = k_i[\text{ClO}_{2(\text{aq})}][S_i] \qquad (11\text{-}18)$$

where:

k_i = observed pseudo-first order rate constant for substrate i
$[\text{ClO}_2]$ = instantaneous aqueous ClO_2 concentration
$[S_i]$ = concentration of substrate i

The solution pH does not appear to play much of a role in the rate process for chlorine dioxide. It has been demonstrated that rates of reaction are constant over a pH range of 4.0 to 9.5, and most reactions are carried out in the neutral pH range (Rav-Acha and Blits 1985; Rav-Acha and Chosen 1987; Amor et al. 1984).

EVALUATION OF CHEMICAL OXIDATION

Total Oxidant Demand

Chemical oxidants are not specific for particular organic compounds, or even for organic rather than inorganic compounds. In mixed-wastewater systems, the potential total oxidant demand (TOD) depends on the sum of all organic (including biodegradable components) and reduced inorganic constituents in the wastewater. Therefore, the TOD may be expressed as:

$$\text{TOD} = \text{ID} + \text{BD} + \text{RD} \qquad (11\text{-}19)$$

where:

ID = inorganic oxidant demand, from reduced inorganic species
BD = biocompatible organic oxidant demand
RD = refractory, toxic, or inhibitory (nonbiocompatible) organic oxidant demand.

It is convenient to consider the organic compounds as fractionated between the "biocompatible" components and the "nonbiocompatible" components.

Eq. (11-19) may be rearranged as:

$$TOD = ID + RD/(1 - f_b) \qquad (11\text{-}20)$$

where f_b is the "biocompatible" fraction of the organic components.

It may not be economical to treat a wastewater that has a high "biocompatible" fraction, i.e., more oxidant will be used to treat readily biodegradable components than the refractory, toxic or inhibitory components. However, this will depend upon the total concentrations, TOD, as well as the fraction of each component and the susceptibility of the various components oxidation (some specificity does, in fact, exist). To illustrate the influence of the "biocompatible" fraction, TOD/RD has been plotted in Fig. 11-2 as a function of f_b. As can be seen from Fig. 11-2, the relative oxidant demand increases exponentially as f_b approaches 1.0, compared to the requirement for the "nonbiocompatible" components alone. The TOD plays a major role in the acceptance of chemical

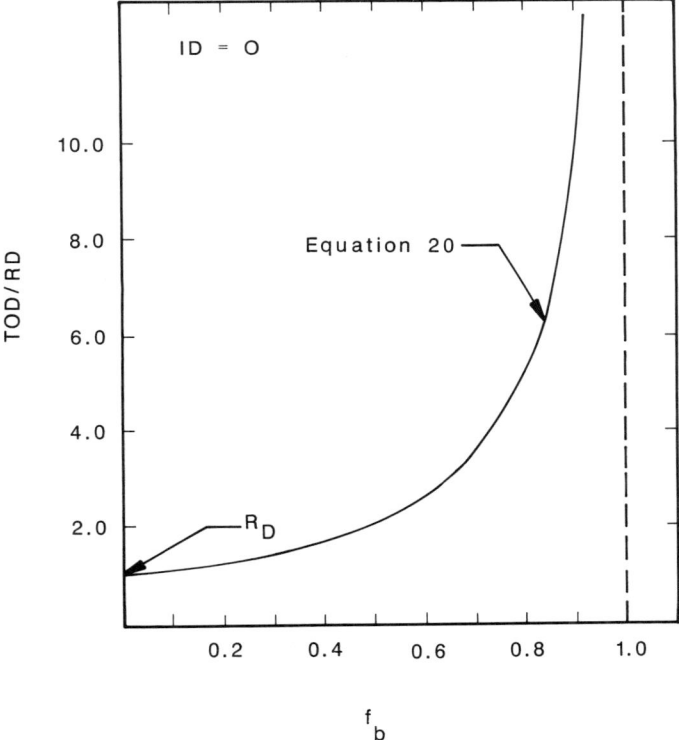

Fig. 11-2. TOD/RD as a function of the biocompatible fraction of the wastewater components, f_b.

oxidation as a valid treatment process, since the cost of applying oxidants is directly proportional to the TOD. In addition, the "biocompatible fraction" along with the magnitude of toxicity and inhibition induced by the "nonbiocompatible" fraction of the waste components will determine whether a pre- or post-oxidation scheme will be required.

Preoxidation is mandatory when sufficient toxicity is present in the wastewater to retard biological growth and prevent sufficient treatment in a biological treatment process. However, when refractory compounds result in little or no bad effect on the biological treatment process, it may be preferred to use chemical oxidation as a tertiary polishing step prior to final effluent discharge (postoxidation). This makes use of the biological treatment stage to reduce the chemical oxidant demand by removing the "biocompatible" fraction of the waste.

The TOD may also indicate the need for better waste management practices within the wastewater collection system. Segregation and preoxidation of "nonbiocompatible" wastes prior to mixing with "biocompatible" or reduced inorganic compounds could result in substantial savings in chemical costs. The potential effectiveness of waste segregation is illustrated in Fig. 11-3 which

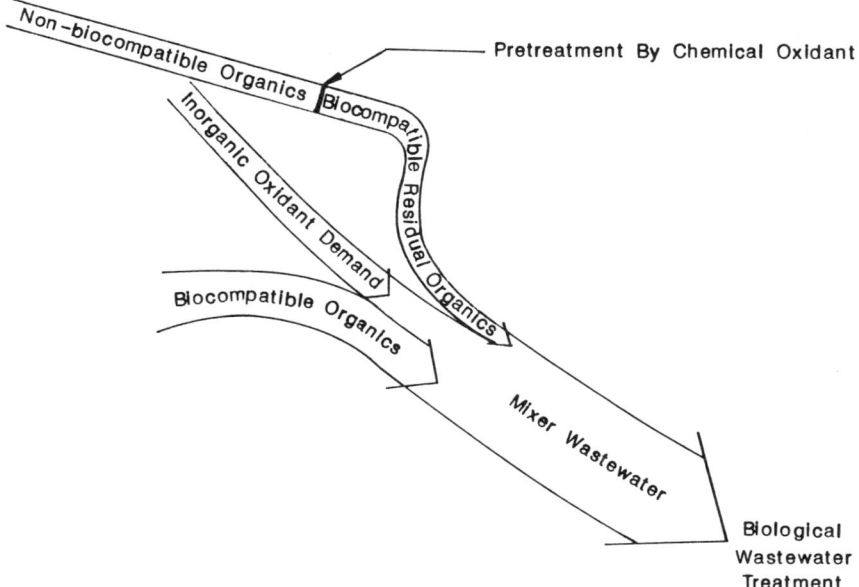

Fig. 11-3. Potential mixed wastewater makeup based on oxidant demands. Pretreatment of segregated non-biocompatible components is illustrated to reduce the volume of waste oxidized and the total oxidant demand.

shows a definite reduction in waste volume and total oxidant demand. Waste segregation is critical, since the major costs associated with chemical oxidation are not capital costs, but the chemical and/or energy costs, which accrue on a daily basis.

Laboratory Testing

The evaluation of the efficacy of chemical oxidation for waste treatment must be determined in the laboratory. The complete lack of a uniform theory or design procedure for chemical oxidation of wastewaters is not surprising considering the complexity and variability of wastewaters. The primary parameters of importance when evaluating wastewaters—initially, during, and after oxidation—may be outlined as follows:

- Concentrations of original toxic or refractory compounds (if possible)
- Total organic carbon (TOC)
- Chemical oxygen demand (COD)
- Toxicity
- Biodegradability

Monitoring the course of reaction during oxidation may be difficult. Due to the stepwise transformation of materials, the total organic carbon (TOC) may remain constant until ultimate degradation is achieved. On the other hand, particular compounds, if known and readily measurable, are direct indicators of primary degradation. The chemical oxygen demand (COD) will also change during the various degradation stages. However, in mixed-wastewater systems (containing ''biocompatible'' and ''non-biocompatible'' organics), there is no way to predict which components will react. The course of reaction with respect to typical reaction parameters is shown in Fig. 11-4. This illustrates the loss of original organics and reduction in COD while the TOC remains unchanged, i.e., no ultimate degradation. Points A, B, and C, as indicated in Fig. 11-4, are significant points for the comparison of toxicity and biodegradability.

In the event that the initial specific organic components are unknown, TOC and COD should be monitored to prevent the reaction from proceeding too far. In addition, the average oxidation state of the organic carbon in the waste mixture may be found from the COD and TOC, or (Bowers et al. 1989):

$$\text{Oxidation State} = 4(\text{TOC} - \text{COD})/\text{TOC} \qquad (11\text{-}21)$$

where COD and TOC are in molar units.

A comparison of the mean oxidation states of the original compounds, Point A in Fig. 11-4, and the mean oxidation states of the final oxidation products, Point C in Fig. 11-4, are shown in Fig. 11-5, where hydrogen peroxide was the oxidant. These data indicate the shift in the organic carbon from reduced

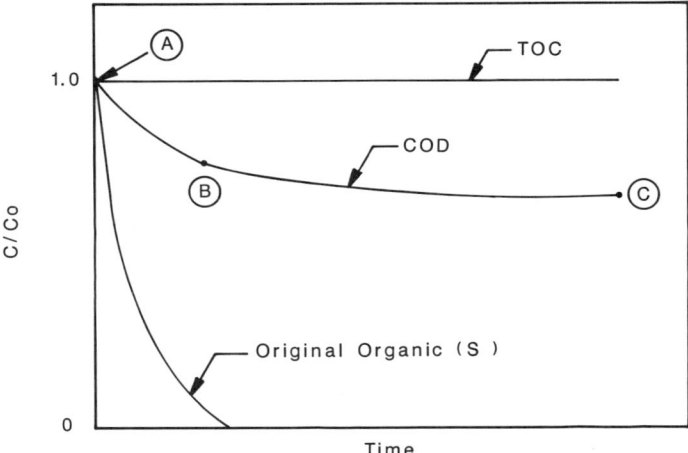

Fig. 11-4. Idealized reaction of oxidant with a specific organic compound(s). Change in original organic and COD while TOC remains constant implies only primary and/or acceptable degradations occurring. Note: A = initial components before oxidation, B = extent of reaction at which no original substrate(s) remain, C = reaction completion.

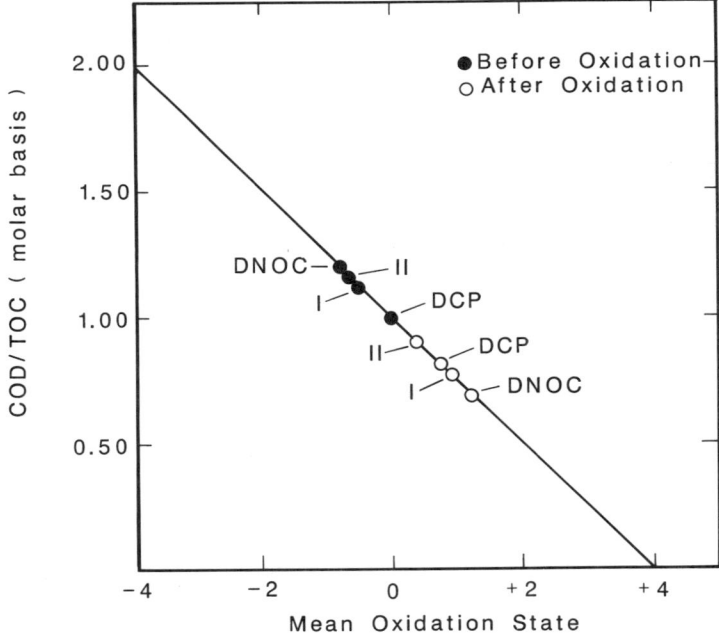

Fig. 11-5. COD/TOC versus mean oxidation states of soluble carbon before and after oxidation. Note: H_2O_2 was the oxidant, DCP = dichlorophenol, DNOC = dinitroortho-cresol, I and II = wastewaters from specialty chemical industries. (*Source:* Bowers et al. 1989.)

265

organic carbon (oxidation state < 0) to more oxidized organic carbon (oxidation state > 0; *note*: inorganic carbon $= +4$). This happens when the COD changes disproportionately with respect to the TOC, i.e., primary and acceptable degradation of original compounds.

The relation of the extent of reaction and reaction products to the "biocompatibility" of the wastewater is more difficult to quantify. Biochemical oxygen demand (BOD) is not a valid tool for measuring the original "nonbiocompatible" components because of their toxic or refractory nature. Upon reaction with oxidants, however, these original species may be degraded into more "biocompatible" components. The BOD test calls for an acclimated seed, and in fact the final oxidation products may consist primarily of common organic acids. However, failure to exhibit a significant increase in BOD after reaction does not rule out the success of the chemical oxidation process. Instead, a continuous biological treatment test may be required, using activated sludge which has been acclimated to the pretreated wastewater.

Toxicity of the wastewater may be evaluated before and after chemical oxidation by any of a variety of standard techniques. Typically, one would compare effluents based on the toxicity tests required by their discharge permit (if any). However, tests based on any accepted organism (Microtox®, *Daphnia magna, Mysidopsis bahia*, etc.) will adequately reflect the effects of chemical oxidation on toxicity.

A complete analysis of chemical oxidation must include the analysis of many parameters. A schematic of a general batch test procedure is provided in Fig. 11-6. This schematic exhibits the main points in the design of experiments for evaluating the effectiveness of chemical oxidation including:

1. A complete characterization of the initial wastewater (identity of organics, COD, TOC, toxicity, and biodegradability)
2. Chemical oxidation with various oxidants, catalysts, pHs, oxidant : organic ratios and oxidant : catalyst ratios
3. Quenching the oxidant
4. Chemical analysis of the oxidation products

Quenching the reaction refers to stopping the oxidation reaction by adding a reducing agent to destroy the residual oxidant(s). For peroxide, some form of enzyme catalase may be used for quenching, while a reduced form of sulfur, such as sodium bisulfite, may be used for most other oxidants. The quenching agent must not interfere with later testing procedures, i.e., add toxicity or affect biodegradability.

The next step is to determine the toxicity of the oxidized wastes and to evaluate the biodegradability of the oxidation products. Biodegradability may be evaluated under aerobic or anaerobic conditions depending upon the final biological treatment scheme which will be used. The effectiveness of chemical

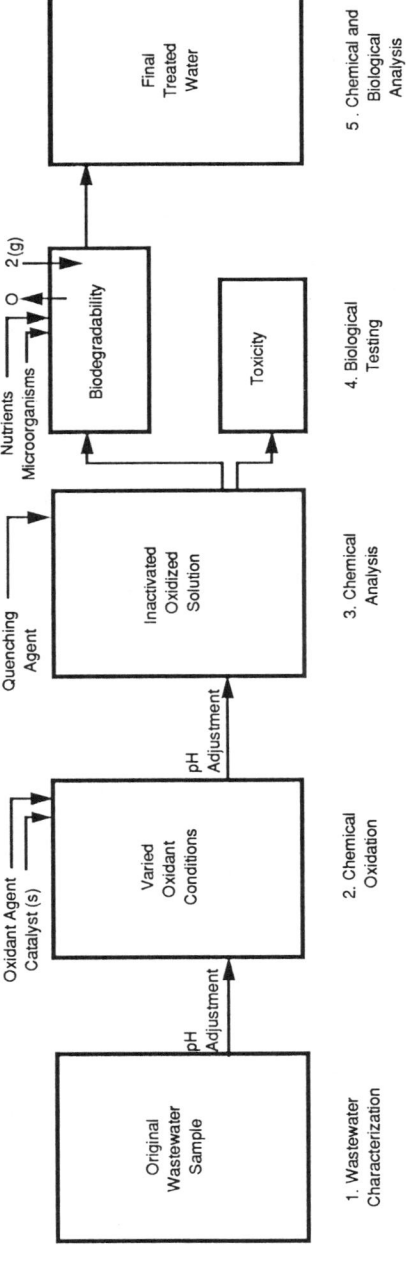

Fig. 11-6. A generalized scheme for evaluating chemical oxidation as a technique for reducing toxicity and/or improving biodegradability.

oxidation may be evaluated by comparing the toxicity, biodegradability and residual organics of the original wastewater and the oxidized wastewater. Improvements in biodegradability may appear as a decrease in the acclimation period, a faster rate of biodegradation, less residual refractory organics, and/or decreased toxicity. Evaluation procedures are provided in Chapters 4 and 7 for aerobic and anaerobic biodegradation, respectively.

It is important to note that blanks should always be run to determine whether any compounds have been lost due to volatility. Even when gas transfer is not used, ample opportunities exist for the loss of volatile compounds into the atmosphere.

CASE STUDIES

A variety of data are available in the literature, which compare the character of oxidation products to the original compounds. Bowers et al. (1989a and 1989b) evaluated a hydrogen peroxide oxidation for several aromatic compounds. An improvement was found in rates of microbial uptake of organics (based on TOC), acclimation time, toxicity (based on Microtox®), and residual refractory organics *after* aerobic biodegradation. A summary of the data collected is shown in Tables 11-4 and 11-5.

Both sets of data presented indicate a dramatic reduction in toxicity in nearly all cases, and in most instances, no detectable toxicity was observed after oxidation with H_2O_2; for example, 11 out of 18 compounds in Table 11-5 showed no toxicity after oxidation. However, 2,6-DCP showed only a threefold improvement in toxicity, due to a change in the oxidation mechanism. This is

Table 11-4. Performance Data with Hydrogen Peroxide.[a]

Wastewater	Percent Phenolics Converted	Percent[b] Ultimate Conversion	EC_{50}, %[c]		q_{max},[d] mg TOC/g VSS-hr	
			Initial	After Oxidation	Initial	After Oxidation
DCP[e]	95	36.1	0.32	2.53	0	3.47
DNOC[f]	—	30.2	8.96	>100	—	—
I[g]	94	12.1	1.05	83.1	3.89	5.10
II[g]	86	22.4	1.30	15.2	4.17	6.28

Source: Adapted from Bowers, et al. (1989a and 1989b).
[a]Based on an initial 4 : 1 molar ratio of H_2O_2 to organics, data compare initial wastewater to 4-hour oxidation products; 10 to 100 mg/l Fe(II) as catalyst.
[b]Based on TOC reduction.
[c]Expressed as percent volume of waste sample to induce a 50 percent reduction in light production of Microtox organisms.
[d]Maximum specific uptake rate, from FBR test, see Chapter 4.
[e]2,4-Dichlorophenol.
[f]Dinitro-*ortho*-cresol.
[g]Industrial wastewaters from specialty chemical manufacturer.

Table 11-5. Hydrogen Peroxide Oxidation of Aromatic Organics.

Compound[a]	Initial Concentration mg/l	Percent Reduction After H_2O_2 Oxidation[b,c] COD	Percent Reduction After H_2O_2 Oxidation[b,c] TOC	EC_{50}, % Before Oxidation	EC_{50}, % After Oxidation	Biological COD Reduction After 2 Days Reaction Before Oxidation	Biological COD Reduction After 2 Days Reaction After Oxidation
Phenol	471	76	44	6.1	NT[d]	41	47
Benzoic Acid	611	76	48	24.0	> 100[e]	69	32
Nitrobenzene	616	72	38	6.0	76.2 (47.2)[f]	59	31
Aniline	466	77	43	35.7	NT	0	40
o-Cresol	541	75	56	2.5	NT	16	51
m-Cresol	541	73	38	1.3	NT	0	51
p-Cresol	541	72	40	0.4	NT	65	47
o-Chlorophenol	625	75	48	5.1	NT	18	37
m-Chlorophenol	625	75	41	1.8	NT	0[g]	39
p-Chlorophenol	625	76	22	0.3	NT	0[g]	39
2,3-DCP	815	70	53	1.0	NT	12	31
2,4-DCP	815	69	50	0.6	> 100	9	32
2,5-DCP	815	74	42	1.9	NT	14	38
2,6-DCP	815	61	33	5.7	17.3 (11.5)[f]	0	9
3,5-DCP	815	69	49	0.5	NT	0[g]	9
2,3-DNP	921	80	51	6.3	85.6 (41.9)[f]	0[g]	19
2,4-DNP	921	73	51	2.0	100	0[g]	49
2,4,6-TCP	800	47	44	2.8	52.2 (29.2)[f]	0	39
Average of Above		72	44			17	36

Source: Bowers et al. 1989b
[a]DCP = dichlorophenol, TCP = trichlorophenol, DNP = dinitrophenol.
[b]Conditions—stoichiometric dosage of H_2O_2, pH 3.5, 50 mg/l, Fe^{++}, 24 hour reaction time.
[c]100 mg/l initial COD, 50 mg/l initial MLVSS, open to atmosphere.
[d]NT = Not Toxic.
[e]> 100 implies some light reduction at no dilution, but not a 50 percent reduction.
[f]Numbers in parenthesis reflect the actual toxicity of the organics, i.e., adjusted to account for TOC removed, or, numbers represent fraction of TOC remaining multiplied by the EC_{50}.
[g]Not biodegraded at all, even after 14 days.

due to the positioning of both chlorines adjacent to the $-OH$ group, which causes a blockage of the primary reaction sites. The only other similar compound, 2,4,6-TCP, showed the same trend, where blockage of both sites adjacent to the $-OH$ group resulted in a reduced toxicity improvement compared to the other compounds tested.

Other data, Table 11-4, indicate that the rate of biodegradation (using an unacclimated activated sludge) was improved after oxidation with hydrogen peroxide; for example, 2,4-DCP was not removed at all, while its oxidation products were removed at a rate of 3.47 mg TOC/g VSS-hr. This removal rate after oxidation is superior to the removal rate of 2,4-DCP using a well acclimated sludge of 1.44 mg TOC/g VSS-hr (Bowers et al. 1989a). All of these data except for three compounds indicate an acceptable chemical degradation of the aromatic compounds tested, and that the oxidation byproducts are readily degradable substrates that require little or no acclimation by typical municipal activated sludge.

Additionally, the residual refractory compounds were studied by Bowers et al. (1989b). For the 18 compounds in Table 11-5, the average residual refractory COD (after 14 days of biodegradation) was 29 percent for the original compounds (excluding nonbiodegradables) and 13 percent for the oxidation products (overall removal based on ultimate chemical oxidation to CO_2 + biodegradation).

Wang et al., (1989a; 1989b) studied the oxidation of 2,5-dichlorophenol and *ortho*-cresol using ozone. They evaluated the toxicity and biodegradability of these compounds by methanogenic bacteria under anaerobic conditions. A summary of their results is presented in Table 11-6. The data in Table 11-6 indicate an acceptable degradation of *o*-cresol and 2,5-dichlorophenol to simple organic acids.

In the case of *o*-cresol, ozonation at a pH of 9.0 resulted in an improvement in methane production while pH 3.0 yielded no particular improvement. While not necessarily expected, the mechanism of oxidation at pH 3.0 would be primarily by means of direct reaction with O_3, while at pH 9.0 reaction with the hydroxyl radical, $\cdot OH$, is the primary mechanism (Mehta et al. 1989). Therefore, it is expected that the oxidation byproducts will differ (as also shown in Table 11-6), and there is no guarantee that both sets of byproducts will be acceptably less toxic.

Other studies indicates that permanganate, chlorine dioxide and ferrate are capable of oxidizing a variety of toxic, refractory and inhibitory organics. Chlorine dioxide has been shown to react with polyaromatic hydrocarbons, alkenes

Table 11-6. Ozonation of *o*-Cresol and 2,5-Dichlorophenol.

	o-Cresol		2,5-Dichlorophenol	
	pH = 3.0[a]	pH = 9.0[b]	pH = 3.0[a,c]	pH = 9.0[b,d]
Major ozonation byproducts[a]	glyoxylic acid acetic acid formic acid oxalic acid	oxalic acid formic acid acetic acid	oxalic acid formic acid acetic acid	oxalic acid
Improved methane production	no[f]	yes[g]	yes[h]	yes[h]

Source: Wang et al (1989a, 1989b).
[a]pH = 3.0 implies direct oxidation by O_3.
[b]pH = 9.0 implies a free radical mechanism, $\cdot OH$.
[c]Identified about 50 percent of the remaining TOC during ozonation.
[d]Identified about 35 percent of the remaining TOC during ozonation.
[e]Listed in order of concentration, some other minor components given by authors.
[f]Early ozonation products were more toxic than 0-cresol.
[g]After 60 percent COD reduction or 32 percent reduction in TOC.
[h]Reduced toxicity to methane production for acetic acid and phenol principal organics, using 2,5-DCP as an added toxin.

and phenolic compounds (Masschelein 1984; Rav-Acha and Blits 1985; Rav-Acha and Chosen 1987; Amor et al. 1984). For phenolics, it has been demonstrated that up to 50 percent of the TOC may be removed by ClO_2, and that the common byproducts ultimately are organic acids, such as oxalic, malic, and formic acid (Amor et al. 1984).

Deluca et al. (1981) found that potassium ferrate could effectively remove naphthalene at ferrate/napthtalene molar ratios of greater than 20:1. Ferrate has been shown to be reactive with phenol and other organics (Waite and Gilbert 1979).

Permanganate has long been used in drinking water treatment for removal of tastes and odors, and currently for the removal of THM precursors. However, its use in treatment of toxic and refractory compounds is surprisingly not well documented. There is some evidence that permanganate reacts with a variety of phenolics, and may yield decreases in toxicity and improvements in biodegradability (Throop 1977; Reides et al. in press).

REFERENCES

Amor, H. B., J. DeLaat and M. Dore. 1984. Chlorine Dioxide Consumption and Reactions on Phenolic Compounds. *Water Research*, 1545–1560.

Baral, S., C. Lume-Pereira, E. Janata and A. Henglein. 1985. Chemistry of Colloidal Manganese Dioxide. 2. Reaction with O_2 and H_2O_2. *J. Phys. Chem.*, **89:** 5779.

Bowers, A. R., P. Gaddipati, W. W. Eckenfelder, Jr. and R. M. Monsen. 1989a. Treatment of Toxic or Refractory Wastewaters with Hydrogen Peroxide. *Water Sci. Technol.*, **21:** 477–486.

Bowers, A. R., W. S. Kong, W. W. Eckenfelder, Jr. and R. M. Monsen. 1989b. Treatment of Aromatic Organic Compounds with Hydrogen Peroxide. Industrial Waste Symposium, 62nd Annual Conference of the Water Pollution Control Federation, San Francisco, CA.

Brandes, R., T. Wall, B. Newton and H. M. Biswas. 1987. *Permit Writer's Guide to Water Quality-Based Permitting for Toxic Pollutants.* EPA-440/4-87-005, Washington, DC. U.S. Environmental Protection Agency.

Carus Chemical Co., Inc. 1979. *The Cairox® Method for Water Treatment.* LaSalle, Illinois: Carus Chemical Co.

Deluca, S., A. C. Chao, and C. Smallwood, Jr. 1981. Removal of Selected Pollutants with Potassium Ferrate. *Proc. Thirteenth Mid Atlantic Industrial Waste Conference* (C. P. Huang, ed.). Ann Arbor, MI: Ann Arbor Science, Inc.

Eckenfelder, W. W., Jr. 1989. *Industrial Water Pollution Control*, 2nd edition. New York: McGraw-Hill, Inc.

Eisenhauer, H. R. 1968. Ozonation of Phenolic Wastes. *J. WPCF*, **40:** 1887–1899.

Fenton, H. S. H. 1894. Oxidation of Tartaric Acid in the Presence of Iron. *J. Chem. Soc.*, **65:** 899.

Francis, K. C., D. Cummins, and J. Oakes. 1985. Kinetic and Structural Investigations of [FeIII(EDTA)]-[EDTA] Catalyzed Decomposition of Hydrogen Peroxide. *J. Chem. Soc.*, **3:** 493.

Froment, G. F. and K. B. Bischoff. 1979. *Chemical Reactor Analysis and Design.* New York: McGraw-Hill, Inc.

Gould, J. P. and W. J. Weber. 1976. Oxidation of Phenols by Ozone. *J. WPCF*, **48:**47–60.

Grady, C. P. L. 1986. Biodegradation of Hazardous Wastes by Conventional Biological Treatment. *Hazardous Wastes Hazardous Materials*, **3:** 333.

Gurol, M. D. 1985. Factors Controlling the Removal of Organic Pollutants in Ozone Reactors. *J. Amer. Water Works Association*, **77**: 77.

Gurol, M. D. and S. Nekouinaini. 1984. Kinetic Behavior Ozone in Aqueous Solutions of Substituted Phenols. *Ind. Eng. Chem. Fund.*, **23**: 54–60.

Gurol, M. D. and S. Nekouinaini. 1985. Effect of Organic Substances on Mass Transfer in Bubble Aeration. *J. WPCF*, **57**: 235–240.

Hoffman, J. M. and P. C. Moser. 1985. *Chemical Process Hazard Review*, ACS Symposium Series No. 274. Washington, DC: American Chemical Society.

Klibanov, A. M., B. N. Alberti, E. D. Morris, and L. M. Felshin. 1980. Enzymatic Removal of Toxic Phenols and Anilines from Wastewaters. *J. Applied Biochem.*, **2**: 414.

Lim, P. K., J. A. Cha, and B. S. Fagg. 1984. Use of Manganese(II)–Polyol Complexes to Accelerate High-pH Peroxide Oxidation Reactions. *Ind. Eng. Chem. Fund.*, **23**: 29.

Lyman, W. J., W. F. Reehl and D. H. Rosenblatt. 1982. *Handbook of Chemical Property Estimation Methods*. New York: McGraw-Hill, Inc.

Masschelein, W. J. 1984. Experience with Chlorine Dioxide in Brussels: Generation of Chlorine Dioxide. *J. Amer. Water Works Assoc.*, **76**: 70–76.

Mehta, Y. M., C. E. George, and C. H. Kuos. 1989. Mass Transfer and Selectivity of Ozone Reactions. *Canad. J. Chem. Eng.*, **67**: 118–126.

Merz, J. H. and W. A. Waters. 1949. Some Oxidations Involving the Free Hydroxyl Radical. *J. Chem. Soc.*, Part V: 515.

Rav-Acha, Ch. and R. Blits. 1985. The Different Reaction Mechanisms by Which Chlorine and Chlorine Dioxide React with Polycyclic Aromatic Hydrocarbons (PAH) in Water. *Water Research*, **19**: 1273–1281.

Rav-Acha, Ch. and E. Chosen. 1987. Aqueous Reactions of Chlorine Dioxide with Hydrocarbons. *Environ. Sci. Technol.*, **21**: 1069–1074.

Reides, A. H., J. E. Bold, and W. M. Joyce. In press. Potassium Permangate. In *Handbook of Chemical Oxidation in Wastewater Treatment*. Lancaster, PA: Technomics, Inc.

Rice, R. G. and A. Netzer, editors. 1982. *Handbook of Ozone Technology and Applications*. Ann Arbor, MI: Ann Arbor Science, Inc.

Roberts, P. V., C. Munz, D. Dandliker and C. Matter-Muller. 1984. Volumization of Organic Pollutants in Wastewater Treatment: Model Studies. U.S. EPA: Cincinnati, OH, EPI.8912; 60015 2-84-047.

Roth, J. A., W. L. Moench, Jr., and K. A. Debelak. 1982. Kinetic Modeling of the Ozonation of Phenol in Water. *J. WPCF*, **54**: 135–139.

Roth, J. A. and D. E. Sullivan. 1981. Solubility of Ozone in Water. *Ind. Eng. Chem. Fund.*, **20**: 137–140.

Stumm, W. W. and J. J. Morgan. 1981. *Aquatic Chemistry*. New York: John Wiley & Sons, Inc.

Sullivan, D. E. and J. A. Roth. 1980. Kinetics of Ozone Self-Decomposition in Aqueous Solution. *Water-1979*, **76**(197): 142–149.

Throop, W. M. 1977. Alternative Methods of Phenol Wastewater Control. *J. Hazard Mater.*, **1**: 319–329.

Waite, T. D. and M. Gilbert. 1979. Oxidation Destruction of Phenol and Other Organic Water Residuals by Iron(VI) Ferrate. *J. WPCF*, **50**: 543–551.

Wang, Y. T., P. C. Pai, and J. L. Lutchaw. 1989a. Effects of Preozonation on Anaerobic Biodegradability *J. Environ. Eng.*, **115**: 336–347.

Wang, Y. T., P. C. Pai, and J. L. Latchaw. 1989b. Effects of Preozonation on the Methanogenic Toxicity of 2,5 Dichlorophenol. *J. WPCF*, **61**: 320–326.

Weast, R. C. 1987/88. *CRC Handbook of Chemistry and Physics*, 68th edition. Boca Raton, FL: CRC Press, Inc.

Wood, R. H. 1952. The Heat, Free Energy and Entropy of the Ferrante(VI) Ion. *J. Amer. Chem. Soc.*, **80**: 2036.

12

Management of Sludges from Treatment of Toxic Wastewaters

Jeffrey L. Pintenich

INTRODUCTION

Several of the unit operations utilized to reduce the aquatic toxicity of industrial effluents generate semi-solid residues or sludges. In some systems, the inherent toxicity of the effluent is concentrated almost totally in the sludge, such as heavy metals removed by precipitation. In other systems, such as biological activated sludge processes, little or none of the original toxicity of the effluent may be present as such in the waste sludge. As depicted in Fig. 12-1, the sludge management strategy chosen in a specific application depends principally upon three interrelated factors: (1) the physical and chemical characteristics of the sludge; (2) the susceptibility of the sludge to various treatment processes, i.e., its "treatability"; and (3) the final disposal option for the sludge. Geographic and economic constraints can also be important.

In the U.S., especially since 1980, owners and operators of industrial wastewater treatment facilities generating sludges have been subject to a pervasive set of government regulations, at both federal and state levels, which place a number of constraints upon the management of wastes containing toxic materials. The most restrictive of these regulations deal with ultimate disposal practices.

In addition, there are proposals for even further regulation of solid and hazardous wastes. As a result, for many wastewater treatment systems, one-half or more of the total cost of treatment is now attributable to sludge management costs.

REGULATORY CONSIDERATIONS

The environmental statute which has had the greatest impact upon sludge management practices in the U.S. is the Resource Conservation and Recovery Act

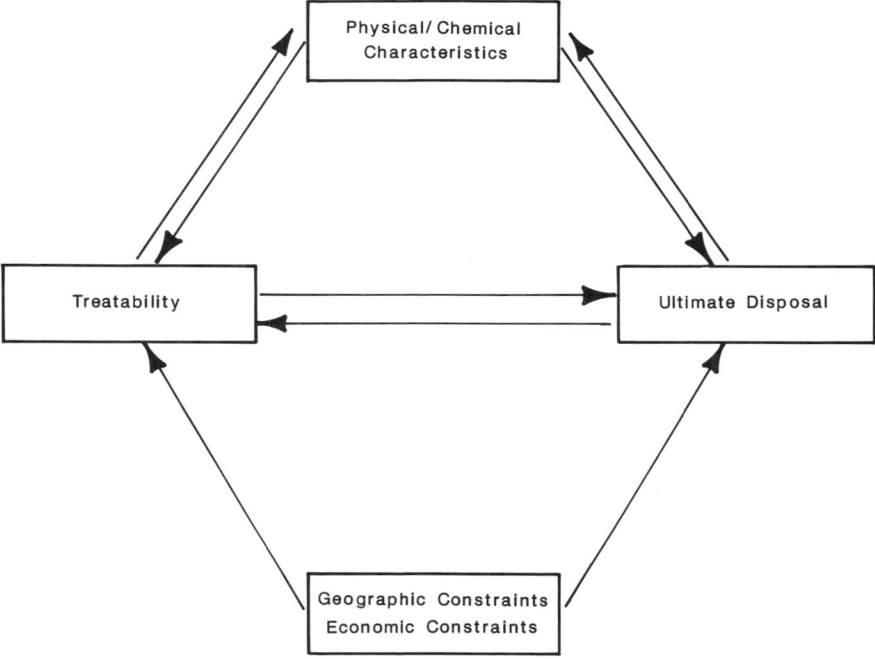

Fig. 12-1. Key factors in the selection of a sludge management strategy.

(RCRA), enacted by Congress in 1976. RCRA directed the U.S. Environmental Protection Agency (U.S. EPA) to develop and implement a national regulatory program for solid and hazardous wastes. The first major rulemaking activities under RCRA took place in 1980, with the promulgation of a comprehensive set of hazardous waste management (HWM) regulations (40 *CFR*, Subtitle C). The enforcement of the HWM regulations has been delegated, in most cases, to the various state environmental agencies. Several of the states have promulgated regulations which are even more stringent than those of the U.S. EPA. The federal HWM regulations have been in a constant state of evolution over the last decade. Recently, U.S. EPA has proposed regulations to govern municipal solid waste landfills (for nonhazardous waste) and, under the federal Clean Water Act, has proposed a comprehensive set of restrictions for the treatment and disposal of municipal sewage sludge. It is anticipated that these may ultimately be extended to non-hazardous industrial wastes as well.

Hazardous Wastes

By definition, a *solid waste* is any discarded material, whether it be a solid, liquid, semi-solid, or contained gas. Solid wastes include municipal refuse, sludges, many industrial wastes, and other wastes that are not excluded by regulation or excluded by a variance to the regulations. Wastes which are excluded by regulation include sanitary wastewater or mixtures of sanitary wastewater and other wastewaters that flow through sewer systems to publicly owned treatment works (POTWs); point source discharges, including treated industrial effluents, that are regulated under the Clean Water Act; nuclear wastes; and several others. U.S. EPA defines a *sludge* as any solid, semi-solid, or liquid waste generated by a wastewater treatment facility, water supply treatment plant, or air pollution control facility, exclusive of the treated effluents from a wastewater facility. It is clear, therefore, that industrial wastewater treatment sludges are solid wastes, unless a specific variance has been obtained.

Hazardous wastes are those solid wastes which could cause or contribute to an increase in human mortality or serious illness, or could pose a hazard to human health or the environment when improperly managed. Hazardous wastes have been specifically identified by U.S. EPA and listed in the regulations ("listed wastes"), or they are solid wastes which possess one or more of the characteristics of hazardous wastes ("characteristic wastes"). In deciding whether or not a waste should be listed, U.S. EPA has determined that the waste possesses one of the four characteristics (ignitability, corrosivity, reactivity, EP toxicity) or that it fits the definition of "acutely hazardous" or "toxic." Title 40 of the *Code of Federal Regulations* identifies hazardous wastes at Part 261 as shown in Table 12-1.

A solid waste exhibits the characteristic of *ignitability* if a representative sample of the waste has any of the following properties (see 40 *CFR* 261.21):

- It is a liquid, other than an aqueous solution containing less than 24 percent alcohol by volume, and has a flash point less than 60°C (140°F), as determined by a Pensky-Martens Closed Cup Tester, using the test method specified in ASTM Standard D-93-79 or D-93-80 (incorporated by reference, see Section 260.11), or a Setaflash Closed Cup Tester, using the test method specified in ASTM Standard D-3278-78 (incorporated by reference, see Section 260.11), or as determined by an equivalent test method

Table 12-1.   U.S. EPA Hazardous Waste Designations.

Designation	Characteristic	Hazard Code
Ignitable	Yes	I
Corrosive	Yes	C
Reactive	Yes	R
EP toxic	Yes	E
Acutely hazardous	No	H
Toxic	No	T

approved by the Administrator of the U.S. EPA under procedures set forth in Sections 260.20 and 260.21.

- It is not a liquid and is capable, under standard temperature and pressure, of causing fire through friction, absorption of moisture, or spontaneous chemical changes and, when ignited, burns so vigorously and persistently that it creates a hazard.
- It is an ignitable compressed gas as defined in 49 *CFR* 173.300 and as determined by the test methods described in that regulation or equivalent test methods approved by the Administrator under Sections 260.20 and 260.21.
- It is an oxidizer as defined in 49 *CFR* 173.151.

A solid waste that exhibits the characteristic of ignitability (Hazardous Code I), but is not listed as a hazardous waste in 40 *CFR* 261, Subpart D, has the U.S. EPA Hazardous Waste Number of D001.

A solid waste exhibits the characteristic of *corrosivity* if a representative sample of the waste has either of the following properties (40 *CFR* 261.22):

- It is aqueous and has a pH less than or equal to 2 or greater than or equal to 12.5, as determined by a pH meter using either a U.S. EPA test method or an equivalent test method approved by the Administrator under the procedures set forth in Sections 260.20 and 260.21. The U.S. EPA test method for pH is specified as Method 5.2 in "Test Methods for the Evaluation of Solid Waste, Physical/Chemical Methods" (incorporated by reference, see Section 260.11).
- It is a liquid and corrodes steel (SAE 1020) at a rate of greater than 6.35 millimeters (0.250 inches) per year at a test temperature of 55°C (130°F) as determined by the test method specified in NACE (National Association of Corrosion Engineers) Standard TM-01-69 as standardized in "Test Methods for the Evaluation of Solid Waste, Physical/Chemical Methods" (incorporated by reference, see Section 260.11) or an equivalent test method approved by the Administrator under the procedures set forth in Sections 260.20 and 260.21.

A solid waste that exhibits the characteristic of corrosivity (Hazard Code C), but is not listed as a hazardous waste in 40 CFR 261 Subpart D, has the U.S. EPA Hazardous Waste Number of D002.

A solid waste exhibits the characteristic of *reactivity* if a representative sample of the waste has any of the following properties (40 *CFR* 261.23):

- It is normally unstable and readily undergoes violent change without detonating.
- It reacts violently with water.
- It forms potentially explosive mixtures with water.

- When mixed with water, it generates toxic gases, vapors, or fumes in a quantity sufficient to present a danger to human health or the environment.
- It is a cyanide or sulfide bearing water which, when exposed to pH conditions between 2 and 12.5, can generate toxic gases, vapors, or fumes in a quantity sufficient to present a danger to human health or the environment.
- It is capable of detonation or explosive reaction if it is subjected to a strong initiating source or if heated under confinement.
- It is readily capable of detonation of explosive decomposition or reaction at standard temperature and pressure.
- It is a forbidden explosive as defined in 49 *CFR* 173.51, a Class A explosive as defined in 49 *CFR* 173.53, or a Class B explosive as defined in 49 *CFR* 173.88

A solid waste that exhibits the characteristic of reactivity (Hazard Code R), but is not listed as a hazardous waste in Subpart D, has the U.S. EPA Hazardous Waste Number of D003.

A solid waste exhibits the characteristic of *EP (Extraction Procedure) toxicity* if, using test methods approved by the U.S. EPA Administrator, the extract from a representative sample of the waste contains any of the contaminants listed at concentrations greater than or equal to the values presented in Table 12-2. Where the waste contains less than 0.5 percent filterable solids, the waste itself, after filtering, is considered to be the extract (see 40 *CFR* 261.24).

Table 12-2. Threshold Values for the Characteristic of EP Toxicity.

EPA Hazardous Waste Number	Contaminant	Maximum Contaminant Concentration, mg/l
D004	Arsenic	5.0
D005	Barium	100.0
D006	Cadmium	1.0
D007	Chromium	5.0
D008	Lead	5.0
D009	Mercury	0.2
D010	Selenium	1.0
D011	Silver	5.0
D012	Endrin (1,2,3,4,10-hexachloro-1,7,-epoxy-1,4,4a,5,6,7,8,8a-octahydro-1,4-endo-5,8-dimethanonaphthalene)	0.02
D013	Lindane (1,2,3,4,5,6-hexachlorocyclohexane, gamma isomer)	0.4
D014	Methoxychlor (1,1,1-trichloro-2,2-*bis*[*p*-methoxyphenyl] ethane)	10.0
D015	Toxaphene ($C_{10}H_{10}Cl_8$, technical chlorinated camphene, 67–69 percent chlorine)	0.5
D016	2,4-D (2,4-dichlorophenoxyacetic acid)	10.0
D017	2,4,5-TP Silvex (2,4,5-trichlorophenoxypropionic acid)	1.0

A solid waste is *acutely hazardous* (40 *CFR* 261.11) if a representative sample of the waste has been found to be fatal to humans in low doses or, in the absence of data on human toxicity, it has been shown in studies to have an oral LD_{50} toxicity (rat) of less than 50 milligrams per kilogram, an inhalation LC_{50} toxicity (rat) of less than 2 milligrams per liter, or a dermal LD_{50} toxicity (rabbit) of less than 200 milligrams per kilogram, or is otherwise capable of causing or significantly contributing to an increase in serious irreversible, or incapacitating reversible, illness.

A solid waste is *toxic* (see 40 *CFR* 261.11) if a representative sample of the waste contains any of the toxic constituents listed in Appendix VIII of 40 *CFR* 261 unless, after considering several specific factors, the Administrator concludes that the waste is not capable of posing a substantial present or potential hazard to human health or the environment when improperly treated, stored, transported, or disposed of, or otherwise managed. Substances listed in Appendix VIII have been shown in scientific studies to have toxic, carcinogenic, mutagenic, or teratogenic effects on humans or other life forms.

Fig. 12-2 summarizes the general types of hazardous wastes subject to regulation.

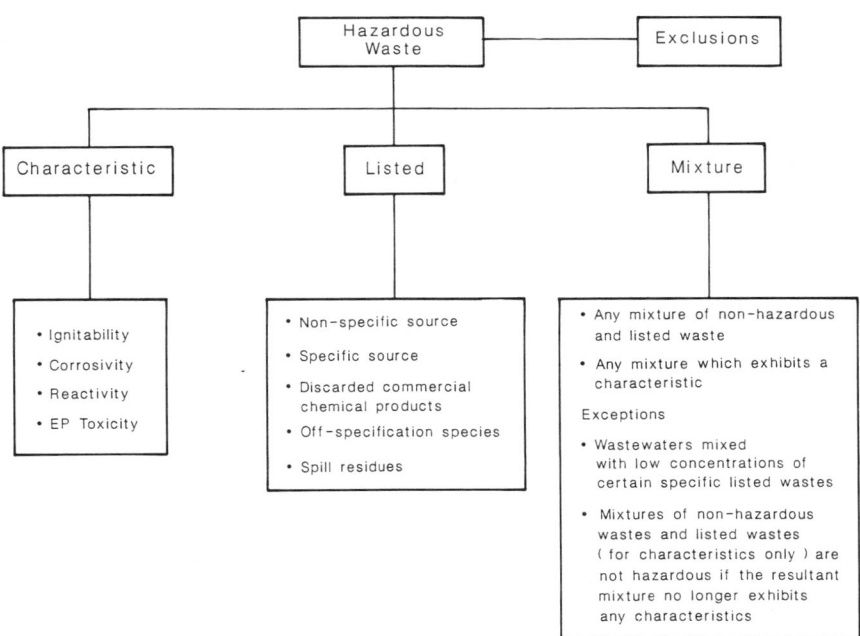

Fig. 12-2. Types of hazardous wastes.

Listed Hazardous Wastes

A variety of process wastewaters and wastewater treatment sludges are designated as listed hazardous wastes, as shown in Table 12-3. For the most part, these wastes contain one or more toxic hazardous constituents as indicated in Table 12-4. It should be noted that unless a specific exclusion is obtained, residuals from the treatment of a listed waste are also considered to be hazardous wastes.

Table 12-3.   Process Hazardous Wastewaters and Treatment Sludges Identified as Listed Hazardous Wastes.[a]

Industry and EPA Hazardous Waste No.	Hazardous Waste	Hazard Code
Non-Specific Sources:		
F001	The following spent halogenated solvents used in degreasing: tetrachloroethylene, trichloroethylene, methylene chloride, 1,1,1-trichloroethane, carbon tetrachloride, and chlorinated fluorocarbons; all spent solvent mixtures/blends used in degreasing containing, before use, a total of 10 percent or more (by volume) of one or more of the above halogenated solvents or those solvents listed in F002, F004, and F005; and still bottoms from the recovery of these spent solvents and spent solvent mixtures.	(T)
F002	The following spent halogenated solvents: tetrachloroethylene, methylene chloride, trichloroethylene, 1,1,1-trichloroethane, chlorobenzene, 1,1,2-trichloro-1,2,2-trifluoroethane, ortho-dichlorobenzene, trichlorofluoromethane, and 1,1,2-trichloroethane; all spent solvent mixtures/blends containing, before use, a total of 10 percent or more (by volume) of one or more of the above halogenated solvents or those listed in F001, F004, or F005; and still bottoms from the recovery of these spent solvents and spent solvent mixtures.	(T)
F003	The following spent nonhalogenated solvents: xylene, acetone, ethyl acetate, ethyl benzene, ethyl ether, methyl isobutyl ketone, *n*-butyl alcohol, cyclohexanone, and methanol; all spent solvent mixtures/blends containing before use, only the above spent nonhalogenated solvents; and all spent solvent mixtures/blends containing, before use, one or more of the above nonhalogenated solvents, and a total of 10 percent or more (by volume) of one or more of those solvents listed in F001, F002, F004, and F005; and still bottoms from the recovery of these spent solvents and spent solvent mixtures.	(I)

Table 12-3. (*Continued*)

Industry and EPA Hazardous Waste No.	Hazardous Waste	Hazard Code
Non-Specific Sources:		
F004	The following spent nonhalogenated solvents: cresols and cresylic acid, and nitrobenzene; all spent solvent mixtures/ blends containing, before use, a total of 10 percent or more (by volume) of one or more of the above nonhalogenated solvents or those solvents listed in F001, F002, and F005; and still bottoms from the recovery of these spent solvents and spent solvent mixtures.	(T)
F005	The following spent nonhalogenated solvents: toluene, methylethylketone, carbon disulfide, isobutanol, pyridine, benzene, 2-ethoxyethanol, and 2-nitropropane; all spent solvent mixtures/blends containing, before use, a total of 10 percent or more (by volume) of one or more of the above non-halogenated solvents or those solvents listed in F001, F002, or F004; and still bottoms from the recovery of these spent solvents and spent solvent mixtures.	(I, T)
F006	Wastewater treatment sludges from electroplating operations except from the following processes: (1) sulfuric acid anodizing of alumnium; (2) tin plating on carbon steel; (3) zinc plating (segregated basis) on carbon steel; (4) aluminum or zinc-aluminum plating on carbon steel; (5) cleaning/ stripping associated with tin, zinc, and aluminum plating on carbon steel; and (6) chemical etching and milling of aluminum.	(T)
F007	Spent cyanide plating bath solutions from electroplating operations.	(R, T)
F008	Plating bath residues from the bottom of plating baths from electroplating operations where cyanides are used in the process.	(R, T)
F009	Spent stripping and cleaning bath solutions from electroplating operations where cyanindes are used in the process.	(R, T)
F010	Quenching bath residues from oil baths from metal heat treating operations where cyanides are used in the process.	(R, T)
F011	Spent cyanide solutions from salt bath pot cleaning from heat treating operations.	(R, T)
F012	Quenching wastewater treatment sludges from metal heat treating operations where cyanides are used in the process.	(T)
Wood Preservation:		
K001	Bottom sediment sludge from the treatment of wastewaters from wood preserving processes that use creosote and/or pentachlorophenol.	(T)

Table 12-3.   (*Continued*)

Industry and EPA Hazardous Waste No.	Hazardous Waste	Hazard Code
Inorganic Pigments:		
K002	Wastewater treatment sludge from the production of chrome yellow and orange pigments.	(T)
K003	Wastewater treatment sludge from the production of molybdate orange pigments.	(T)
K004	Wastewater treatment sludge from the production of zinc yellow pigments.	(T)
K005	Wastewater treatment sludge from the production of chrome green pigments.	(T)
K006	Wastewater treatment sludge from the production of chrome oxide green pigments (anhydrous and hydrated).	(T)
K007	Wastewater treatment sludge from the production of iron blue pigments.	(T)
Organic Chemicals:		
K104	Combined wastewater streams generated from nitrobenzene/ aniline production.	(T)
Inorganic Chemicals:		
K106	Wastewater treatment sludge from the mercury cell process in chlorine production.	(T)
Pesticides:		
K032	Wastewater treatment sludge from the production of chlordane.	(T)
K033	Wastewater and scrub water from the chlorination of cyclopentadiene in the production of chlordane.	(T)
K035	Wastewater treatment sludges generated in the production of creosote.	(T)
K037	Wastewater treatment sludges from the production of disulfoton.	(T)
K038	Wastewater from the washing and stripping of phorate production.	(T)
K040	Wastewater treatment sludges from the production of phorate.	(T)
K041	Wastewater treatment sludges from the production of toxaphene.	(T)
K098	Untreated process wastewater from the production of toxaphene.	(T)
K099	Untreated wastewater from the production of 2,4-D.	(T)
K123	Process wastewater (including supernates, filtrates, and washwaters) from the production of ethylene*bis*dithiocarbamic acid and its salt.	(T)
Explosives:		
K044	Wastewater treatment sludges from the manufacturing and processing of explosives.	(R)

Table 12-3. (Continued)

Industry and EPA Hazardous Waste No.	Hazardous Waste	Hazard Code
Explosives:		
K045	Spent carbon from the treatment of wastewater containing explosives.	(R)
K046	Wastewater treatment sludges from the manufacturing, formulation, and loading of lead-based initiating compounds.	(T)
K047	Pink/red water from TNT operations.	(R)
Petroleum Refining:		
K048	Dissolved air flotation (DAF) float from the petroleum refining industry.	(T)
K049	Slop oil emulsion solids from the petroleum refining industry.	(T)
K051	API separator sludge from the petroleum refining industry.	(T)
Iron and Steel:		
K062	Spent pickle liquor generated by steel finishing operations of facilities within the iron and steel industry (SIC Codes 331 and 332).	(C, T)
Primary Lead:		
K065	Surface impoundment solids contained in and dredged from surface impoundments at primary lead smelting facilities.	(T)
Primary Zinc:		
K066	Sludge from treatment of process wastewater and/or acid plant blowdown from primary zinc production.	(T)
Veterinary Pharmaceuticals:		
K084	Wastewater treatment sludges generated during the production of veterinary pharmaceuticals from arsenic or organic-arsenic compounds.	(T)
Ink Formulation:		
K086	Solvent washes and sludges, caustic washes and sludges, or water washes and sludges from cleaning tubs and equipment used in the formulation of ink from pigments, driers, soaps, and stabilizers containing chromium and lead.	(T)

[a]The reader is cautioned that a case-by-case evaluation is necessary. Other wastewaters and sludges may be hazardous wastes due to the mixture provisions and/or based upon the characteristics of the wastes. Furthermore, this list is accurate as of June 1989, and the U.S. EPA may have added or deleted listed wastes subsequent to this date.

Table 12-4.   Basis for Listing Hazardous Waste.

EPA Hazardous Waste No.	Hazardous Constituents for Which Listed
F001	Tetrachloroethylene, methylene chloride, trichloroethylene, 1,1,1-trichloroethane, carbon tetrachloride, chlorinated fluorocarbons.
F002	Tetrachloroethylene, methylene chloride, trichloroethylene, 1,1,1-trichloroethane, 1,1,2-trichloroethane, chlorobenzene, 1,1,2-trichloro-1,2,2-trifluoroethane, *ortho*-dichlorobenzene, trichlorofluoromethane.
F003	NA[a]
F004	Cresols and cresylic acid, nitrobenzene.
F005	Toluene, methyl ethyl ketone, carbon disulfide, isobutanol, pyridine, 2-ethoxyethanol, benzene, 2-nitropropane.
F006	Cadmium, hexavalent chromium, nickel, cyanide (complexed).
F007	Cyanide (salts).
F008	Cyanide (salts).
F009	Cyanide (salts).
F010	Cyanide (salts).
F011	Cyanide (salts).
F012	Cyanide (complexed).
K001	Pentachlorophenol, phenol, 2-chlorophenol, *p*-chloro-*m*-cresol, 2,4-dimethylphenyl, 2,4-dinitrophenol, trichlorophenols, tetrachlorophenols, 2,4-dinitrophenol, creosote, chrysene, naphthalene, fluoranthene, benzo(b)fluoranthene, benzo(a)pyrene, indeno(1,2,3-cd)pyrene, benzo(a)anthracene, dibenz(a)anthracene, acenaphthalene.
K002	Hexavalent chromium, lead.
K003	Hexavalent chromium, lead.
K004	Hexavalent chromium.
K005	Hexavalent chromium, lead.
K006	Hexavalent chromium.
K007	Cyanide (complexed), hexavalent chromium.
K032	Hexachlorocyclopentadiene.
K033	Hexachlorocyclopentadiene.
K035	Creosote, chrysene, naphthalene, fluoranthene, benzo(b)fluoranthene, benzo(a)pyrene, indeno(1,2,3-cd)pyrene, benzo(a)anthracene, dibenzo(a)anthracene, acenaphthalene.
K037	Toluene, phosphorodithioic and phosphorothioic acid esters.
K038	Phorate, formaldehyde, phosphorodithioic and phosphorothioic acid esters.
K040	Phorate, formaldehyde, phosphorodithioic and phosphorothioic acid esters.
K041	Toxaphene.
K044	NA
K045	NA
K046	Lead.
K047	NA
K048	Hexavalent chromium, lead.

Table 12-4. (*Continued*)

EPA Hazardous Waste No.	Hazardous Constituents for Which Listed
K049	Hexavalent chromium, lead.
K051	Hexavalent chromium, lead.
K062	Hexavalent chromium, lead.
K065	Lead, cadmium.
K066	Lead, cadmium.
K084	Arsenic.
K086	Lead, hexavalent chromium.
K098	Toxaphene.
K099	2,4-dichlorophenol, 2,4,6-trichlorophenol.
K104	Aniline, benzene, diphenylamine, nitrobenzene, phenylenediamine.
K016	Mercury.
K123	Ethylene thiourea.

[a]NA denotes that waste is a listed hazardous waste because it fails the test for the characteristic of ignitability, corrosivity, or reactivity.

Overview of Hazardous Waste Regulatory Program

The federal HWM regulations provide for control over all aspects of the management of the wastes, from their point of generation to their ultimate disposal; hence, they are said to provide a "cradle-to-grave" system. The major sections of the regulations in Title 40 of the *Code of Federal Regulations* are:

40 *CFR* 260	Definitions
40 *CFR* 261	Identification of hazardous waste
40 *CFR* 262	Standards for generators
40 *CFR* 263	Standards for transporters
40 *CFR* 264, 265	Standards for treatment, storage, and disposal facilities
40 *CFR* 268	Land disposal regulations
40 *CFR* 270	Permit requirements

Land Disposal Restrictions (LDRs)

In Congress' 1984 amendments to RCRA, the U.S. EPA was directed to evaluate and, if necessary, prohibit the land disposal of certain types of hazardous waste. The so-called "land ban" regulations were promulgated at 40 *CFR* 268 and their restrictions went into effect in late 1986. Land disposal means any placement of hazardous waste in a landfill, surface impoundment, waste pile, injection well, land treatment facility, salt dome formation, salt bed formation, underground mine or cave, or concrete bunker or vault.

U.S. EPA has established or will establish treatment standards for land disposal of seven groups of RCRA hazardous wastes by specific dates, as listed in Table 12-5.

The first category of wastes includes the F001–F005 spent solvent-containing wastes and the F020–F023 and F026–F028 dioxin-containing wastes. The second category, the "California List" wastes, is a distinct category of RCRA hazardous wastes described below. The three categories of scheduled wastes (i.e., First Third, Second Third, Third Third wastes) include all listed and characteristic hazardous wastes identified as of November 8, 1984 (excluding the solvent and dioxin wastes mentioned above). U.S. EPA ranked the scheduled wastes based on their toxicity and volume and placed the highest toxicity/volume wastes in the "First Third." Soil and debris contaminated with spent solvent- or dioxin-containing and "California List" wastes generated during CERCLA response and RCRA corrective actions were given a separate statutory deadline. Wastes newly identified or listed after 1984 must have standards set within 6 months of their identification or listing as a hazardous waste.

To be classified as a "California List" waste, three conditions must be met:

1. The waste must be a RCRA listed or characteristic waste
2. The waste must be a liquid (i.e., it fails Method 9095 Paint Filter Liquids Test [PFLT]), except for halogenated organic compounds (HOCs), which may be liquid or nonliquid
3. The waste must exceed statutory prohibition levels for specified constituents

The types of wastes that may be "California List" wastes are: free cyanides, certain metals, corrosive wastes, PCBs, and HOCs. HOCs are compounds containing carbon and a halogen, such as fluorine, chlorine, bromine, iodine, and

Table 12-5.   LDR Statutory Deadlines.

Waste	Date
Spent solvent and dioxin-containing wastes	November 9, 1986
"California List" wastes	July 8, 1987
First third wastes	August 8, 1988
Spent solvent, dioxin-containing, and "California List" soil and debris from CERCLA/RCRA corrective actions	November 8, 1988
Second third wastes	June 8, 1989
Third third wastes	May 8, 1990
Newly identified wastes	Within 6 months of identification as a hazardous waste

astatine, in their molecular formulas. The U.S. EPA has limited the restricted HOCs to approximately 100 HOCs listed in Appendix III to 40 *CFR* Part 268. These restricted HOCs include solvents, pesticides, PCBs, and dioxins. These hazardous wastes are referred to as "California List" wastes because the State of California developed regulations to restrict the land disposal of wastes containing these constituents, and Congress incorporated these provisions into the 1984 amendments to RCRA.

The types of LDRs available to U.S. EPA include setting treatment standards, specifying minimum technology requirements, or complete prohibition of all land disposal practices. Waste generators may seek treatability variances, "no migration" petitions, or delistings for a particular waste, but otherwise must fully comply with the LDRs. The LDRs are relatively complex and the reader is cautioned to examine each waste on a case-by-case basis to see which restrictions apply.

Proposed Regulations

U.S. EPA is constantly evaluating possible changes to the HWM regulations. Perhaps the most significant of the changes being considered in mid 1989 is the use of a new extraction procedure called the toxicity characteristic leaching procedure (TCLP). The TCLP would replace the EP toxicity protocol. Not only would the procedure change, but the U.S. EPA is also considering an expansion of the list of regulated chemicals in the extract or leachate from the test. If promulgated, the number of wastes defined as hazardous would greatly increase. The TCLP and EP toxicity tests are compared in Table 12-6. The proposed TCLP regulatory limits are summarized in Table 12-7.

Sludges Containing PCBs

The disposal of materials containing polychlorinated biphenyls (PCBs) in the U.S. is regulated by U.S. EPA pursuant to the Toxic Substances Control Act. The PCB disposal regulations are found at 40 CFR 761. Allowable disposal methods for PCB wastes are summarized in Table 12-8.

Other Regulations

Several other types of regulations may be of importance to sludge management practices. Examples include those statutes relating to air emissions, spills, transportation, and worker health/safety. A thorough assessment of potentially applicable regulations should be conducted for each project.

Table 12-6. Differences Between EP Toxicity and TCLP Tests (40 *CFR* 261 and 268).

Item	EP Toxicity	TCLP
Leaching media	0.5N acetic acid added to distilled deionized water to a pH of 5 with 400 ml maximum addition. Continual pH adjustment.	0.1N pH 2.9 acetic acid solution for moderate to high alkaline wastes and 0.1N pH 4.9 acetate buffer for other wastes.
Liquid/solid separation	0.45 μm filtration to 75 psi in 10 psi increments. Unspecified filter type.	0.6 to 0.8 μm glass fiber filter filtration to 50 psi.
Monolithic material particle size reduction	Use of structural integrity procedure or grinding and milling.	Grinding or milling only. Structural integrity procedure not used.
Extraction vessels	Unspecified design. Blade/stirrer vessel acceptable.	Zero-headspace vessel required for volatiles. Bottles used for nonvolatiles. Blade/stirrer vessel not used.
Agitation	Prose definition of acceptable agitation.	Rotary agitation only in an end-over-end fashion at 30 ± 2 rpm.
Extraction time	24 hours	18 hours
Quality control requirements	Standard additions required. One blank per sample batch.	Standard additions required in some cases. One blank per 10 extractions and every new batch of extract. Analysis specific to analyte.

Note: All other attributes between the two tests are generally the same, although there are some minor differences. Note also that, while the EP toxicity only addresses those species for which National Interim Primary Drinking Water Standards (NIPDWS) exist, the TCLP can be applied to other toxicants.

**Table 12-7.   Toxicity Characteristic Leaching
Procedure (TCLP). Proposed Toxicity Characteristic
Contaminants and Levels (June 1986).**

Contaminant	Regulatory Level, mg/l
Acrylonitrile	5.0
Arsenic	5.0
Barium	100
Benzene	0.07
Bis(2-chloroethyl)ether	0.05
Cadmium	1.0
Carbon disulfide	14.4
Carbon tetrachloride	0.07
Chlordane	0.03
Chlorobenzene	1.4
Chloroform	0.07
Chromium	5.0
o-Cresol	10.0
m-Cresol	10.0
p-Cresol	10.0
2,4-D	1.4
1,2-Dichlorobenzene	4.3
1,4-Dichlorobenzene	10.8
1,2-Dichloroethane	0.40
1,1-Dichloroethylene	0.1
2,4-dinitrotoluene	0.13
Endrin	0.003
Heptachlor (and hydroxide)	0.001
Hexachlorobenzene	0.13
Hexachlorobutadiene	0.72
Hexachloroethane	4.3
Isobutanol	25
Lead	5.0
Lindane	0.06
Mercury	0.2
Methoxyclor	1.4
Methylene chloride	8.6
Methyl ethyl ketone	7.2
Nitrobenzene	0.13
Pentachlorophenol	3.6
Phenol	14.4
Pyridine	5.0
Selenium	1.0
Silver	5.0
1,1,1,2-Tetrachloroethane	10.0
1,1,2,2-Tetrachloroethane	1.3
Tetrachloroethylene	0.1
2,3,4,6-Tetrachlorophenol	1.5
Toluene	14.4

Table 12-7.   (*Continued*)

Contaminant	Regulatory Level, mg/l
Toxaphene	0.07
1,1,1-Trichloroethane	25
1,1,2-Trichloroethane	1.2
Trichloroethylene	0.07
2,4,5-Trichlorophenol	5.8
2,4,6-Trichlorophenol	0.30
2,4,5-TP (Silvex)	0.14
Vinyl chloride	0.05

Table 12-8.   Allowable Disposal Methods for PCB Wastes in the U.S.[a,b]

	PCB Concentration	Incinerator	Chemical Waste Landfill	High Efficiency Boiler	Other Approved Method
Mineral oil dielectric fluid	≥ 50 < 500	X	X[c]	X	X
Other liquids	≥ 50 < 500	X	X[c]	X	X
Liquids	≥ 500	X			
Non-liquid	> 50	X	X		
Dredged materials and municipal sewage treatment sludges	> 50	X	X[c]		
Other	> 50	X			

[a]Facilities must be U.S. EPA-approved.
[b]See 40 *CFR* 761.60 for specific requirements for PCB transformers.
[c]Prohibited for liquids which are hazardous wastes since July 8, 1987 per 40 *CFR* 268.32(a)(2).

SLUDGE CHARACTERISTICS

Knowledge of the characteristics of a sludge is essential to selection of appropriate treatment and disposal options.

Primary sludges are generated in front-end wastewater treatment processes, which are typically physical-chemical technologies, such as sedimentation or oil/water separation. In many cases, these sludges contain readily settleable solids which also thicken and dewater easily. They may be inorganic or organic in nature, or a mixture of the two types of substances.

Chemical sludges result from unit operations in which chemicals are added to wastewater, such as coagulation or metals precipitation. Although in most cases they are predominantly inorganic in composition, organics may be entrained in the sludge matrix in particulate or soluble forms. Some chemical sludges, such as lime sludge, are relatively easy to handle, whereas others, including iron and aluminum hydroxides, are gelatinous materials which do not dewater easily.

Biological sludges are formed from biological treatment processes such as activated sludge, trickling filters, and rotating biological contactors. These sludges are principally organic in nature. The degree to which the organics have been stabilized (i.e., oxidized) is dependent upon the nature of the process (e.g., high rate activated sludge vs. extended aeration processes). Depending upon the type of upstream processing, these sludges may contain some or all of the suspended matter present in the raw wastewater.

Table 12-9 provides a list of most of the sludge characteristics which are important to the selection and design of management programs. Some of these characteristics are concerned with defining the proper regulatory classification, while others are needed for the selection of appropriate treatment processes. The utility of most of the parameters is apparent; however, several deserve explanation.

Viscosity and/or consistency are directly relevant to conveyance or movement of sludges. Table 12-10 contains a group of generic consistency classifications. Particle size distribution can be a key parameter when considering stabilization/solidification processes, as fine particles can delay curing or coat larger particles and prevent bonding. Alkalinity/acidity can dictate requirements for neutralization to prevent, for example, corrosive attack of refractory materials in incinerators. Specific resistance and capillary suction time are useful indicators of the dewaterability of a sludge. The heating value of a sludge is an indication of combustibility.

It should be recognized that other parameters may be useful in defining processes for a particular sludge.

SELECTION OF APPROPRIATE TECHNOLOGIES

For some sludges, the selection of the optimum treatment and disposal technologies is relatively straightforward. Usually, however, several combinations could be employed in a sludge management strategy. Experimental evaluations, regulatory considerations, and economic comparisons are needed to make final

Table 12-9.   Important Sludge Characteristics for the Selection and Design of Management Program.

Characteristic	Regulatory Classification	Physical Processes	Chemical Processes	Biological Processes	Thermal Processes	Landfill
Generation rate		X	X	X	X	X
Solids content		X	X	X	X	X
Ash content		X	X	X	X	
Volatile solids content		X	X	X	X	
Density		X	X	X	X	X
Viscosity/consistency		X	X	X	X	
Particle size distribution		X	X		X	
Free liquid content	X	X				X
pH	X	X	X	X	X	
Alkalinity/acidity		X	X	X	X	
Heating value					X	
Specific resistance		X				
Capillary suction time		X				
Explosivity	X				X	
Oil and grease		X	X	X	X	
Halogen Content (Cl, F, Br)			X	X	X	
Nitrogen				X	X	
Phosphorus				X	X	
Sulfate					X	
Alkali metals	X				X	
Heavy metals	X			X	X	
Cyanide	X			X	X	
Sulfide	X			X	X	
Specific organics	X		X		X	
Hazardous constituents	X					X
Flashpoint	X	X				
EP Toxicity	X		X			
TCLP	X		X			

Table 12-10. Waste Consistency Classifications.

Description	Characteristics
1. Liquid waste	< 1% suspended solids,[a] pumpable liquid, generally too dilute for sludge dewatering operation.
2. Pumpable waste	< 10% suspended solids,[a] pumpable liquid, generally suitable for sludge dewatering.
3. Flowable waste	> 10% suspended solids,[a] not pumpable, will flow or release free liquid, will not support heavy equipment, may support high flotation equipment, will undergo extensive primary consolidation.
4. Nonflowable waste	Solid characteristics, will not flow or release free liquids, will support heavy equipment, may be 100% saturated, may undergo primary and secondary consolidation.

[a]Suspended solids ranges are approximate.

selections. Fig. 12-3 depicts the most common types of sludge management technologies in use in recent years.

Physical processes have as their goal the reduction of the volume and weight of the sludge, accomplished through the removal of free liquids and a portion of the bound liquids in interstitial pores. In *chemical processes*, chemicals are added to the sludge and through a reaction the desired modification of sludge properties is accomplished. *Biological processes* are used to biochemically oxidize organic materials, while in *thermal processes*, the applied energy is used to dewater, destruct, and/or detoxify the sludge. Table 12-11 is a technology selection matrix which may be of assistance as a starting point for evaluating appropriate technologies.

PHYSICAL PROCESSES

The primary role of physical processes in sludge management is to reduce the volume and weight of material to be handled. The volume of a sludge varies inversely with the solids content; for example, if a wastewater sludge is thickened from 1 percent solids to 4 percent solids, the resultant volume will be only 25 percent of the original value. Thickening and dewatering are the two general categories of physical processes for sludge treatment. Volume reduction can greatly reduce the expense involved with subsequent thermal processes or land disposal, however, it may not be desirable or necessary prior to biological or chemical processes. Chemical conditioning may be employed to facilitate these processes. Eckenfelder (1980) provides a description of the most commonly employed process design approaches.

Thickening typically effects a two- to fourfold increase in the solids concentration of sludges with initial solids concentrations less than 2 to 3 percent.

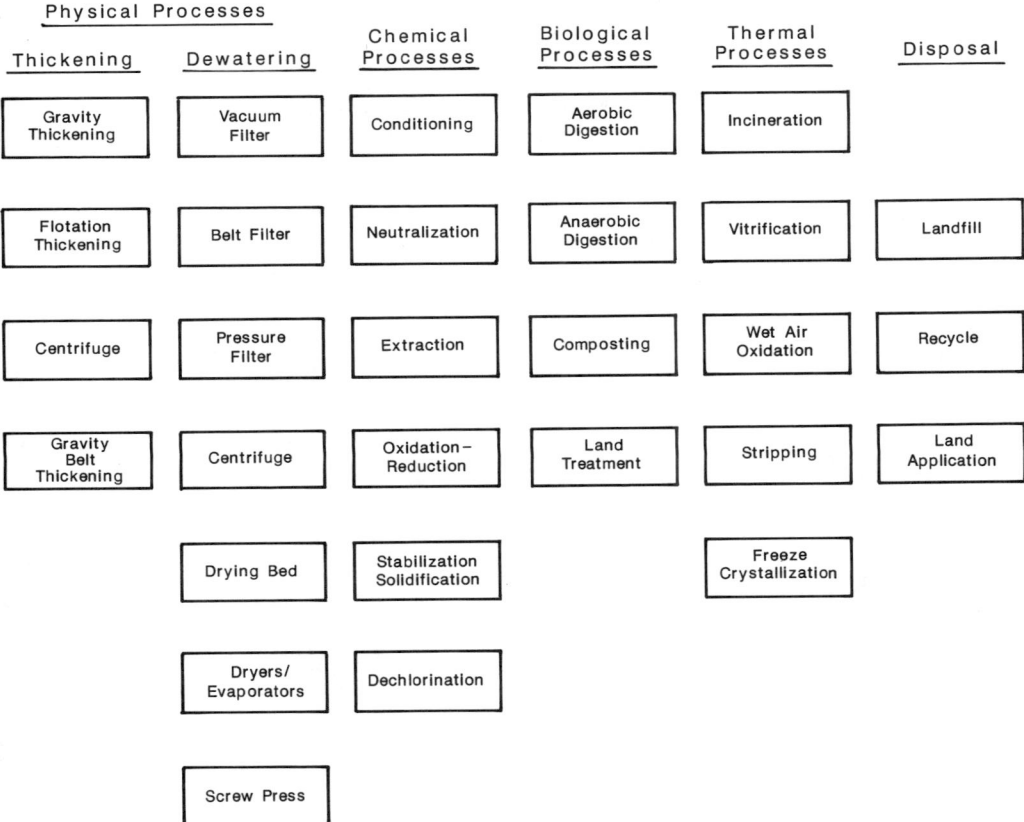

Fig. 12-3. Alternative technologies.

Sludges are usually thickened prior to mechanical dewatering processes. Gravity thickening, flotation thickening, centrifugal thickening, and gravity belt thickening are the four options typically considered by the process engineer. Thickened sludges generally contain 5 to 10 percent solids. Chemical conditioners can be used to assist the thickening process.

Dewatering of sludges is commonly accomplished with vacuum filters, belt filters, pressure filters, centrifuges, drying beds, dryers or evaporators, and screw presses. These processes result in a greater degree of volume reduction than do the thickening operations, changing the sludge from a liquid slurry to a semi-solid or solid waste. The solids content of a dewatered sludge can generally be expected to range from 10 to 60 percent. Chemical conditioning of the sludge is usually required prior to mechanical dewatering.

Table 12-11. Technology Selection Matrix.

Process	General	Inorganic Sludges	Heavy Metals	Halogenated Organics	Nonhalogenated Organics	PCBs
				Type of Sludge		
Gravity thickener	X	X				
Flotation thickener						
Centrifuge	X	X				
Gravity belt thickener	X	X				
Vacuum filter	X	X				
Belt filter	X	X				
Pressure filter	X	X				
Drying bed	X	X				
Dryers/evaporators	X	X				
Screw press	X	X				
Conditioning	X	X	X			
Neutralization		X	X			
Extraction				X	X	X
Oxidation-reduction		X	X			
Stabilization/solidification		X	X			
Dechlorination				X		X
Aerobic digestion					X	
Anaerobic digestion					X	X
Composting				X	X	
Land treatment				X	X	
Incineration				X	X	X
Vitrification		X	X	X	X	X
Wet air oxidation				X	X	X
Freeze crystallization		X	X			

Gravity Thickening

This process is conducted in a tank with a slowly rotating rake mechanism which gently disrupts the bridging between sludge particles, allowing for increased settling and compaction. The objective is to provide a concentrated underflow stream. The supernatant is usually recycled back to the liquid stream processing sequence. The key process design parameter is the areal solids loading rate (lb/sq ft-day).

Flotation Thickening

Dissolved air flotation (DAF) is quite commonly utilized to thicken waste activated sludge prior to subsequent handling steps. In this process, a portion of the DAF effluent or plant utility water is pressurized, and then aerated to 70 to 85 percent of the saturation concentration of oxygen and nitrogen. The pressurized liquid is mixed with the influent sludge stream and the mixture is released to atmospheric conditions in the flotation tank. The liberated microscopic air bubbles attach to and become enmeshed in the flocculated sludge causing it to rise to the surface to be skimmed off. The clarified effluent (less the recycle stream for pressurization) is routed back to the liquid stream process train. Important design parameters for DAF thickening are the initial solids concentration, solids loading rate, hydraulic loading rate, air requirement (air to solids ratio) and recycle flow or ratio.

Centrifugal Thickening

Centrifuges have also been employed for the thickening and dewatering of wastewater sludges. In the centrifugation process, the sludge is introduced into a rotating vessel. The centrifugal force pushes the solid matter to the walls of the vessel. The process is merely multiplicative gravity sedimentation. Basket-type centrifuges and disc-nozzle centrifuges have been used for sludge thickening.

Gravity Belt Thickening

Gravity belt thickening involves gravity drainage of water from a sludge and typically requires polymer conditioning. Metal salts conditioning is not typically used due to filter media blinding. This equipment can be an integral part of a complete dewatering system. Gravity belt thickeners are energy efficient, however, the system requires a spray wash of approximately 10 percent or more of the sludge feed rate.

Vacuum Filters

The most common type of vacuum filter that has been historically employed to dewater wastewater treatment sludges is the rotary drum unit. In this operation, a cylindrical drum, covered with a filter fabric, rotates through the sludge slurry in which it is partially submerged while a vacuum is drawn across the filter media. The screened sludge is held on the surface of the filter media and water is released as the drum rotates through the drying cycle where air is drawn through the sludge cake. A scraper knife releases the dewatered sludge from the drum before it is spray washed and rotates back into the slurry tank.

Of obvious importance in determining the relative success of this process are the characteristics of the sludge being handled, such as solids content, viscosity, compressibility, etc. The designer can evaluate the effect of vacuum, drum submergence, rotational velocity, chemical conditioning, and type of filter media. Vacuum filter loading rates for wastewater sludges generally range from 2 to 20 lb/sq ft-hr.

Belt Filters

Belt filter presses have become relatively popular in the last decade. These units dewater the sludge by passing it between two moving belts with rollers. Most models include a gravity drainage zone, a pressure zone, and a shear zone. Process variables commonly available to the designer include type/amount of chemical conditioner, belt speed, belt width, and pressure.

Pressure Filters

Plate-and-frame or recessed plate filter presses can achieve a relatively high cake solids content by the application of hydraulic pressure in the range of 50 to 200 psi. In these machines, the sludge is pumped into the space between two plates, each of which is covered with a filter cloth. The liquid passes through the cloth, depositing a sludge cake between the plates. When the chambers are completely filled and dewatered, the plates are separated and the sludge cake falls free or is removed through bumping or air pulsing. The filter media may be covered with a precoat to facilitate dewatering and/or cake release, if needed. In addition, the process designer can vary the type/amount of conditioning agent, chamber volume, pressure, and cycle time to optimize performance.

Centrifuges

Solid bowl centrifuges are often used for dewatering sludges. In these devices, a solid cylindrical-conical bowl rotates at high speeds. Sludge is forced by the centrifugal energy to the wall of the bowl, and an internal rotating scroll conveyor moves the dewatered sludge to discharge at the conical section, or beach, of the bowl. The released liquid, or centrate, is skimmed at the feed end. Key process variables include type/amount of conditioner, bowl speed, conveyor speed, pool volume, and feed rate.

Drying Beds

Sludges can also be dewatered by applying them to sand beds, where drying occurs by percolation and evaporation. The beds may be open or covered; rainfall and evaporation have a significant impact on the rate of drying. Generally, the bed configuration consists of 4 to 9 inches of sand underlain by 8 to 18 inches of graded gravel with lateral underdrain pipes. Wet sludge is usually applied in lifts of 8 to 12 inches thickness which initially drain to 1 to 2 inches. The sludge is removed when it cracks indicating it is dry enough for disposal. In practice, application rates range from 5 to 30 lb dry solids/sq ft-yr.

Dryers/Evaporators

In sludge dryers, the water contained in the sludge is vaporized. This is accomplished by turbulent contact with heated air, as in flash dryers, or by directly and indirectly heated rotary dryers. Internal operating temperatures for sludge dryers are typically in the range of 650° to 700°F. Sludge dryers are commonly preceded by mechanical dewatering devices. To evaporate water from sludges, the Carver-Greenfield process mixes the sludge with heated oil, and then evaporates the moisture in a multi-effect evaporator.

Screw Presses

Screw presses may involve the injection of steam in conjunction with a multiple stage enclosed screw mechanism. The technique is particularly applicable to high fibre sludges; high proportions of biological sludge may preclude its use. High feed solids levels are required, in the range of 10 percent solids. Advantages include the potential for extremely high solids levels, low energy requirements, and no spray wash requirements. Process variables include feed solids content, steam usage, screw speed, chemical conditioning, and fibre content.

Performance data for the most common thickening and dewatering processes are summarized in Table 12-12.

CHEMICAL PROCESSES

Chemical processes for sludge treatment are used to alter the properties for subsequent handling (e.g., conditioning) or to reduce or immobilize toxic ingredients (e.g., extraction or stabilization/solidification).

Table 12-12. Performance of Thickening and Dewatering Processes.

Process	Expected Range of Final Solids Concentration, %
Gravity thickening	2 to 9
Flotation thickening	3 to 6
Centrifugal thickening	4 to 13
Gravity belt thickening	3 to 6
Vacuum filter	15 to 35
Belt filter	16 to 35
Pressure filter	30 to 45
Centrifugal dewatering	15 to 35
Drying bed	10 to 30
Screw presses	35 to 65

Conditioning

Chemical conditioning of most wastewater sludges is necessary to facilitate effective, economical thickening and dewatering. Conditioners that are in widespread use include lime, ferric chloride or sulfate, alum, fly ash, and polyelectrolytes. These conditioners cause sludge particles to coagulate, which frees some of the interstitial liquids. Mixing is necessary to successfully intermix the conditioners, but the mixing intensity should be controlled to avoid floc shear. Eckenfelder (1989) describes specific resistance tests which can be used to optimize chemical conditioning for dewatering processes.

Neutralization

The neutralization of extremely acidic or alkaline sludges may reduce chemical conditioner dosages or reduce the leachability of waste constituents. Aqueous solutions of lime, sodium hydroxide, sulfuric acid, and hydrochloric acid are the most frequently used neutralizing agents.

Extraction

Extraction is used to partition sludge constituents into a solvent in which they are relatively soluble. Several types of extraction processes are available, although some of these have been applied to waste sludges only in recent years.

Elutriation has been used in municipal wastewater treatment plants to extract excess alkalinity from anaerobically digested biological sludge. In this case, the solvent used is water which is mixed with the sludge; the mixture is then separated into the washed sludge and spent washwater components. The washwater is recycled in some multi-stage systems. The removal of the excess alkalinity reduces chemical conditioning requirements for subsequent dewatering processes.

The basic extraction sludge treatment (B.E.S.T.™) process, developed by Resources Conservation Company, has reportedly been successful for sludges containing oils, pesticides, and PCBs. A secondary or tertiary amine, usually triethylamine, is mixed with the sludge at cool temperatures. This solvent extracts the organics from the mixture which is then centrifuged or filtered to separate waste solids. Fig. 12-4 provides a schematic of the process. The separated solids are dried to allow for solvent recovery. The solvent/oil/water solution is heated, decanted, and then distilled and stripped in two or three steps to result in aqueous affluent, waste oil/heavies, and solvent streams. The solvent is recycled, but the other two streams and the solids require disposal, possibly incineration for the waste oil/heavies mixture.

Values greater than or equal to pH 10 are required in the process, so some sludges need pH adjustment. The presence of large amounts of volatile organics

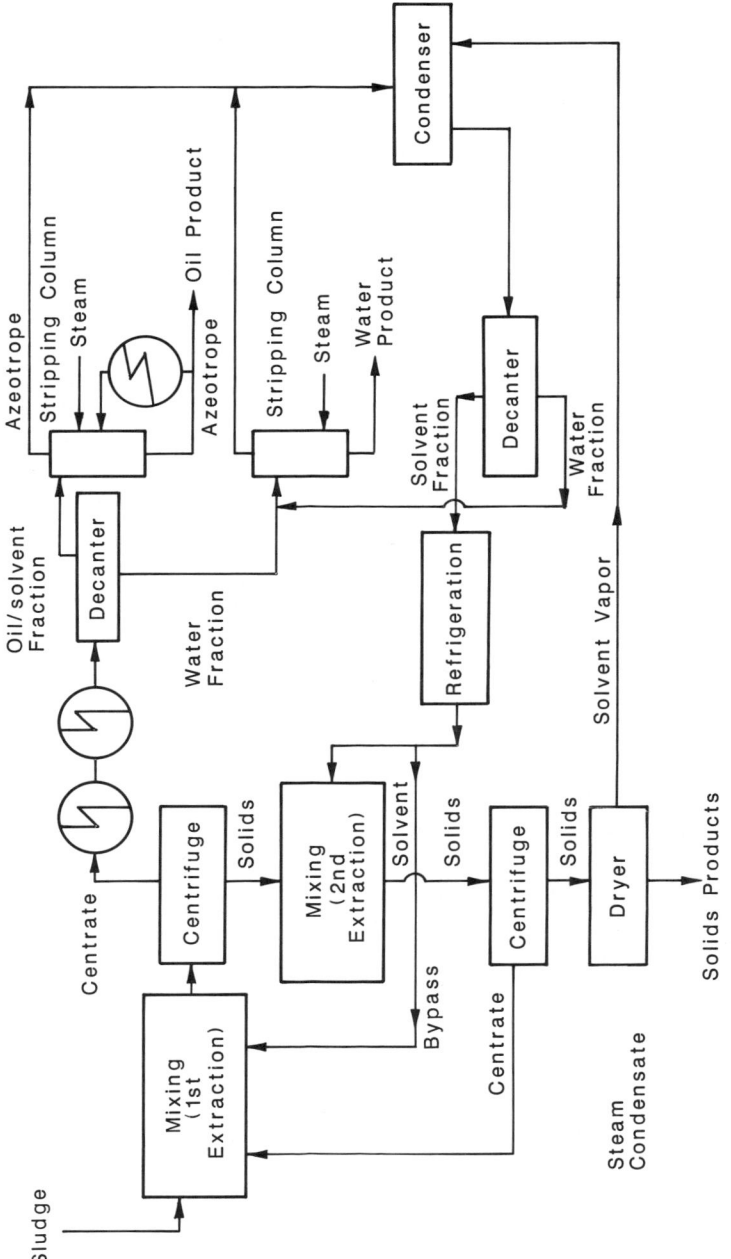

Fig. 12-4. BEST™ process. (*Source:* U.S. EPA 1988a.)

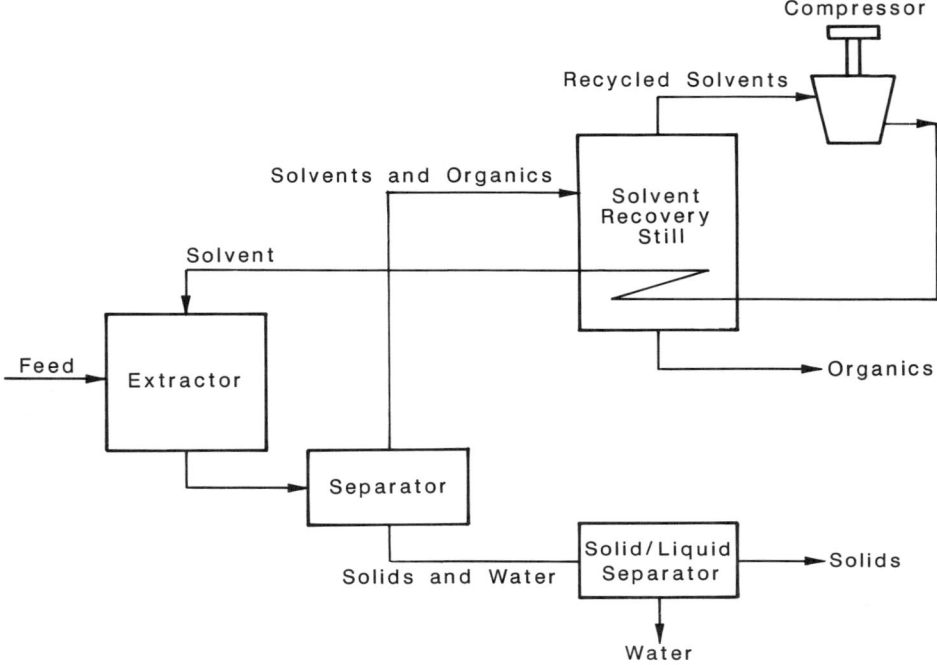

Fig. 12-5. Critical fluid extraction process. (*Source:* U.S. EPA 1988b.)

makes solvent recovery troublesome. Emulsifiers can hinder oil/water separation. Heavy metals are not removed to any appreciable extent.

In critical fluid extraction, a liquified gas such as propane is used as the extracting solvent to remove hydrocarbons from the sludge. As shown in Fig. 12-5, the sludge and compressed, liquefied solvent are both fed into a countercurrent, mixed extraction vessel. Treated sludge is removed from the extraction vessel, while the solvent/extracted hydrocarbon mixture flows into a separator through a pressure reducing valve. The liquefied gas is vaporized in the separator and then compressed for reuse. The extracted organics are removed from the bottom of the separator for subsequent treatment. CF Systems Corporation, the developer of the process, reports success with extraction of oils and other organics.

Wilson and Clarke (1988) have reported success with a biodegradable, relatively low cost surfactant, sodium dodecyl sulfate (SDS), for extraction of oils, PCBs, and chlorinated organics from soils or sludges. The SDS can be recovered for reuse by gentle extraction with a nonpolar solvent, such as mineral spirits. Oils/heavies can be separated from the solvent by distillation.

Oxidation-Reduction

Chemical oxidation-reduction may be employed to destruct toxic constituents of some sludges or to convert constituents to less toxic forms. In these processes, a chemical reactant is added to the sludge which causes the oxidation state of one reactant to increase with a coincident decrease in the oxidation state of the other. Sludges with high solids concentrations are difficult to treat via oxidation-reduction due to lengthy reaction times, sometimes resulting in elevated dosage requirements.

At low pH conditions, sulfur dioxide or other reducing agents may be used to reduce hexavalent chromium (Cr^{+6}) to the less toxic and less soluble trivalent form (Cr^{+3}). Sodium borohydride may be used to reduce other metals, such as mercury and silver, to elemental forms for recovery. Organics, such as cyanides, can be chemically oxidized to innocuous end products. However, oxidant dosages may be quite high for sludge. In addition, nonselective oxidation may occur, resulting in the formation of more toxic compounds.

Stabilization/Solidification

The major concerns with land disposal of sludges are leaching of chemicals and ease of handling. Several methods have been developed to convert wastes into a low permeability material to greatly reduce leaching potential and provide physical stability. The process by which this is accomplished is known by several names: stabilization, solidification, and fixation. As shown schematically in Fig. 12-6, the process is principally a blending operation. During stabilization/solidification/fixation, the mobility of hazardous constituents is reduced through binding into a solid mass with low permeability that resists leaching.

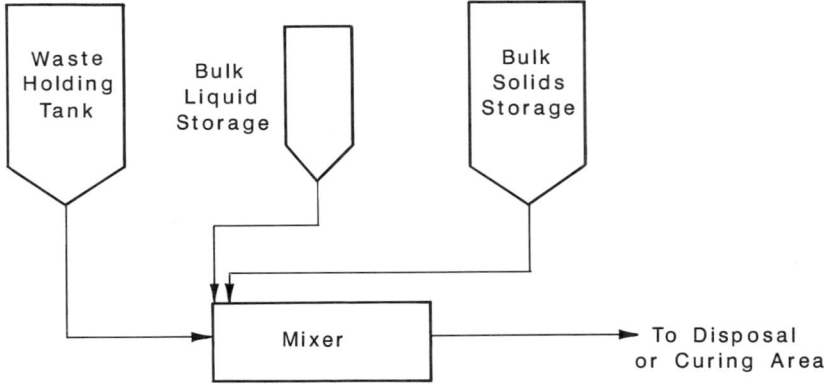

Fig. 12-6. Stabilization/solidification process. (*Source:* U.S. EPA 1988b.)

The actual mechanisms of binding can be categorized by the primary stabilizing agent used: cement-based, pozzolanic, silicate-based, thermoplastic-based, organic polymer-based, or a combination of these. The pH of the waste may be adjusted prior to stabilization to precipitate heavy metals. Gas monitoring, collection, and treatment may be necessary with wastes containing ammonium or volatile organics. The alkalinity associated with some stabilizing agents and the heat generated during stabilization may cause the release of both these constituents.

The efficacy of the process can be markedly affected by the presence of organic materials, fine solids, or excess water. High concentrations of organic materials may interfere with bonding of waste constituents. Fine particles can delay setting and curing. Extremely dilute sludges can require excessive quantities of stabilization reagents.

Dechlorination

Dechlorination of sludges with potassium polyethylene glycolate (KPEG) is an innovative technology recently developed by General Electric (U.S. EPA 1988b). This process has been demonstrated to dechlorinate PCB-containing wastes. KPEG is mixed with the waste in a heated, mixed reactor. Excess KPEG is decanted and the waste is then washed with water to remove the reaction products and residual KPEG. The washwater and KPEG can be recycled to treat additional waste. Galson Research Corporation (U.S. EPA 1988b) has developed a similar process using alkaline polyethylene glycolate (APEG) catalyzed by dimethyl sulfoxide. Blowdown (excess) washwater containing the reaction products from these processes requires further treatment. Sludges with high organic, moisture, and aluminum contents require large volumes of dechlorination reagent.

BIOLOGICAL PROCESSES

Several biological oxidation processes are in widespread use for organic-containing wastewater treatment sludges. These include aerobic digestion, anaerobic digestion, composting, and land treatment. Table 12-13 summarizes some of the conditions which impact the feasibility of the processes.

Aerobic Digestion

When biological solids are treated in an aerobic digester, the biomass is aerated to retain the endogenous state, and cellular biodegradation occurs. Since limited BOD is available, the biomass may also degrade residual organics that resisted prior biodegradation processes. Also, because cellular material is degraded,

Table 12-13.   Biological Sludge Treatment Considerations.

Characteristics Impacting Process Feasibility	Reason for Potential Impact
Variable waste composition	Inconsistent biodegradation caused by variation in biological activity.
Water solubility	Contaminants with low solubility are harder to biodegrade.
Biodegradability	Low biodegradability reduces process efficiency.
Temperature outside 15 to 70°C range	Larger, more diverse microbial production present in this range
Nutrient deficiency	Lack of adequate nutrients for biological activity (although nutrient supplements may be added).
Oxygen deficiency (aerobic process)	Lack of oxygen is rate limiting.
Moisture content	A moisture content below 40 percent severely inhibits bacterial activity.
pH outside 4.5 to 8.5 range	Inhibition of biological activity.
Lack of microbial population	If indigenous microorganisms are not present, cultured strains can be added.
Water and air emissions and discharges (composting only)	Potential environmental and/or health impacts (control achieved through air scrubbing, carbon filtration, forced aeration, impermeable liner).
Compaction of compost (composting only)	Particles tend to coalesce and form an amorphous mass that is not easily maintained in an aerobic environment (wood chips or shredded tires may be added as bulking agents).
Nonuniform particles (composting only)	Waste mixtures must be of uniform particle size.
Presence of elevated levels of: • Heavy metals • Highly chlorinated organics • Some pesticides, herbicides • Inorganic salts	Can be highly toxic to microorganisms.

Source: U.S. EPA (1988b).

those chemicals that sorbed or were retained within the biomass may be released back to the water phase and be recycled to the liquid stream treatment train. Due to the relatively long hydraulic retention time (normally 10 to 30 days) in the aerobic digestion system, some volatile organics may be stripped. Eckenfelder (1989) presents a process design approach.

Anaerobic Digestion

Anaerobic microorganisms function in the absence of oxygen, utilizing hydro-carbons as their energy source. Anaerobic biochemical reactions occur in two steps, as shown in Fig. 12-7; first, acid-producing bacteria degrade organic compounds in the sludge to volatile acids, and second, methane-forming bac-teria convert the acids to methane and carbon dioxide. A proper balance must be maintained between the rates of the two reactions for successful operation. The process, shown schematically in Figure 12-8, operates with long residence times and at mesophilic temperatures (90 to 100°F). In unmixed digesters, solids residence times of 30 to 60 days are typical, while high-rate mixed units

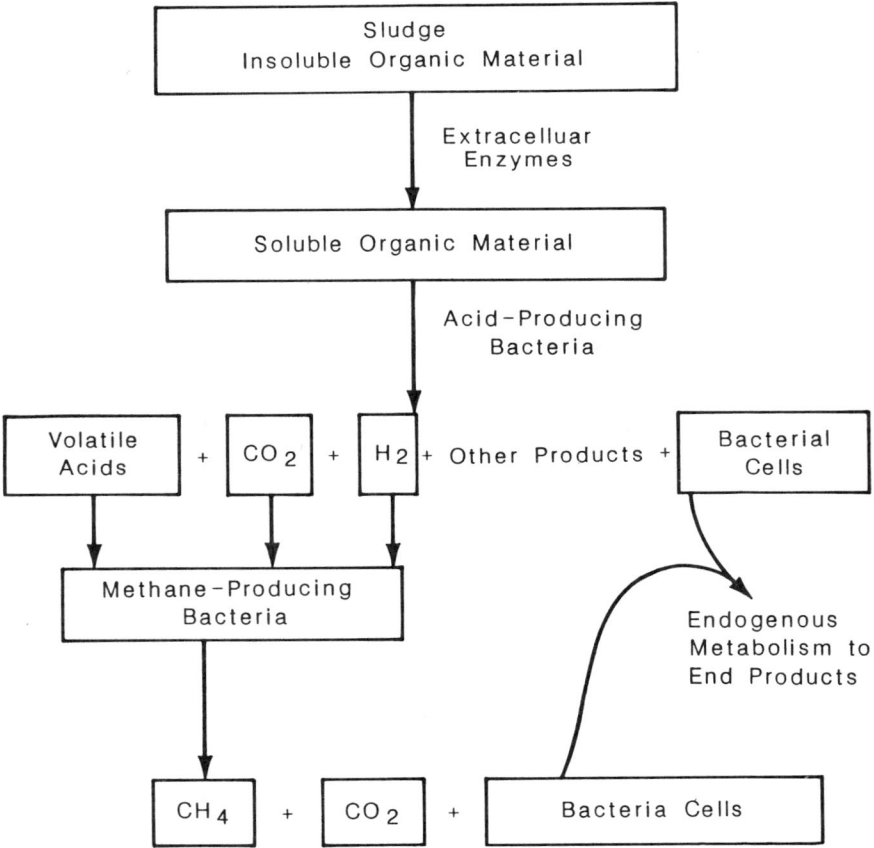

Fig. 12-7. Mechanism of anaerobic sludge digestion. (*Source:* Eckenfelder 1980.)

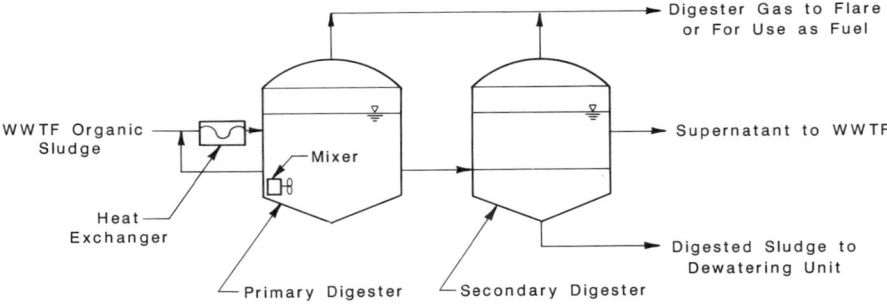

Fig. 12-8. Anaerobic solids digestion process.

may effectively provide stabilization in 10 to 20 days. Eckenfelder (1980) and Metcalf and Eddy (1972) provide useful design and performance information.

Composting

Dewatered sludges and mixtures of sludge and solid waste may be biodegraded by composting. The humus material formed is commonly used as a soil conditioner. The composting process employs aerobic thermophilic degradation of organics. The moisture content of the composting mass generally should not exceed 60 percent. Open windrow systems, static windrow systems, and rotating vessels have been used for composting. Open windrows are aerated by periodically disturbing and then reforming the stacks of material. Forced air is used to aerate static windrows. Rotating vessels are aerated and mixed by tumbling, stripping, and forced aeration. Between one and three weeks are required to complete the process. Runoff and leachate from composting operations must be collected and treated.

Land Treatment

Land treatment is both a treatment and disposal process in which the sludge is mixed with surficial soils and constituents are degraded and/or immobilized. The process is also called land application, landfarming, and sludge spreading. Agricultural discing equipment is used to mix and aerate the sludge-soil mixture so that soil microorganisms thrive. Nutrient addition (fertilizer) may be necessary. Heavy metals and other components can limit application rates. There are specific design and operational standards for hazardous wastes which are subjected to land treatment (see 40 *CFR* 264). With proper consideration to siting, stormwater runoff, and loading rates, land treatment can be an effective way to deal with wastewater sludges.

THERMAL PROCESSES

Thermal treatment processes for wastewater sludges use high or low temperature conditions to destruct or detoxify waste constituents. Major categories of thermal processes include incineration, vitrification, wet air oxidation, stripping, and freeze crystallation. In addition to volume reduction and detoxification, thermal processes may provide energy recovery and material recovery. Auxiliary fuel and oxygen (air) are needed to initiate and sustain combustion.

Incineration

Incineration processes accomplish high temperature combustion of organic components in sludges. Multiple-hearth incinerators, as depicted in Fig. 12-9 are often used to incinerate municipal treatment sludges. These units consist of

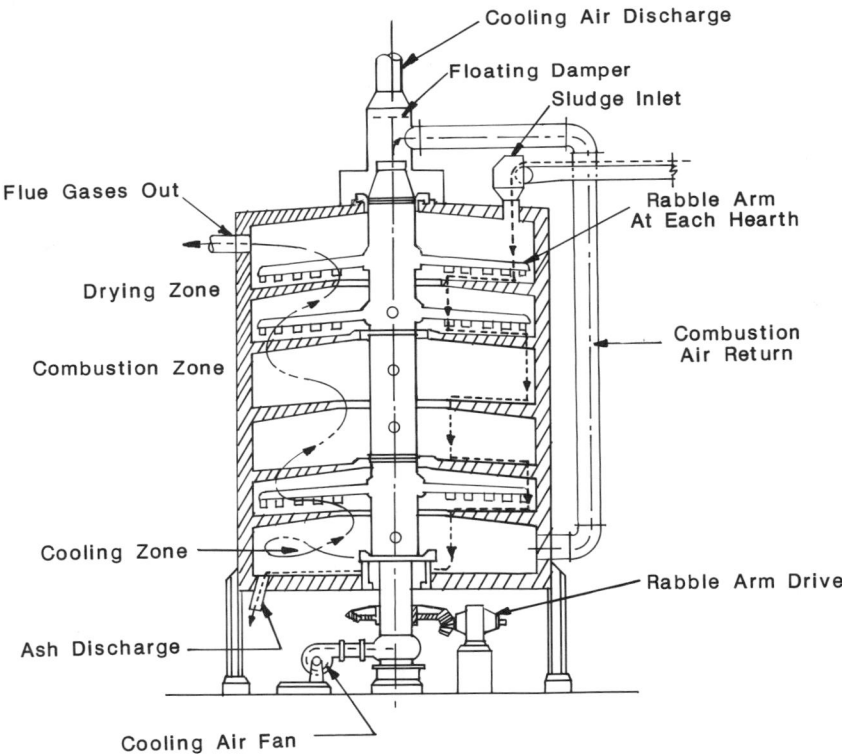

Fig. 12-9. Cross section of a typical multiple-hearth incinerator, (*Source:* Eckenfelder 1980.)

a series of vertical hearths with a rabble arm assembly. Drying occurs on the upper hearths and combustion on the intermediate hearths. Internal temperatures range from 600 to 1,000°F. The ash residue requires further handling. Air emissions are controlled with baghouses, electrostatic precipitators, and scrubbers.

Rotary kiln incinerators are slightly inclined, refractory-lined cylinders. Wastes and auxiliary fuel are introduced at the upper end of the kiln, and the wastes slowly tumble through the combustion chamber. Combustion times vary from several minutes to an hour or more. As shown in Fig. 12-10, ash is removed at the lower end of the kiln and flue gases pass through a secondary combustion chamber and then through treatment devices for particulate and acid gas removal.

Incineration of sludges with high inorganic salt content is discouraged due to problems associated with degradation of the refractory lining and slagging of ash, unless the kiln is designed to operate in the "slagging" mode. Incineration of sludges with a high toxic metal content is also discouraged due to the difficulty in removing associated air pollutants. The residuals from this process are ash/slag, treated combustion gases, and solutions from ash quench and wet scrubbers.

Rotary kiln incinerators have several advantages over other combustion technologies. They are capable of handling liquids, slurries, sludges, bulk solids of varying size, contaminated soils and materials, and containerized wastes, all in large quantities. This type of incinerator is capable of providing effective, thorough destruction of virtually any type of solid or liquid hazardous waste that is combustible. An assortment of different types of wastes can be incinerated simultaneously in a rotary kiln because of its ability to achieve high operating temperatures and to obtain a gentle, continuous mixing of the wastes, both of which facilitate complete waste combustion. A slagging rotary kiln produces incineration residues (slag) similar to glass-like vitrified residues that normally do not have hazardous properties. Mobile hazardous constituents, which are sometimes associated with ash residues from an ashing type kiln, are physically and chemically bound into the vitreous matrix of slag residue.

The two principal disadvantages of rotary kilns include high capital and operating costs and the generation of fine particulates and inorganic oxides in the exhaust gases. The high operating costs are primarily due to the frequent replacement of refractory lining if abrasive or corrosive conditions exist within the kiln. The formation of fine particulates requires the installation of expensive air emissions control equipment. The impact of these disadvantages can be minimized with the application of effective engineering designs, controls, and operating practices.

There are two basic types of rotary kilns for the incineration of wastes:

- Ashing type. Inorganic materials are not melted and the kiln walls are dry. Most rotary kiln systems currently operating in the U.S. for waste incin-

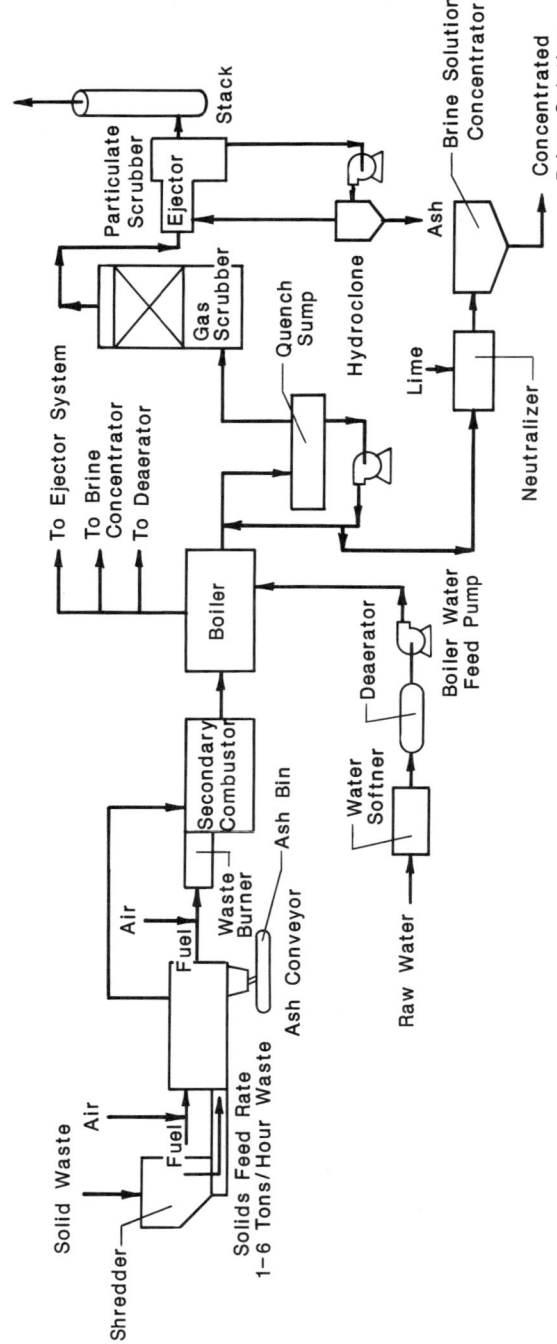

Fig. 12-10. Rotary kiln incineration process. (*Source:* U.S. EPA 1988b.)

eration are the ashing type. Ashing type incinerators are typically operated at temperatures ranging from 1,400 to 1,650°F.

- Slagging type. Many inorganic materials are melted and higher melting point inorganic components are oxidized and combined into the slag with the lower melting point materials. The kiln walls are wet with viscous slag. Slagging type incinerators are typically operated at temperatures ranging from 1,300 to 2,400°F. Some may operate at temperatures up to 3,000°F.

For disposal of hazardous waste, the slagging rotary kiln system with its integral secondary combustion chamber is thought by many to be the most modern and economical technology available. The secondary combustion chamber is designed to provide the necessary residence time for complete combustion and destruction of hazardous wastes.

The fluidized bed incinerator consists of a refractory-lined vessel containing a bed of sized, crushed refractory. The process flow diagram is presented in Fig. 12-11. Wastes are injected into the bed or at its surface. Combustion air is forced upward through the bed causing its fluidization. Ash resulting from combustion is removed from the bottom of the reaction chamber. Constant temperature is maintained in the reaction chamber through auxiliary fuel addition, thorough waste mixing, and heat conservation by the bed material. A secondary reaction chamber is employed to permit adequate retention time (at least 2 seconds) for combustion of volatiles. Combustion gases are drawn out of the secondary reaction chamber and treated for removal of particulates and sometimes acid gas.

A variation of the fluidized bed incinerator, called the circulating bed com-

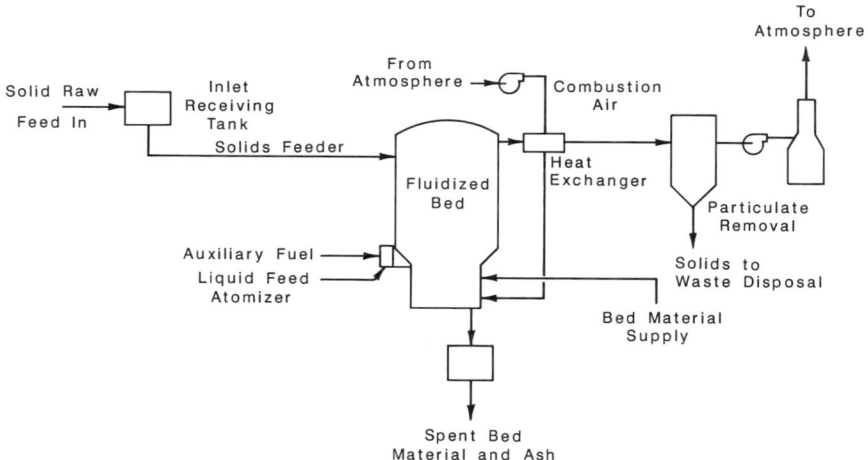

Fig. 12-11. Fluidized bed incineration process. (*Source:* U.S. EPA 1988b.)

bustor (CBC), has been developed by Ogden Environmental Services. The CBC uses higher air velocity and circulating solids to create a larger and highly turbulent combustion zone for the efficient destruction of toxic chemicals and the retention of resultant acid vapor. A schematic of the CBC incineration process is shown in Fig. 12-12. Solids, liquids, or sludges are burned along the height of the combustion section. Dry limestone, added to the feed, reacts in the combustion zone and captures acid gases avoiding the need for wet scrubbers. The high turbulence, staged combustion, and long residence time in CBCs allow incineration of the waste at lower temperature (1,500 to 1,600°F), thus eliminating ash agglomeration and reducing nitrous oxide (NO_x) emissions. The entrained solids are separated from off-gases by an integral cyclone and recycled to the combustor through a nonmechanical seal. The flue gases are cooled in an off-gas cooler by the heating of water, steam, or combustion air. Any remaining particulates in the cooled off-gas are separated in a baghouse filter, and the clean off-gas is vented to the atmosphere.

Infrared thermal treatment consists of passing sludge through an infrared chamber on a conveyor belt. Silicon carbide elements are used to generate the infrared radiation. The primary process variables are temperature (760°C to 870°C for the incineration mode and 430°C for the pyrolysis mode), residence

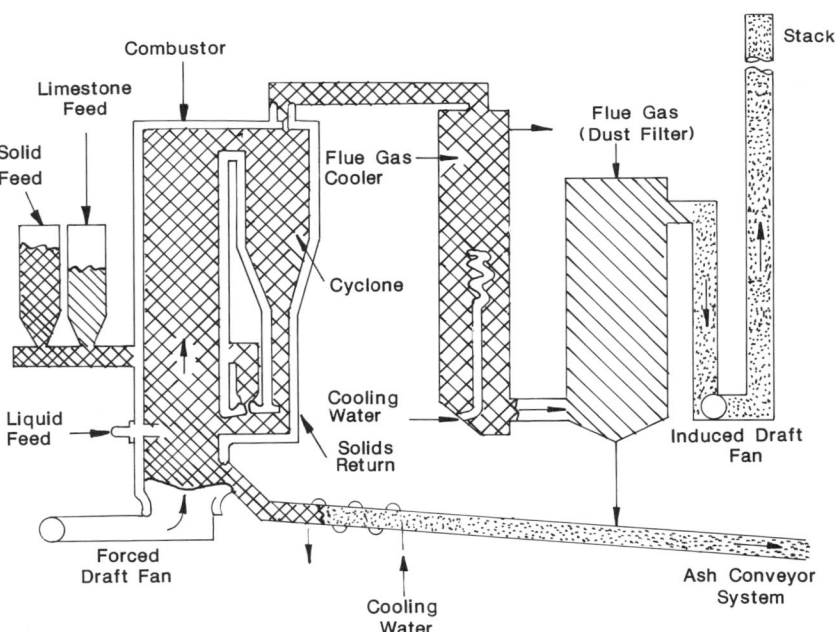

Fig. 12-12. Circulating bed combustor. (*Source:* Ogden Environmental Services.)

time 5 to 50 minutes), material layer thickness (optimum 2 inches), and combustion air flow. As shown in Fig. 12-13, off gases pass through a secondary reaction chamber for further irradiation and retention time and then through treatment devices for particulate and acid gas removal. The residues from this process are ash, treated combustion gases, and wet scrubber water.

Vitrification

Vitrification consists of thermally treating wastes such that the treated residues are immobilized into a vitreous mass. In the upper section of the refractory-lined reaction zone, the wastes are combusted at temperatures greater than 1,300°C. The off-gas is withdrawn by an induction fan and treated. The lower section of the reaction zone contains the two layer molten zone for the melts of the metal and siliceous components.

The advantages of vitrification over other thermal processes are the reduced oxidation products and reduced air emissions (due to high operating temperatures), and reduced leachability of inorganic materials. The residues from this process are treated combustion gases, molten mass, and scrubber water.

Wet Air Oxidation

The wet air oxidation (WAO) treatment process involves mixing wastes with compressed air in a high pressure (150 to 3,000 psig), high temperature (350 to 600°F) reactor as shown in Fig. 12-14. The reaction temperature is reached initially through addition of steam and maintained by the exothermic oxidation which occurs. The residuals from this process are treated oxidation by-product gases, wet scrubber water, oxidized sludge, and residues from decant water treatment. Off-gases are treated for acid gas removal. The solids residual undergoes dewatering prior to disposal. The waste liquid from the reaction may be treated biologically (activated sludge, fixed film aerobic or anaerobic treatment).

Stripping

Low temperature thermal stripping is normally applied to those sludges that have relatively high organic and solids contents. Chemical Waste Management has developed a system called (X* TRAX™), which was designed for treating sludges containing low boiling point (less than 430°C) organics at less than 10 percent by weight, and more than 40 percent solids.

Low temperature thermal stripping is being applied to relatively dry solids (as noted above for the X* TRAX™ process). Hot air is contacted with the sludge in a pug mill or rotary drum system. Volatile organics are stripped from

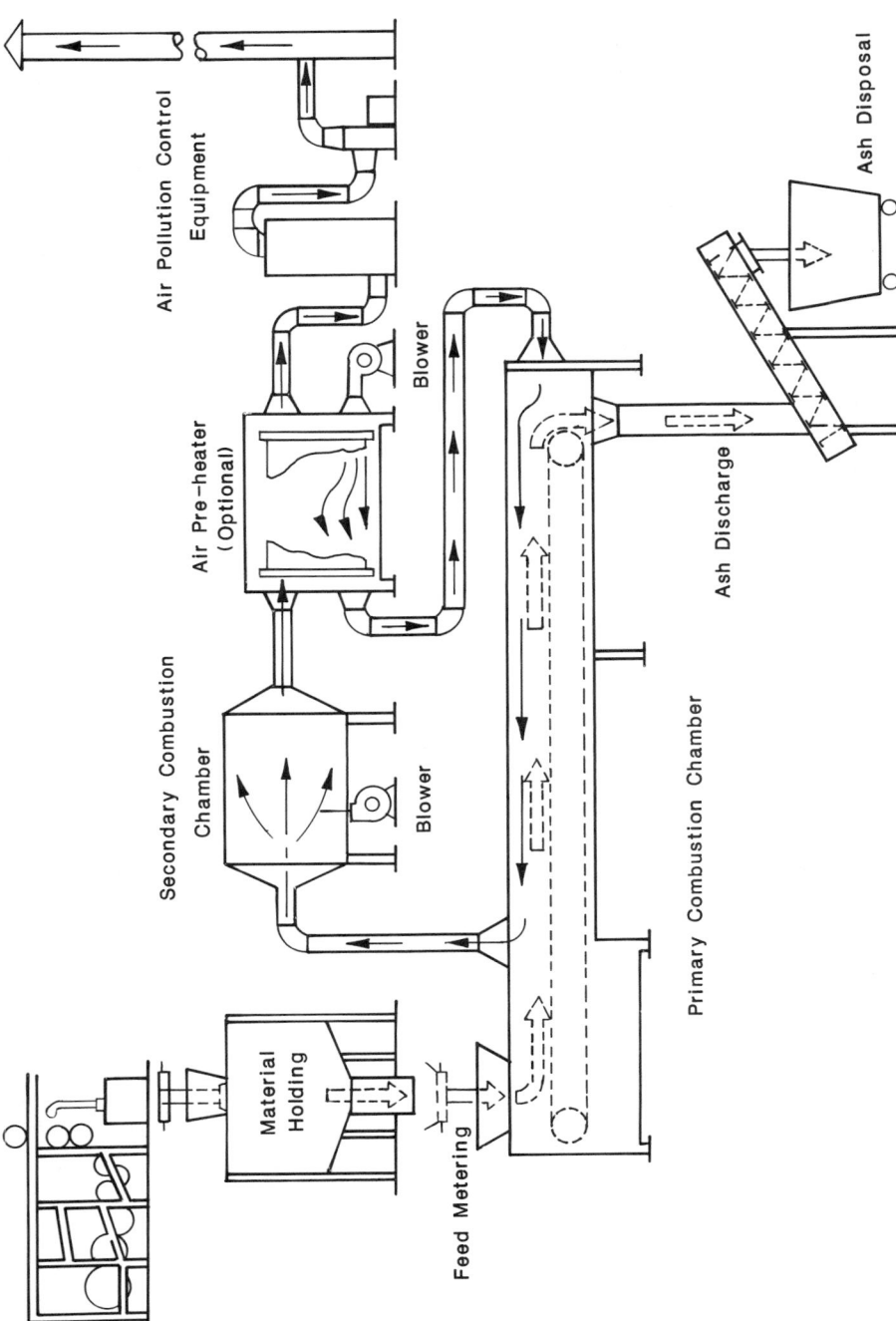

Fig. 12-13. Infrared thermal treatment process. (*Source:* U.S. EPA 1988b.)

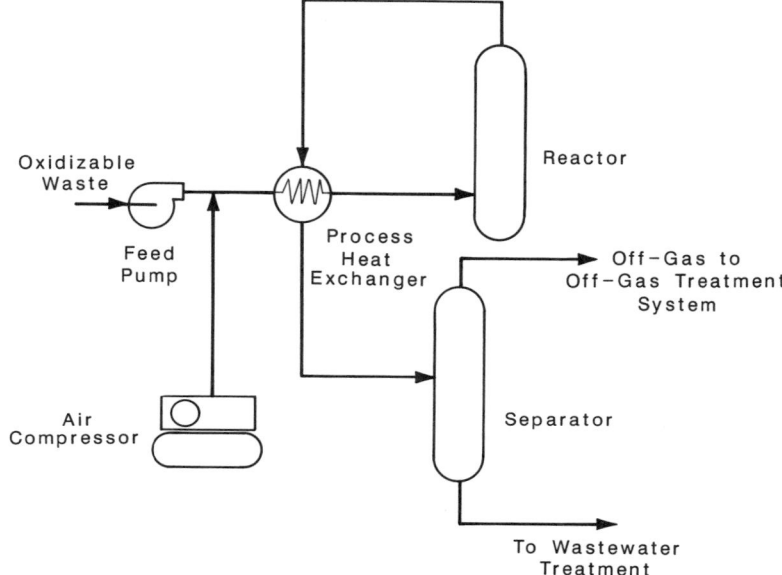

Fig. 12-14. Wet air oxidation process. (*Source:* U.S. EPA 1988b.)

the sludge and either sorbed onto activated carbon or incinerated. An example process flow diagram for a low temperature thermal stripper is shown in Fig. 12-15. This system uses a hot oil system to maintain the temperature in the mill and an afterburner to treat the air-stripped organics. The corrosion characteristics of the sludge are especially important in evaluating this type of system. The residuals from this process are combustion gases and treated sludge.

Freeze Crystallization

Freeze crystallation is a developing process which has the claimed ability to concentrate both volatile and nonvolatile components. Dissolved solids and organics are concentrated and separated in the process to produce a contaminant-free product water stream.

The use of freezing as a concentration process is based on the concept that when ice is crystallized from an aqueous solution, the ice crystal is pure water, with all the impurities originally present in the solution excluded from the crystalline phase. Thus, given a suitable means to separate the ice from the mother liquor, freezing is adaptable to the production of very high waste concentrations. In many cases, this concentration is to the point of wet contaminant solids. At the same time, melting the pure ice yields a reusable water product of exceptionally high quality.

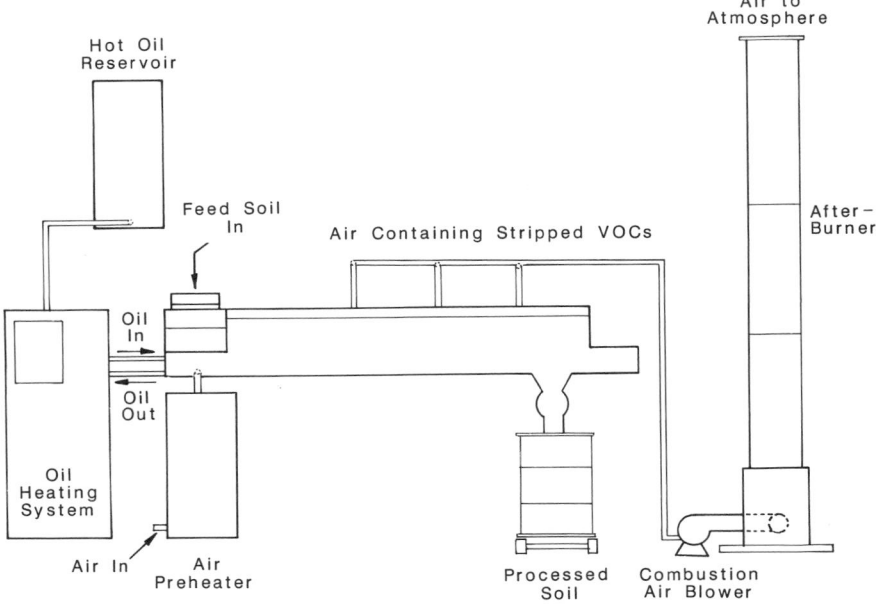

Fig. 12-15. Low temperature thermal stripping process. (*Source:* U.S. EPA 1988b.)

The freeze crystallization process consists of four main unit operations. Heat transfer to bring the solution to its freezing point is accomplished by indirect cooling heat exchangers, direct cooling by compressed refrigerants, or by combinations of the two. As the cooled liquid reaches its freezing point, a slurry of small ice crystals is formed in the concentrated liquor. This slurry is pumped to the first physical separation step where gravity or induced density separation of the crystals and mother liquor is accomplished. The solid crystals are then washed and melted to provide the product stream. The usual processing scheme utilizes internal heat exchange to maximize the efficiency of heat transfer between the warm influent and cold effluent streams. If solid precipitate forms in the process from exceeding the solubility product of various chemical compounds, the solids and ice crystals separate physically due to density effects in the crystallizer mother liquor.

ULTIMATE DISPOSAL

The most common, viable methods for ultimate disposal of wastewater treatment residues include:

- Landfilling

- Soil conditioning
- Land application
- Recycling

Landfilling is becoming much more restrictive, as discussed earlier. Numerous municipalities distribute and market their stabilized wastewater sludges as soil conditioners. Sludges are applied to agricultural or non-agricultural lands in many instances. Inorganic sludges may be recycled for chemical recovery (e.g., aluminum hydroxide sludges and ferric hydroxide sludges), or in the case of ash, incorporated into construction aggregate and landfill intermittent cover. Lagooning and ocean disposal are still practiced to some degree, although the outlook is not favorable for continuation of these methods.

REFERENCES

Eckenfelder, W. W., Jr. 1980. *Principles of Water Quality Management*. Boston, MA: CBI Publishing.

Eckenfelder, W. W., Jr. 1989. *Industrial Water Pollution Control*, 2nd edition. New York: McGraw-Hill, Inc.

51 *FR* 21648, June 13, 1986.

40 *CFR*, Parts 260 through 270.

40 *CFR*, Part 761.

Metcalf and Eddy, Inc. 1972. *Wastewater Engineering*. New York: McGraw-Hill, Inc.

U.S. EPA 1982. *Closure of Hazardous Waste Surface Impoundments*, SW-873, Washington, DC.

U.S. EPA. 1988a. *The Superfund Innovative Technology Evaluation Program: Technology Profiles*. EPA/540/5-88/003. Washington, DC.

U.S. EPA. 1988b. *Technology Screening Guide for Treatment of CERCLA Soils and Sludges*. EPA/540/2-88/004. Washington, DC.

Wilson, D. J., and Clarke, A. N. 1988. Soil Clean-Up by In Situ Surfactant Flushing I. Mathematical Modeling and Lab Scale Results. *Proc. DOE Model Conference*, Oak Ridge, TN.

13

Economics of Toxicity Reduction

Kevin D. Torrens

INTRODUCTION

Economics play a critical and often decisive role in final process selection. In many cases, there are several different treatment technologies which can achieve the required effluent quality, but there will typically be significant differences in capital and operating costs. Rational determination of these cost differences, followed by development of accurate final cost estimates, is critical to successful project completion.

ALTERNATIVE ESTIMATES

Generally, cost estimates are conducted to either determine the most economical treatment technology(ies) for a given wastewater or for developing estimated project costs for internal budgeting or project bidding. The level of detail and accuracy required in the estimating process varies according to the purpose. Comparison and selection of alternatives generally requires less accuracy than budgetary cost estimates. In developing either type of cost estimate, both capital and operation and maintenance (O&M) costs must be evaluated due to the potentially large operating costs associated with certain technologies relative to capital cost. Examples include technologies such as chemical oxidation or carbon adsorption. In addition, a present value (PV) comparison which accounts for capital and O&M costs should be conducted for cost comparisons over the expected life of the facility. Due to the impact of operating costs on the PV, there may be significant and unexpected cost differentials between system alternatives.

SOURCE OF DATA

The source used for developing the cost estimate can vary according to the accuracy desired. Sources include published cost curves, previous in-house cost estimates, computerized cost estimating programs and vendor quotes. When using any source, with the exception of recent vendor quotes, the data obtained must be adjusted to reflect current costs and actual design conditions.

Cost estimates may be required at different stages of a project for initial technology comparison and selection, specific treatment component selection or for final budget estimating purposes. The required level of accuracy of the estimate would be dependent on the expected use. Technology comparison and selection would typically require less accuracy and detail than budget level estimates for funding requests. Published cost curves may be sufficiently accurate for initial technology comparisons and screening. More detailed and accurate estimates typically obtained from equipment suppliers are often required for final process comparison, process selection and for budgeting purposes.

This chapter will present resources and methods for preparing cost estimates as well as a comparison of the costs of various technologies for toxicity reduction of industrial wastewaters. The intent is to provide a guide in preparing comparative or budget level cost estimates and to provide a means of comparison of alternatives.

COST ESTIMATE DEVELOPMENT PROCEDURE

In general, development of cost estimates requires a knowledge of the following items:

- Wastewater characterization data
- Design parameters for applicable treatment processes
- Effluent requirements
- Unit cost information of treatment processes under consideration

Preliminary process screening cost estimates may be based on assumed information or historical data while final process selection and budgeting estimates must be based on reliable operating or treatability data to insure cost estimate accuracy.

A general procedure for development of treatment cost estimates which applies to most industrial situations has been developed. Modifications of this procedure may be applied to cases requiring special treatment and to allow for different degrees of cost estimate accuracy. The general estimating procedure is outlined in sequence below with a brief description of each step.

Step 1. Wastewater Characterization

Characterization data for the wastewater to be treated are necessary for development of process design information. The completeness of the available characterization data will significantly affect the accuracy of the cost estimate.

Wastewater characterization data should be collected from the actual waste-stream(s) under investigation where feasible. It this is not possible, alternate sources including published data for similar industries or synthesis of waste-water samples must be utilized. Typical parameters of interest, depending on the unit process, include flow, BOD, TOC, COD, TSS, pH, VOC's and heavy metals. If significant changes in plant production practices are anticipated over the life of the treatment plant, these must be factored into the system basis of design.

Step 2. Treatment Process Selection

The methodology for treatment process selection for cost estimation will be dependent on the intent of the cost estimate; i.e., preliminary process screening or budgetary estimating. If the intent of the cost estimate is to assist in prelim-inary process screening and selection at the start of a project, preliminary pro-cess design information must be developed based on knowledge of wastewater characteristics, typical process design criteria and the effluent quality targets for specific parameters. Order of magnitude cost estimates can then be prepared based on this preliminary information to determine if a process would be cost-effective and should be considered for further process testing and definition work.

If the intent of the estimate is final process selection or project budgeting, more detailed process definition information, based on treatability studies, is necessary. Based on the process design information, accurate cost estimates can be developed based on vendor estimates and equipment take-offs.

Other factors affecting process selection beyond system cost include:

• Volume of waste to be treated
• Character of the wastewater
• Effluent requirements
• Treatment process efficiency and reliability
• Potential for recovery of salable byproducts
• Ease of operation
• System flexibility in responding to changing conditions.

Step 3. Selection of Plant Design Capacity

The design capacity of each process must be determined by considering the characteristics of the wastestream. If increases in plant flow and loading are anticipated, the treatment plant must be designed to accommodate these antic-ipated increases. Typically,

• Hydraulic structures are designed to handle the flow not exceeding 90 per-cent of the time (90 percentile)

- Tankage volumes are sized on the 50 percentile flow
- Oxygen requirements for biological systems are based on the 90 percentile oxygen demand
- Chemical feed systems should accommodate the 90 percentile demand.

Additionally, sufficient redundancy must be provided to allow operation of the plant with key system components off line due to equipment failure. The required accuracy of the cost estimate will determine the extent to which these factors are incorporated in developing the process designs.

Step 4. Selection of Cost Estimating Source

The source of the cost estimate will typically be dependent on the degree of accuracy required. For instance, preliminary process comparison and selection may only require order of magnitude estimates which can be obtained from published cost curves while more detailed estimates will require equipment supplier quotes and/or bids. Cost estimating resources include the following:

- U.S. EPA "Innovative and Alternative Technology Assessment Manual" (U.S. EPA 1980)
- U.S. EPA "Treatability Manual, Volume IV, Cost Estimating" (U.S. EPA 1983)
- "CAPDET" cost estimating program (United States Army Corps of Engineers 1976)
- "Water Cost" cost estimating program (Culp and Culp 1986)
- "ECAS Expert System" cost estimating program (Van Nostrand Reinhold Company, Inc. 1986)

Cost data obtained from dated resources such as cost curves or cost estimating programs must be updated to reflect current cost indices due to continual changes in the cost of goods and services. Different resources utilize different cost indices to provide benchmarks for cost adjustments. The most common indices include the *Engineering News Record* (ENR) construction cost index (ENR 1989), United States Environmental Protection Agency (U.S. EPA) Sewage Treatment Plant (STP) index and various Bureau of Labor Statistics (BLS) indices for categories such as specific types of labor and materials. These resources are all readily available and must be utilized to update any cost estimates which rely on old cost data.

Step 5. Development of Treatment Cost Estimates

As described previously, the type of cost estimate to be conducted is dependent on the phase of the project and intent of the estimate. Table 13-1 presents a summary of the various types of cost estimates typically used, the degree of

Table 13-1. Types of Cost Estimates.

Phase	Estimate Range	Accuracy Range	Basis	Decisions/Uses
Process screening	Order-of-Magnitude	+50% to −30%	Capacity, waste type and strength	Elimination of non-cost-effective processes Order-of-magnitude funding needs Fund authorization for experimental programs
Process selection	Budget	+30% to −15%	Equipment specifications and preliminary design based on pilot-plant data base	Process selection Financial planning Fund authorization for definitive estimate
Project authorization and construction	Definitive	+15% to −5%	Full scope, equipment and building design, piping, electrical, etc.—preliminary design	Construction fund authorization Construction project cost control

accuracy of the estimate, the basis of the estimate and the type of decision for which it is used; i.e., process selection or budgeting.

The accuracy range of plus 50 percent to minus 30 percent for order-of-magnitude estimates represents the cost range which might normally be expected between the estimate and the final plant cost. In order to have confidence in the accuracy of comparative estimates, the following guidelines should be followed:

- The estimate must be done by a single estimator.
- Cost data should be developed from a common resource; i.e., cost curves, vendor estimates, etc.
- Comparable design assumptions should be used for all individual process components.
- The scope of each estimate should be as similar as possible.

If the individual process estimates prepared using these guidelines differ by more than 25 percent, the differences should be considered significant. If the estimates for various options are within 25 percent, other factors must be utilized in process selection such as reliability, ease of operation, flexibility, etc. More accurate estimates may also be needed to determine actual cost differences.

Other capital cost factors, such as contingencies, engineering, piping, electrical, instrumentation and site preparation must be added to the unit process cost data to provide total capital costs. Operation and maintenance costs must also be determined for the proposed system and should include labor, electricity, sludge disposal, chemicals, maintenance and any other applicable expenses.

Present Value Determination

Final process costs should be compared on a present value (PV) basis to account for the effects of capital and operating costs over the life of the plant. This method of comparison provides a rational basis for process selection from estimates of long-term system costs.

Present value analysis converts all costs for the life of the system to today's value based upon a discount rate selected to correspond with the owners estimated cost of funds. Constant O&M costs can be converted back to present value by multiplying the cost times the present value factor found in Table 13-2 for the appropriate discount rate and project economic life. The same can be done with anticipated replacement costs or major one time costs through use of the factors found in Table 13-3. Following this conversion, these present value costs can be summed and added to the initial capital cost to determine total present value. The present value of each alternative can then be compared and a selection made.

Table 13-2. Present Value Factors for Constant Annual Expenditures.[a]

N Year	8%	10%	11%	12%	13%	14%	15%
1	.926	.909	.901	.893	.885	.877	.870
2	1.783	1.736	1.713	1.690	1.668	1.647	1.626
3	2.577	2.487	2.443	2.402	2.361	2.322	2.283
4	3.312	3.170	3.102	3.037	2.975	2.914	2.855
5	3.993	3.791	3.696	3.605	3.517	3.433	3.352
6	4.623	4.355	4.231	4.111	3.998	3.889	3.785
7	5.206	4.868	4.712	4.564	4.423	4.288	4.160
8	5.747	5.335	5.146	4.968	4.799	4.639	4.487
9	6.247	5.759	5.537	5.328	5.132	4.946	4.772
10	6.710	6.145	5.889	5.650	5.426	5.216	5.019

[a]Present value of $1 expended each year for N years at various discount rates. $PV = f \times$ $ expended.

Table 13-3. Present Value Factors for One-Time Expenditures.[a]

N Year	8%	10%	11%	12%	13%	14%	15%
1	.926	.909	.901	.893	.885	.877	.870
2	.857	.826	.812	.797	.783	.769	.756
3	.794	.751	.731	.712	.693	.675	.658
4	.735	.683	.659	.635	.613	.592	.572
5	.681	.621	.593	.567	.543	.519	.497
6	.630	.564	.535	.507	.480	.456	.432
7	.583	.513	.482	.452	.425	.400	.376
8	.540	.467	.434	.404	.376	.351	.327
9	.500	.424	.391	.361	.333	.308	.284
10	.463	.386	.352	.322	.295	.270	.247

[a]Present value of $1 expended in the Nth year at various discount rate $PV = f \times$ $ expected.

TECHNOLOGIES CONSIDERED

As discussed in preceding chapters, there are numerous and different technologies applicable to toxicity reduction of industrial effluents. Of major interest are the following:

- Biological treatment—aerobic and anaerobic processes
- Carbon adsorption—PACT® and GAC
- Chemical oxidation—hydrogen peroxide and ozone
- Precipitation
- Stripping
- Sludge handling methods

Interrelationships

Although these candidate processes are broadly stated, it is important to note that each may be composed of several individual treatment components which often employ different removal mechanisms and are combined in variable sequence to achieve the required effluent quality. Examples include:

- Filtration preceding GAC columns for removal of suspended solids to prevent clogging
- Filtration following PACT® or activated sludge to reduce effluent suspended solids and the associated organics
- Chemical oxidation preceding biological treatment to improve biodegradability
- pH adjustment preceding chemical oxidation to achieve optimum conditions for chemical reaction
- Filtration following precipitation for enhanced removal of suspended solids
- Precipitation preceding stripping for removal of iron to prevent fouling
- Sludge conditioning preceding dewatering and disposal

Seldom does any unit process operate alone, therefore, the impact of each unit process on other system components must be considered in design and included in the cost estimating function.

Process Cost Sensitivity

Capital and operation and maintenance (O&M) costs for various treatment technologies are sensitive to different variables; i.e., flow, constituent concentrations, pH, and type of wastewater constituents. For example, the capital cost of a GAC adsorption system is most strongly influenced by wastewater flow rate, while operating costs of GAC are most sensitive to contaminant concentration and adsorbability.

Table 13-4 presents a summary of the major toxicity reduction treatment technologies of interest and the cost sensitivity factors associated with each. The factors are presented in decreasing order of importance for the particular cost estimate and technology.

DESCRIPTION OF PROCESS UNITS

Development of treatment system cost estimates is typically conducted using a building block approach where costs for individual components are added together. Other associated costs such as piping, instrumentation, electrical, contingencies, engineering and profit are then customarily added as a percent of the total installed equipment costs. Outlined below are brief descriptions of

Table 13-4. **Treatment Technology Cost Sensitivity Parameters.**

Treatment Technology	Capital Costs	O&M Costs
Precipitation	Flow	Acidity
Air stripping	Flow	Off-gas treatment
	Constituents	Pretreatment
	Percent removal	Air requirements
	Off-gas treatment	
Chemical oxidation	Flow	Chemical requirements
	Chemical, i.e., peroxide or ozone	
Activated sludge	Flow	Organic loading
	Organic concentration	Sludge disposal
		Aeration costs
PACT®	Flow	Carbon requirements
		Sludge disposal
GAC	Flow	Carbon requirements
		Carbon regeneration

typical treatment systems with a discussion of other technologies which are often associated with that technology. Various equipment items associated with the technology of concern are also provided. These should be included in any cost estimates.

Activated Sludge

The activated sludge process includes numerous operational modes and configurations including continuous flow, sequencing batch reactor (SBR), high rate, low rate or extended aeration. In addition, actual construction features can take various forms including above ground or below ground; concrete, steel or lined lagoon; circular or rectangular; and numerous types of aeration devices. Additional treatment components often associated with an activated sludge system include equalization, pH adjustment, and primary clarification. Costs of these unit processes must be included in the complete project estimate. Costs presented in Tables 13-5 and 13-6 include the activated sludge system, sludge handling and dewatering and sludge disposal cost items as discussed below. Major cost items for the activated sludge system include the following: concrete tank, aeration system, blowers, clarifier, return sludge pumps, sludge wastage pumps, instrumentation, piping, and electrical.

Sludge Disposal

Sludge disposal can often be one of the largest single cost items associated with wastewater treatment system operation, particularly if the sludge is considered

Table 13-5.   Incremental Capital Costs for Toxicity Reduction Technologies (as a percent of base system cost in 1989 dollars).

	Flow,[a] MGD						
	0.01	0.10	0.50	1.0	5.0	10.0	Average
Activated sludge (base cost, $\times 10^6$)[b]	1.0 (0.60)	1.0 (1.60)	1.0 (3.30)	1.0 (5.5)	1.0 (13.5)	1.0 (19.7)	1.0
Air stripping	0.12	0.13	0.17	0.20	0.20	0.26	0.18
Equalization	0.22	0.15	0.22	0.17	0.15	0.18	0.18
Neutralization	0.14	0.11	0.10	0.08	0.07	0.07	0.09
Chemical precipitation	0.38	0.40	0.42	0.40	0.40	0.43	0.41
Filtration	0.19	0.20	0.15	0.13	0.12	0.11	0.15
Chemical oxidation (peroxide)	0.20	0.17	0.17	0.17	0.19	0.22	0.19
Chemical oxidation (ozone)	0.39	0.30	0.31	0.30	0.26	0.29	0.31
GAC columns	0.37	0.28	0.25	0.22	0.25	0.23	0.27
PACT®	0.12	0.09	0.13	0.10	0.08	0.07	0.10
PAC regeneration	NA	NA	0.44	0.42	0.48	0.47	0.44
Anaerobic biological treatment	0.60	0.85	0.90	0.90	0.92	0.91	0.85

[a]Wastewater, 1,000 mg/l COD, 500 mg/l BOD.
[b]Includes cost of sludge handling and dewatering.
[c]Based on carbon utilization rate of 2 lb carbon/1,000 gal.
[d]Based on 250 mg/l carbon dose.
ENR Index = 4,570.
NA = Not Applicable.

Table 13-6. Incremental Operation and Maintenance Costs for Toxicity Reduction Technologies (as a percent of base system cost in 1989 dollars).

	Flow,[a] MGD						
	0.01	0.10	0.50	1.0	5.0	10.0	Average
Activated sludge (base cost, $ \times 10^6$)[b]	1.0 (0.07)	1.0 (0.16)	1.0 (0.50)	1.0 (0.65)	1.0 (1.60)	1.0 (2.42)	1.0
Air stripping	0.15	0.25	0.2	0.17	0.18	0.17	0.18
Equalization	0.015	0.03	0.03	0.06	0.06	0.07	0.045
Neutralization	0.07	0.12	0.16	0.2	0.32	0.34	0.20
Chemical precipitation	0.16	0.2	0.2	0.24	0.33	0.35	0.31
Filtration	0.015	0.07	0.07	0.1	0.08	0.14	0.08
Chemical oxidation (peroxide)	0.15	0.25	0.22	0.29	0.43	0.45	0.30
Chemical oxidation (ozone)	0.05	0.09	0.15	0.23	0.45	0.60	0.26
GAC columns	0.10	0.12	0.10	0.1	0.09	0.09	0.10
GAC regeneration	0.14	0.12	0.16	0.18	0.31	0.31	0.21
PACT®[c]	0.35[d]	0.43[d]	0.07[e]	0.11[e]	0.09[e]	0.08[e]	0.39[d]
							0.09[e]
PAC regeneration	NA	NA	0.22	0.21	0.16	0.17	0.19
Anaerobic biological treatment	0.14	0.19	0.3	0.34	0.37	0.36	0.28

[a]Wastewater, 1,000 mg/l COD, 500 mg/l BOD.
[b]Includes sludge handling, dewatering and disposal.
[c]PAC dose of 250 mg/l.
[d]Includes cost of makeup carbon with no regeneration.
[e]Assumes on-site PAC regeneration as a separate cost.
ENR Index = 4,570.
NA = Not Applicable.

hazardous. Disposal options for nonhazardous sludges include landfilling, land application and incineration. Disposal options for hazardous sludges include disposal in permitted hazardous waste landfills, and incineration in a permitted incinerator. Costs for nonhazardous sludge disposal range from $10 to $150/ton with hazardous sludge disposal costs ranging from $200 to $1,000/ton of dry solids. Cost for sludge disposal are highly site and waste specific and must be considered during any cost estimating procedure.

Sludge Handling and Dewatering

Sludge handling and dewatering accounts for a substantial capital and O&M cost for wastewater treatment. Sludges are generated from physical-chemical processes, primary clarification, and biological treatment operations. Typical processes associated with sludge handling include aerobic or anaerobic digestion, gravity and flotation thickening, chemical conditioning and dewatering using various devices such as belt filter presses, plate and frame filter presses, vacuum filters or sand drying beds. The actual system configuration will be based on the type and quantity of sludge produced and the ultimate point of disposal; i.e., landfill, land application, incineration or wet air oxidation. Costs for sludge handling and dewatering include: sludge conditioning tanks, chemical storage and feed systems, thickening and dewatering equipment.

Anaerobic Biological Treatment

Anaerobic process alternatives include complete mix, fixed media, suspended upflow media and low, medium and high rate designs. Other components often associated with anaerobic treatment systems include pH adjustment, solids handling and off-gas handling. Costs for these additional unit processes must be included in the project estimate. Major cost items include the following: anaerobic reactor, agitators, off-gas control system, instrumentation, pumps and degasifier and sludge handling, dewatering and disposal.

Air Stripping

Stripping is used as a treatment or pretreatment step for removal of volatile compounds. Air stripping can be used as either a stand-alone treatment process or in conjunction with other treatment components, typically activated sludge or GAC. Recent air emission regulations typically require off-gas treatment for removal of volatile compounds from the off-gas prior to release. Off gas treatment may include vapor phase carbon adsorption or incineration. Costs include the following items: stripping tower, blower, mist eliminator, packing media, pumps, and feed tank. Air emission control equipment is at additional (and often substantial) cost.

Equalization

Equalization of wastewater flow and contaminant concentration is often required in industrial facilities due to the high variability of wastewater characteristics associated with industrial facilities. Equalization is rarely used as a stand-alone treatment method except as a pretreatment step prior to discharge to a municipal facility. In most instances, equalization is used to reduce the size of subsequent treatment processes and improve system performance by reducing the variability of influent wastewater flow and concentration. Equalization design is generally either constant volume/variable flow or variable volume/constant flow operation and should be well mixed. Major costs items include the following: piping carbon steel tank, concrete foundation, agitator(s), insulation, pumps, valves, and instrumentation.

Neutralization

Neutralization is rarely used as a stand-alone treatment process prior to direct discharge, however, it is commonly used as a pretreatment step prior to discharge to a municipal facility or on-site treatment facility. Neutralization consists of pH adjustment to near neutral conditions with acid or base, depending on the initial pH of the wastewater. Neutralization is typically one of the first unit processes of a treatment system and may precede or follow equalization, depending on pH and wastestream variability. Major cost items include the following: mixing reaction vessel(s), attenuation vessel, acid/base storage and feed system, agitators.

Chemical Precipitation

The chemical precipitation process is widely used for removal of heavy metals and involves pH adjustment, coagulant addition and flocculation and sedimentation of metal hydroxide or metal sulfide precipitates. Each heavy metal has an optimum pH at which minimum solubility is achieved. If several metals are present, either a compromise pH (providing some removal of each metal) is used or a multi-stage system is designed to provide optimum removal of each metal of concern. The compromise design is often used for relatively low metals concentrations while the latter is typically utilized as a primary treatment process for metal finishing wastewaters. Sludge handling is required with either process alternative. Major cost items include the following equipment items: mixing and flocculation vessels, chemical and polymer storage, day tank and feed pumps, agitators, clarifier and mechanical dewatering equipment (if sludge must be handled separately). Sludge handling, dewatering and disposal costs are also included. Costs for sludge handling and dewatering include: sludge conditioning tanks, chemical storage and feed systems, thickening and dewatering equipment.

Filtration

Sand and multi-granular media filtration is typically used in conjunction with metals precipitation, GAC, PACT®, or for polishing of clarified effluents from biological systems. Filtration can serve as either a pre- or post-treatment process. Major cost items include the following components: filter structure and internal components, feed pumps, backwash pumps, air compressor, and backwash holding tank.

Wet Air Oxidation

Wet air oxidation (WAO) is essentially identical to the process used for powdered activated carbon (PAC) regeneration except it is used as a dedicated treatment system for destruction of organic material. The system is typically considered only for low flow, high-strength, refractory or highly toxic wastestreams. Costs for wet air oxidation systems are primarily based on flow. Major cost items include: wet air oxidation system, waste sludge storage tank, feed pumps and instrumentation.

Chemical Oxidation

Chemical oxidation can be utilized as either a pretreatment step prior to biological treatment to improve biodegradability, post treatment to oxidize residual organic materials or as a stand-alone treatment process. pH adjustment is often a required pretreatment step to chemical oxidation. Equipment requirements for chemical oxidation include chemical storage tanks (peroxide), ozone generation equipment, delivery pumps, and a reaction tank with agitation. The major operating costs are typically chemicals (peroxide) or electricity (ozone generation). Major capital cost items include the following for a peroxide system: chemical storage tanks, feed pumps, reaction tank and instrumentation. Major cost items for an ozone system include: ozone generator, oxygen storage tanks and instrumentation.

Granular Activated Carbon

Granular activated carbon (GAC) can be used as a stand-alone treatment process or in conjunction with other processes such as precipitation, biological treatment, air stripping or filtration. Depending on the wastewater characteristics, pretreatment for iron and suspended solids removal is often necessary to maintain GAC operating efficiency and prevent clogging. Costs include the following: steel carbon columns (at least two in series or parallel), virgin and spent carbon holding vessels, backwash holding tank, initial carbon charge, feed and backwash pumps, agitators, and instrumentation.

On-site carbon regeneration, typically by multiple hearth or fluidized bed furnances, is only justified at high carbon usage rates; i.e., greater than 5,000 lb/day. For lower carbon usage rates, off-site regeneration through a carbon supplier is the preferred alternative. Major cost items include the following: fluidized bed furnace, air scrubber, regenerated carbon holding tank, afterburner, and quench tank.

Powdered Activated Carbon Treatment (PACT®)

The PACT® process is a proprietary treatment process (licensed by Zimpro, Inc.) consisting of the addition of powdered carbon to an activated sludge system. As such, license fees are required and are dependent on system size, as shown in Table 13-7. The PACT® process can be constructed as a new system or retrofitted to existing activated sludge systems. Of particular concern in PACT® design are materials of construction due to the abrasive nature of carbon. This pertains primarily to aeration equipment, pumps, piping and valves. Costs for the PACT® system include carbon storage, carbon addition equipment, carbon slurry tank, and license fee. Cost estimates assume an activated sludge system is in place and can be readily converted to a PACT® system.

Carbon can be added to the system as either virgin carbon, regenerated carbon from the carbon supplier, or as regenerated carbon from an on-site wet air oxidation regeneration system. The source of carbon, virgin or regenerated, is typ-

Table 13-7. PACT® License Fees.[a]

Capacity, MGD	Carbon Dose Range, ppm					
	< 125	125–200	201–350	351–600	601–1000	> 1000
< 0.1	10[a]	15	20	25	30	40
0.5	25	35	45	60	75	100
1.0	35	45	60	80	105	140
2.0	50	65	85	115	150	200
5.0	90	115	150	200	270	360
10.0	135	175	230	310	400	540
20.0	200	260	340	460	600	800
50.0	400	520	680	800	800	800
100	670	800	800	800	800	800
100	670	800	800	800	800	800
≥ 150	800	800	800	800	800	800

Source: Zimpro, Passavant, Inc.
[a]Thousands of 1981 dollars. Intermediate capacity fees are determined by interpolation. Effective May 1, 1981.

ically based on system size and carbon utilization rate. Typically, on-site carbon regeneration becomes cost-effective at flows of approximately 1.0 MGD, however, this can vary depending on carbon dosage.

COMPARATIVE COSTS

Accurate comparisons of various toxicity reduction technology costs can only be conducted on a case-by-case basis due to the extreme variability encountered in industrial wastewaters and the varying degree of treatment required. Chemical doses, carbon utilization rates and tankage requirements can vary by orders of magnitude from wastewater to wastewater for any given flow. In addition, other non-cost items such as reliability, simplicity and operability often play a decisive role in technology comparison and selection.

However, with the above considerations in mind, comparative costs for several of the most common treatment technologies; i.e., activated sludge, anaerobic biological treatment, GAC, PACT® precipitation, chemical oxidation and stripping were developed for various wastewater flows ranging from 10,000 to 10 mil gal/day.

The values presented in Table 13-5 are the capital costs for each technology, at the indicated flow rate and wastewater characteristics, provided as a fraction of the cost of a base system consisting of activated sludge (including waste sludge handling and dewatering). This table can be used in conjunction with specific project capital cost estimates to gauge the relative capital costs of various treatment technologies in relation to a basic activated sludge system.

Table 13-6 presents a summary of the relative operation and maintenance costs of the various treatment technologies. The basic activated sludge system, including sludge handling, dewatering and disposal, was also used as the base cost for this comparison. Operation and maintenance costs must be adjusted to reflect differences in wastewater character such as wastewater contaminant concentration and constituency which can impact O&M costs. These items can impact degradability, reactivity and adsorbability which will in turn influence system capital and O&M costs.

In developing the costs, similar wastewater characteristics and treatment objectives were chosen (as available) for the various technologies; extreme characteristics or objectives were avoided. Data were gathered from published cost curves, vendor estimates and previous project experience. The costs presented for the various flows should be used for relative technology comparison only and not for budgeting purposes. The intent of this comparison was to demonstrate the relative cost relationship of the various technologies for a given wastewater.

The slight variability in specific process cost factors at different wastewater flow rates reflects the differences in sensitivity of individual unit treatment processes to wastewater flow. The relative relationship of the technology costs may change with different wastewater characteristics. Each of the individual treatment components exhibit slightly different economies of scale. Therefore, the individual process component cost factors will vary at different flow rates in relation to one another. The individual process component cost factors do not necessarily decrease with increasing flow rates, as is typical for economies of scale, since the cost factors are derived as a percent of the base cost of the activated sludge system at each flow rate which already exhibit economies of scale. Therefore, the variability in the specific process cost factors is due to economies of scale associated directly with that particular treatment process process in relation to economies of scale exhibited by the activated sludge system.

EXAMPLE PRESENT VALUE CALCULATION

An engineer must determine the most cost-effective alternative for a proposed wastewater treatment plant. The two treatment system alternatives provide similar treatment, however, since different technologies are involved, the alternatives have different capital and O&M costs.

ALTERNATIVE A

1. Capital cost, $2,700,000
2. Operation and routine maintenance cost, $230,000/yr
3. Major rework in 5th Year, $800,000
4. No salvage value anticipated

ALTERNATIVE B

1. Capital cost, $2,100,000
2. Operation and maintenance cost, $400,000/yr
3. Major rework in 5th Year, $600,000
4. No. salvage value anticipated

Determine which system will be more cost-effective over a 10-year economic life assuming a 10 percent discount rate.

Solution:

Alternative A.

1. Initial investment	$2,700,000
2. Present value of annual disbursements =	
$230,000 (6.145)	1,413,350
3. Present value of one-time cost in year 5 =	
$800,000 (0.621)	496,800
Total present value	$4,610,150

Alternative B.

1. Initial Investment	$2,100,000
2. Present Value of Annual Disbursements =	
$400,000 (6.145)	2,458,000
3. Present Value of One-Time Cost in Year 5	
= $600,000 (0.621)	372,600
Total Present Value	$4,930,600

Therefore, Alternative A is the preferred choice although it has a higher initial capital cost. This exemplifies the benefit of conducting a relatively simple present value analysis since capital costs which represent short term expenditures are often outweighed by the differences in long-term operating and maintenance costs. This procedure can be expanded to include purchasing of equipment at different time intervals, varying interest rates with time and recovery of salvageable equipment. A discussion of these procedures is beyond the scope of this book and the reader is referred to the numerous engineering economics books available.

EXAMPLE COMPARATIVE COST EVALUATION

An organic chemicals plant expects to change production processes and begin production of a new product. The overall average wastewater flow rate of 5.0 MGD is not expected to change significantly, however, the new product wastewater is expected to be extremely variable in organic concentration. In addition, the new wastestream will increase the peak flow to 7.5 MGD for durations of up to one day. Additionally, biological treatability testing indicates the new wastewater will adversely impact biological treatment effluent quality resulting in exceedence of the plant's effluent toxicity limit. The existing plant consists only of an activated sludge system.

Further toxicity reduction treatability testing indicates there are several alternative technologies capable of improving effluent quality for permit compliance. These include:

Alt A.   Equalization followed by chemical oxidation for improved biodegradability

Alt. B   Add PAC at a dose of 250 mg/l without equalization due to the effect of peak concentrations

Alt. C   Add PAC at a dose of 150 mg/l with the addition of equalization

Alt. D   Add GAC columns following biological treatment

Alt. E   Provide equalization and post-treatment with GAC (50 percent fewer columns due to reduced peak flows)

Neutralization will be required for all alternatives.

Using Tables 13-6 and 13-7, determine which treatment alternative is most cost-effective. Assume conditions used in development of the table are similar to those of the plant. Determine net present value for each alternative for a 10 year economic life at a 10 percent discount rate.

Solution.
Using Tables 13-5 and 13-6, determine the base cost of a 5.0 MGD activated sludge plant. Then apply the appropriate factors for specific technologies to achieve the total system costs. See Tables 13-8, 13-9, and 13-10 for a summary of costs.

Assumptions made in developing the cost figures in the tables include:

- A design flow of 7.5 MGD was required for some components for peak flows without equalization (alternatives B and D).
- A flow of 3.0 MGD was used to determine costs associated with the lower carbon dose of 150 mg/l with equalization (alternatives A, C, and E) since the data in Tables 13-6 and 13-7 are based on a carbon dosage of 250 mg/l ($5.0 \times 150/250 = 3.0$).
- Costs at design flows other than 5.0 MGD were linearly interpolated.
- Neutralization and filtration for removal of residual suspended solids were required for each alternative.
- No significant change in sludge production would occur. In practice, the chemical oxidation alternative may result in a slight increase in waste biological sludge due to improved biodegradability following oxidation.
- Carbon regeneration for powdered and granular systems would occur on site.

Table 13-8. Example Capital Cost Comparision Summary.

	Cost Estimate Flow, MGD		Alternative Costs, $ \times 10^6$				
Unit Process	A,C,E	B,D	A	B	C	D	E
Base Cost for Activated Sludge (Existing)	NA	NA	13.5	13.5	13.5	13.5	13.5
Equalization	5.0	NA	2.03	NA	2.03	NA	2.03
Neutralization	5.0	7.5	0.95	1.17	0.95	1.17	0.95
Chemical oxidation (peroxide)	5.0	NA	2.57	NA	NA	NA	NA
PACT®	3.0	5.0	NA	1.08	0.90	NA	NA
PAC regeneration	3.0	5.0	NA	6.48	4.67	NA	NA
Filtration	5.0	7.5	1.62	2.00	1.62	2.00	1.62
GAC columns	5.0	7.5	NA	NA	NA	4.00	3.38
GAC regeneration	3.0	5.0	NA	NA	NA	6.34	3.61
Total capital cost (not including activated sludge system)			7.17	10.73	10.17	13.51	11.59

Table 13-9.   Example O&M Cost Comparison Summary.

Unit Process	Cost Estimate Flow, MGD A,C,E	B,D	Alternative Costs, $ \times 10^6$ A	B	C	D	E
Base cost for activated sludge (existing)	NA	NA	1.60	1.60	1.60	1.60	1.60
Equalization	5.0	NA	0.09	NA	0.09	NA	0.09
Neutralization	5.0	5.0	0.50	0.50	0.50	0.50	0.50
Chemical oxidation (peroxide)	5.0	NA	0.68	NA	NA	NA	NA
PACT®	3.0	5.0	NA	0.15	0.11	NA	NA
PAC regeneration	3.0	5.0	NA	0.22	0.26	NA	NA
Filtration	5.0	5.0	0.13	0.13	0.13	0.13	0.13
GAC columns	5.0	5.0	NA	NA	NA	0.14	0.14
GAC regeneration	7.0	5.0	NA	NA	NA	0.50	0.32
Total capital cost (not including activated sludge system)			1.40	1.00	1.09	1.27	1.18

Table 13-10.   Example Present Value Comparison.

Cost Item	Alternative A	B	C	D	E
Capital cost, $ \times 10^6$	7.17	10.73	10.17	13.51	11.59
O&M Cost, $ \times 10^6$	1.40	1.00	1.09	1.27	1.18
$ \times 10^6$	15.77	16.88	16.86	21.31	18.84

[a]Calculated at 10 years at 10 percent interest (PV factor = 6.145).

Conclusions

Based on the results of the cost analyses summarized in Table 13-10, there would be no significant cost differential, based on present value, between alternatives A, B, or C, the chemical oxidation or two PACT® alternatives. Alternatives D and E, the GAC alternative result in substantially higher costs.

The importance of developing the present value costs is clear when the capital costs alone are compared for the various alternatives. Based solely on capital costs, chemical oxidation is clearly the least expensive, followed by PACT® with equalization. However, with O&M costs factored into the total present value cost, there is no significant difference between the chemical oxidation and PACT® alternatives. The decision on which alternative to proceed with then must be based on other concerns, such as:

- Process reliability
- Equipment reliability
- System operability
- Level of technology
- Ease of expansion
- Future sludge disposal considerations
- Client preference

Based on a review of all aspects of each alternative, the best system configuration can then be recommended.

REFERENCES

Aguilar, Rodolfo J. 1973. *System Analysis and Design.* Englewood Cliffs, NJ: Prentice-Hall, Inc.

AWARE Incorporated Internal Correspondence. October 1987. Comparison of Anaerobic Treatment Systems.

Culp, Wesner and Irvin Culp. 1986. Water Cost Computer Program for Estimating Water and Wastewater Costs. California

Dietrich, M. and J. Meidl. April 1989. Treatment of Pesticide Production Wastewater Using the PACT® System. Presented at the AIChE Conference, Houston, Texas.

DuPont, Inc. 1981. *PACT® Technology Manual*, Volume 1.

ENR. 1989. ENR First Quarterly Cost Report 222(12)55.

Hagar, Donald G. 1974. Industrial Wastewater Treatment by Granular Activated Carbon. *Industrial Waste Engineering*, January-February Issue.

Meidl, John A. and Allan R. Wilhelmi. January 1986. PACT®: An Economical Solution in Treating Contaminated Groundwater and Leachate. Presented at New England Water Pollution Control Association Conference.

O'Brien, Robert P. and Stenzel, Mark H. December 1984. Combining Granular Activated Carbon and Air Stripping. *Public Works*.

Sehuliger, Wayne G., et al. 1988. Thermal Reactivation of GAC: A Proven Technology. *Waterworld News*, **4**(1).

Shukron, A. J. and Culp, G. L. 1977. Appraisal of Powdered Activated Carbon Processes for Municipal Wastewater Treatment. EPA 600/2-77-156.

United States Army Corps of Engineers. 1976. Computer-Assisted Procedure for the Design and Evaluation of Wastewater Treatment Systems.

U.S. Environmental Protection Agency. February 1980. Innovative and Alternative Technology Assessment Manual. Publication MCD-53, Office of Research and Development, Washington, DC.

U.S. Environmental Protection Agency. April 1983. Treatability Manual, Volume IV, Cost Estimating. Office of Research and Development, Washington, DC.

Van Nostrand Reinhold Company, Inc. 1986. *Environmental Cost Analysis System*. New York: Van Nostrand Reinhold.

Glossary

acetogens.   acetic acid producing bacteria

activated sludge.   aerobic biological process with secondary sedimentation and biological sludge recycle

acute toxicity.   involving a stimulus severe enough to rapidly induce a response, usually in 96 hours or less

aerated lagoons.   aerated aerobic basins without sludge recycle

anion exchange resins.   resins which are used to exchange with anions in a fractionation test or treatment

anthropogenic.   activity such as pollution resulting from the influence of humans

AQUIRE.   aquatic information retrieval toxicity data base (U.S. EPA)

ATP.   adenosine triphosphate

BAT.   best available technology

binomial procedures.   refers to a statistical procedure used to calculate LC_{50} s and confidence intervals (Peltier and Weber 1985)

bioassay.   synonymous with toxicity test, as used here, i.e., the means to determine the toxicity of a chemical or an effluent using living organisms

biomonitoring.   the use of organisms to access the suitability of an effluent for discharge into receiving waters, and to test the quality of such waters downstream from a discharge

BMP.   best management practices

BOD.   biochemical oxygen demand

cation exchange resins.   resins which are used to exchange with cations in a fractionation test, or treatment

causative.   effective agent, or chemical, producing or causing a response such as toxicity

CETIS.   computerized complex effluent data base (U.S. EPA 1987)

chain-of-custody records.   official, signed record indicating the order of possession or custody of a sample from time of collection to receipt and analysis in the laboratory

chelation.   *see* EDTA

CHRIS.   chemical hazard response information system

CMAS.   complete mix activated sludge

COD.   chemical oxygen demand

colormetric test kits.   chemical tests which compare the color associated with a standard chemical addition to a sample for determining concentration

controls. an exposure or test chamber which receives only dilution water samples and is used in conjunction with an effluent bioassay

C_{18}. a resin containing long-chain carbon molecules with 18 carbon atoms; such resins serve as sorbtive surfaces for removal of hydrophobic organic chemicals in fractionation tests

EC_{50}. the toxicant or effluent concentration producing an effect, not mortality, in 50 percent of exposed organisms in a specific time

EDTA. ethylenediaminetetraacetic acid, used as a metal chelating agent

elution. to separate or remove adsorbed materials by washing, such as shaking a sediment sample in water for subsequent toxicity testing

embryo-larval. toxicity tests that are initiated with embryos or larvae of the test species

false negatives. unknown circumstances where no toxicity is observed in a specific wastewater fractionation when there is toxicity

false positives. unknown circumstance when toxicity is observed in a specific wastewater fractionation when there actually is no toxicity

FBR. fed batch reactor for toxicity evaluation

flow-proportional. effluent samples collected proportionally to the volume of flow over a period of time such as 6, 8, or 24 hours

fractionation. screening process where samples of wastewater are chemically and physically separated to determine cause of toxicity by chemical groups such as polar organics, metals, pH, chlorine, etc.

GC/MS. gas chromatograph/mass spectrometry

hardness. calcium and magnesium, combined concentrations, expressed as mg/l calcium carbonate in water or effluent

HPLC. high pressure liquid chromatography

hydrophobic. refers to chemicals which lack affinity for water

IWC. instream waste concentration

LC_1. the toxicant or effluent concentration estimated to kill 1 percent of exposed organisms in a specific time period

LC_5. the toxicant or effluent concentration estimated to kill 5 percent of exposed organisms in a specific time period

LC_{10}. the toxicant or effluent concentration estimated to kill 10 percent of exposed organisms in a specific time period

LC_{50}. the toxicant or effluent concentration estimated to kill 50 percent of exposed organisms in a specific time period

macroreticular resins. resins employed for the removal of specific organics by adsorption

matrix effects. the effects of the water chemistry "matrix" on a specific chemical's toxicity

MCL. maximum contaminant level (U.S. EPA 1986)

MDL. method detection limit

methanogens. methane producing bacteria

microbial acclimation. the process of microbes (mainly bacteria) acclimating to and utilizing the chemicals present in water as an energy source

Microtox®. trademark for a specific bacterial effluent toxicity test

mixing zones. a localized area of surface water, as may be designated, into which wastewater may be discharged without creating nuisance or toxic conditions

MLSS. mixed liquor suspended solids

MLVSS. mixed liquor volatile suspended solids

molecular sieves. filters or ''sieves'' which remove chemicals from solution based on their molecular weight

mono-media sand beds. water or wastewater filter systems employing a single sand media for filtration of suspended solids

Mysidopsis bahia. an estuarine mysid used to test the toxicity of effluents discharged to estuarine or marine waters

Nitex®. trademark for 300 to 500 micrometer mesh netting sometimes used to remove large debris from effluent samples

nonrenewal. refers to toxicity testing procedure where the effluent or toxicant is not renewed during the test duration

NPDES. National Pollutant Discharge Elimination System

OHM-TADS. Oil, Hazardous Materials—Toxic Air Data System

oligotrophic growth. bacterial growth at very low organic substrate levels

organic binders. organic materials applied to glass fiber filters to hold the glass fibers together, but which may bind up chemicals rendering them nontoxic, i.e., not available

PAC. powdered activated carbon

PACT®. powdered activated carbon process patented by DuPont and licensed by Zimpro

persistent toxicity. toxicants that are not volatile or do not degrade readily in effluents

photolysis. degradation of a chemical in water by light

physical/chemical nature. the physical or chemical characteristics of a toxicant being evaluated, such as volatility, salinity, or degradability

polar organics. compounds which have an electric charge and hence stay in solution rather than partitioning onto solids in water

priority pollutants. pollutants listed in 45 *FR* 79318, November 28, 1980; 49 *FR* 5831, February 15, 1984; and 50 *FR* 30784, July 29, 1985

probit. a statistical method for analyzing toxicity data

QA/QC. quality assurance/quality control

QSAR. quantitative structural activity system, used to predict chemical properties including partition and persistence in the environment, bioaccumulation, and toxicity

reference toxicants. standard chemicals used to test sensitivity of test organisms in QA/QC

refractory organics. organics resistant to biodegradation

replicates. independent samples of the same concentration used in toxicity testing

sorptive. sorptive chemicals are those which tend to absorb or adsorb to suspended solids, sediments, or organisms in water

species. a specific chemical compound or the scientific or common name of a test organism, as used here

spikes. known additions of a chemical to a water sample for the purpose of determining percent recovery for QA/QC

SRT. solids retention time

static acute. *see* nonrenewal; also, a nonrenewal acute toxicity test

taxonomic. adjective of taxonomy which is the study of the scientific classification of biota

timed lethality tests. toxicity tests which measure toxicity as a function of the amount of time necessary to elicit an effect or response

TOC. total organic carbon

toxicant. the chemical(s) causing the response observed in aquatic toxicity tests, a toxic agent

TREs. toxicity reduction evaluations

TU. toxic unit(s), the reciprocal of the effluent concentration that causes the acute effect (TU_a) or chronic effect (TU_c) by the end of the test exposure period

unionized. that form of a chemical which is not dissociated into its component ions in water

volatilized. the process by which a chemical is driven out of a solution in a gaseous state

xenobiotics. a chemical (man-made) that is foreign to a living organism

zeolite. a hydrous silicate that is used to remove ammonia, in this case, for fractionation testing

Index

Index